AF361844

Microgrids

Microgrids: The Path to Sustainability

Editor

Amjad Anvari-Moghaddam

MDPI • Basel • Beijing • Wuhan • Barcelona • Belgrade • Manchester • Tokyo • Cluj • Tianjin

Editor
Amjad Anvari-Moghaddam
Aalborg University
Denmark

Editorial Office
MDPI
St. Alban-Anlage 66
4052 Basel, Switzerland

This is a reprint of articles from the Special Issue published online in the open access journal *Sustainability* (ISSN 2071-1050) (available at: https://www.mdpi.com/journal/sustainability/special_issues/Microgrids_Path_Sustainability).

For citation purposes, cite each article independently as indicated on the article page online and as indicated below:

LastName, A.A.; LastName, B.B.; LastName, C.C. Article Title. *Journal Name* **Year**, *Volume Number*, Page Range.

ISBN 978-3-0365-0662-3 (Hbk)
ISBN 978-3-0365-0663-0 (PDF)

© 2021 by the authors. Articles in this book are Open Access and distributed under the Creative Commons Attribution (CC BY) license, which allows users to download, copy and build upon published articles, as long as the author and publisher are properly credited, which ensures maximum dissemination and a wider impact of our publications.

The book as a whole is distributed by MDPI under the terms and conditions of the Creative Commons license CC BY-NC-ND.

Contents

About the Editor

Amjad Anvari-Moghaddam is an Associate Professor at the Department of Energy Technology, Aalborg University, where he is the Head of the Integrated Energy Systems Laboratory (IES-Lab). His research interests include the planning, control, and operation of energy systems, mostly renewable and hybrid power systems with appropriate market mechanisms. Prof. Anvari-Moghaddam is the Vice-Chair of the IEEE-PES Chapter in Denmark and a Technical Committee Member of the IEEE-CTSoc Consumer Power and Energy Technical Stream Committee; IEEE-PES/SBLC P2418.5 WG for Blockchain in Energy; IEEE-IES Renewable Energy Systems (RES); IEEE-IES Energy Storage (ESTC); IEEE-IES Resilience and Security for Industrial Applications (ReSia); IEEE-PEL/SC P2004 Working Group; IEEE-PES Energy Development and Power Generation (ED&PG); IEEE-PES Power System Operation, Planning and Economics (PSOPE), as well as CIGRE WG on Rural Electrification (TOR C6.38) and Distributed Energy Resources Aggregation Platforms for the Provision of Flexibility Services (TOR C6.35). He was ranked among the top 2% of the most highly cited scientists in the world (according to the 2020 Stanford University list) and was the recipient of the 2020 DUO—India Fellowship Award, the DANIDA Research Fellowship Grant from the Ministry of Foreign Affairs of Denmark in 2018, the IEEE-CS Outstanding Leadership Award 2018 (Halifax, Nova Scotia, Canada), and the 2017 IEEE-CS Outstanding Service Award (Exeter-UK). Prof. Anvari-Moghaddam serves as the Associate Editor of *IEEE Transactions on Power Systems, IEEE Access, IEEE Open Access Journal of Power and Energy, IEEE Power Engineering Letters, IET Renewable Power Generation, IET Smart Grid, Applied Sciences*, and *Electronics*.

Preface to "Microgrids: The Path to Sustainability"

Microgrids are a growing segment of the energy industry, representing a paradigm shift from centralized structures toward more localized, autonomous, dynamic, and bi-directional energy networks, especially in cities and communities. The ability to isolate from the larger grid makes microgrids resilient, while their capability of forming scalable energy clusters permits the delivery of services that make the grid more sustainable and competitive. Through an optimal design and management process, microgrids could also provide efficient, low-cost, clean energy and help to improve the operation and stability of regional energy systems. To cover the above-mentioned promising and dynamic areas of research and development, this book, as an edited collection of the papers published in the Special Issue "Microgrids: The Path to Sustainability", was launched to gather contributions on different aspects of microgrids in an aim to impart higher degrees of sustainability and resilience to energy systems while paving the way for developing new planning, integration, and management solutions in the power and energy sector.

The book is structured in 12 chapters and covers a variety of topics, ranging from the control of AC/DC microgrids in different working conditions and in the presence of renewable resources to the hourly/daily dispatch of such systems. The book also provides great insight into the planning aspects of microgrids and local energy systems, which are key points in determining the right setting and sizing of energy sources with regard to different objectives, policies, and constraints.

The first chapter of the book investigates the technical and economic trends and environmental and social aspects of micro-energy systems, such as a desalination system, and then provides an overview of the role of renewable energy technologies in the sustainability of future water systems with an increasing share of such entities. The impact of government subsidy on renewable microgrid investment considering double externalities is discussed in Chapter 2. Here, the optimal level of government subsidy to correct the market failure of microgrids is investigated, and the impacts of regulation on the interaction between a microgrid and a distribution network operator are analyzed comprehensively. The third chapter elaborates on spatial-based integration models for regional-scale solar energy technical potential, where the main purpose is to identify locations within a certain geographical place that are suitable for solar energy. The chapter also evaluates the technical potential of various solar technologies, namely, concentrating solar power (CSP) and photovoltaic (PV) energy in power plant applications and rooftop PV panels and solar water heaters in general applications. On the way toward developing more resilient energy systems, Chapter 4 discusses different methods to enhance the seismic resilience of integrated power and water distribution networks against natural disasters by leveraging microgrids. In this chapter, resilience improvement planning models are initially introduced for integrated energy systems, and their effectiveness and applicability in theory and practice are verified accordingly.

On the operational aspects, the fifth chapter discusses optimization models for the optimal and stochastic operation scheduling of residential microgrids in the form of smart buildings. The aim of this chapter is thus to provide cost-effective solutions for meeting the electricity demand in the presence of intermittent renewable resources. Chapter 6 is also dedicated to the operations management of microgrids by incorporating innovative solution methodologies. To this end, a deep neural network-assisted approach is proposed to enhance the short-term optimal operational scheduling of a microgrid, and accurate aggregated electrical load demand and bulk photovoltaic power generation forecasting are achieved.

Regarding the control aspects, a power flow control strategy for the hybrid control architecture of a DC microgrid system is discussed in Chapter 7 given an unreliable grid connection and the constraint of electricity price. In the same line of research, Chapter 8 describes a novel virtual synchronous generator-based accurate voltage control and reactive power sharing method for islanded microgrids, which has become a promising method in the area of microgrid stability control. Methods for maximum power point tracking of photovoltaic systems are the subject matter in the ninth chapter, where variable gain proportional-integral controllers tuned with fuzzy rules are presented. With the aim of solving the power fluctuation and bus voltage instability problems caused by external environmental variations in renewable-based grid-connected microgrids, Chapter 10 studies a prescribed performance-based adaptive backstepping controller for the system to regulate the bus voltage and the inverter current.

The last two chapters of the book address the application of machine learning techniques in practical engineering problems. Chapter 11 opens up doors to innovative electricity theft detection methods using supervised learning techniques on smart meter data. The proposed methods can be efficiently applied by microgrid operators and utility companies using real electricity consumption data to identify electricity thieves and overcome major revenue losses in the power sector. Finally, Chapter 12 discusses a transparent unsupervised approach based on dimensional reduction to improve the residential load disaggregation problem via a visual and transparent process. This is an especially critical task in the planning and operation of energy systems in which an accurate load model is needed to conduct a true sustainability assessment.

Amjad Anvari-Moghaddam
Editor

Article

The Role of Renewable Energy Resources in Sustainability of Water Desalination as a Potential Fresh-Water Source: An Updated Review

Esmaeil Ahmadi [1,*], Benjamin McLellan [1], Behnam Mohammadi-Ivatloo [2,3,*] and Tetsuo Tezuka [1]

[1] Energy Economics Laboratory, Graduate School of Energy Science, Kyoto University, Kyoto 606-8501, Japan; b-mclellan@energy.kyoto-u.ac.jp (B.M.); tezuka@energy.kyoto-u.ac.jp (T.T.)
[2] Faculty of Electrical and Computer Engineering, University of Tabriz, Tabriz 5157944533, Iran
[3] Institute of Research and Development, Duy Tan University, Da Nang 550000, Vietnam
* Correspondence: esmaeil.ahmadi.53s@st.kyoto-u.ac.jp (E.A.); mohammadi@ieee.org (B.M.-I.)

Received: 30 May 2020; Accepted: 22 June 2020; Published: 28 June 2020

Abstract: Desalination is becoming a practical option to meet water demand in an increasing number of locations that are facing water scarcity. Currently, more than 150 countries in the world are already using desalination technologies, which account for about one percent of the world's drinking water. Although for specific regions, desalination is the only feasible solution to close the supply–demand gap (for example the production of desalinated seawater in the Middle East is predicted to rise almost fourteen-fold by 2040), the sustainability of desalination systems is still remarkably under question. This review aims first to investigate the technical and economic trends and environmental and social aspects of desalination systems and then, in the second stage, to give an overview of the role of renewable energy technologies in the sustainability of the future water systems with an increasing share of desalination.

Keywords: sustainability; desalination; renewable energy; water–energy-nexus

---

1. Introduction

Water is vital to life, society and the economy. A study in 2016 [1] found that around 4 billion of the global population are facing moderate to severe water scarcity, about 66% of which live under conditions of severe water scarcity for at least one month during a year. A United Nations' report [2] predicted that under average economic growth without improvement in efficiency, global freshwater demand could reach 40 percent above the current demand by 2030. The current pace of efficiency improvement is globally too slow to meet future demand. Authors [3] estimated that by considering the ongoing improvements in water efficiency, only 20 percent of the supply–demand gap would be closed. On the other hand, the potential remaining natural water resources, which are sustainable to utilize, are limited. Beyond risks from climate change, supply-side options to meet future water demand face rising costs due to steep marginal costs [4]. This study seeks to identify the state-of-art of desalination-based water provision, considered from a wide variety of perspectives beyond just the techno-economic analysis. It aims to identify the promising advantages of desalination technologies, particularly in connection with the energy–water nexus, and to clarify the identified disadvantages as shown through a critical review of recent studies.

There are non-conventional options, such as desalination and reuse of wastewater, which are the ultimate solution to meet water demand in specific regions. Rapidly progressing desalination technologies and market maturation have led to a significant drop in desalination costs, and the environmental impacts of the desalination process are progressively being mitigated. Even though desalination costs are likely to remain more expensive than other traditional water options, it increasingly will be considered as an option in specific areas due to climate change, natural and physical water scarcity, freshwater resource security, and the need to improve access to clean water (health problems).

Climate change is likely to impose a greater incidence of drought due to more limited and unpredictable rainfall and a higher rate of evaporation due to rising temperatures. This rising temperature could also cause an increase in water demand, exacerbating competition among agriculture, municipal, and industrial users. Desalination is a solution to enhance climate change resilience.

Additionally, desalination is economically and politically important to achieve self-reliance for specific areas. Israel and Singapore are examples where investments in desalination have been made to reduce their dependency on imported water due to their geopolitical situation. Furthermore, with population growth, providing quality water for cities becomes a challenge for policy-makers. Supplying water to dynamic sectors of the economy, namely commercial and industrial users, is an economic priority. Any failure in providing water to these sectors leads to high economic, social, and political costs. Desalination is considered a secure supply with high reliability for these water demands.

As a result, desalination is becoming an economical and practical option to meet water demand in an increasing number of locations. Currently, more than 150 countries in the world are already using desalination technologies, which account for about one percent of the world's drinking water [5]. A multi-criteria analysis in Kuwait [6] showed that among management options and strategic policies to meet future water demand, desalination powered by renewable resources and wastewater reuse ranked the highest. Based on Figure 1c, municipal and industrial sectors account for the main share of desalinated water production worldwide .

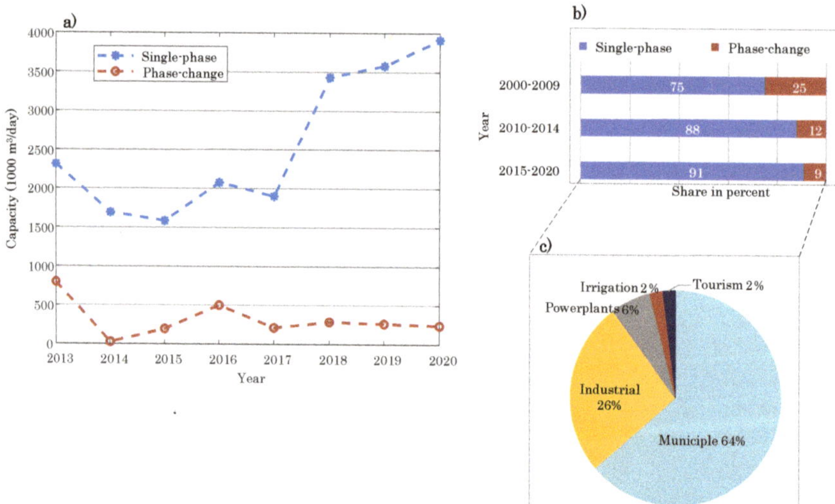

Figure 1. Total worldwide installed desalination: (**a**) capacity by technology; (**b**) technology share; and (**c**) share of each user for worldwide desalinated water [7].

On the other hand, the total world energy consumption has been forecast to increase by 44% from 2006 to 2030, according to a report by the US Department of Energy [8]. Desalination is an energy-intensive process. Energy requirement in commercial desalination processes ranges from a minimum of 1.8 KWh/m^3 for reverse osmosis technology to a maximum of 12.5 KWh/m^3 for multi-stage flash technology [4,9]. On average, desalinating 1000 m^3 of saline water by conventional technologies consumes about 37 barrels of crude oil (utilizing combined cycle power plant and reverse osmosis desalination technology), which causes around 10 tons of CO_2 emissions [10]. According to the World Energy Outlook 2016, in the Middle East, the water sector's share of total electricity consumption is expected to increase from 9% in 2015 to 16% by 2040, because of a rise in desalination capacity [11]. Furthermore, the energy sector is also set to become thirstier over the next decades, with energy-related water consumption increasing by nearly 60% between 2014 and 2040.

Studies have investigated different aspects of the sustainability of water systems and energy systems. The authors of [12] proposed a multi-objective optimization method for the sustainable utilization of energy resources, underground freshwater, and desalinated water. A study [13] explored 12 sample cities in Southeast Europe based on the sustainable development of energy, water, and environment systems. Although this study examined the cities based on a vast range of indicators, the linkage and relations between these indicators have not been evaluated, and the energy and water sectors were studied in isolation.

Based on the technical and economical aspects, other authors [14] applied system dynamics to predict the water and energy demands for a region in China from 2015 to 2030. In the next step, this study investigated the correlation between the risk of availability of energy and water. Another study [15] introduced a simplified method to track the water–energy nexus for a city in China. In this method, the required water for energy generation and required energy for the water sector were calculated in detail and compared with an average, which was obtained from government data. Others [16] developed a tool to calculate the electricity intensity of the water sector in California, USA. The study introduced a bottom-up model to calculate and visualize electricity consumption of the water sector for each person and each cubic meter of water as the energy intensity of the water sector. The result of this study indicated that considering the whole state uniformly for studying electricity intensity of the water sector was not practical because of geographical and climate differences.

Greenhouse gas (GHG) emissions of the water sector are typically the main focal point of studies on environmental impacts. Authors [17] investigated the budget-based plans for CO_2 mitigation and water management at the city-level. The budget-based plans for 13 cities in the USA and the UK were studied. This study proposed recommendations to improve existing and future carbon and water budget programs, indicating that the plans focus primarily on GHG (greenhouse gas) emissions mitigation while ignoring water–energy linkages and challenges due to water resource depletion. The authors of [18] developed a life cycle assessment to study GHG emissions from water resources, including imported water, seawater desalination, brackish groundwater desalination, and recycled water in California, USA. This study found that desalination, coupled with a concentrated solar power (CSP) system, has lower GHG emissions than that of recycled and imported water. Study [19] investigated the energy consumption and CO_2 emissions from the water system in Mexico City, Mexico. The water system in this study was divided into two categories, including water supply and wastewater. Water supplies for this city consist of two groundwater and two surface sources. Due to the long distance from two of the water sources, about 90 percent of the total energy consumption in the water sector was related to water supply. Several solutions proposed to decrease water consumption, the most effective of which are non-revenue water, water pricing reforms, and implementation of rainwater harvesting. Non-revenue water here refers to all water lost to leakage and other losses (such as illegal connections and metering errors), which was reported to be about 40% for Mexico City. In this study, the possible future changes in water demand patterns and energy resources mix (more 90 percent from fossil fuel resources) were not considered.

Authors [20] reviewed quantitative modeling scenarios for sustainable energy pathways and showed a gap in these models, which is not considering the social aspects of sustainability in quantitative technical models. Several models introduced the using of system dynamics, and expressed that taking into account the social aspects (for example, technology adaption by society) makes pathways more realistic. A study [21] highlighted the roles of economic development, technological innovation, and policy-making in shaping national energy transitions. The authors in [22] developed a system dynamics model to study the interactions between urban water and energy systems. The influence of end-use water consumers on the upstream of the water and energy sectors was investigated in London, UK, as a case study. The results indicated that policy targets, such as decarbonization plans, and social aspects, such as quality of life, when incorporated into models, assisted in achieving more realistic models.

All of these studies highlighted the interlinkages between the energy sector and the water sector towards sustainability. Although the deployment rate of desalination systems has accelerated recently and considering recent developments discussed in the literature that reveal that desalination can be affordable, the sustainability of these systems is still under question. Contributions that meaningfully address the technical, economic, environmental, and social issues of desalination are required in this era of water stress in order to achieve sustainable desalination in the future. This review aims to give an overview of the role of renewable energy technologies in the sustainability of future water systems with an increasing share of desalination. In other words, this study highlights the interlinkages and nexus between renewables and desalination.

Different aspects of sustainability are discussed in the following sections. In Section 2, the technical factors and current desalination technologies and their features are clarified. Section 3 highlights the factors which are important in desalination costs and gives an overview of the trend of renewable desalination technologies. Section 4 investigates the environmental impacts. Afterward, in Section 5, the social factors are discussed, and the potential synergy of powering desalination by renewables in order to raise social acceptance is examined. Finally, Section 6 summarizes the main concluding points of this study and proposes some frameworks for future studies.

2. Technical Aspect

2.1. Desalination Technologies

The desalination technologies are divided into two categories: desalination with phase-change or thermal processes and desalination with single-phase or membrane processes. The phase-change desalination technologies include multi-stage flash (MSF), multiple-effect distillation (MED), vapor compression (VC), and freezing. Reverse osmosis (RO), electrodialysis (ED), capacitive deionization (CDI), and membrane distillation (MD) are examples of the single-phase desalination technologies. Reverse osmosis, multi-stage flash, multiple-effect distillation, electrodialysis, and hybrid technologies are commercially viable and commonly used desalination technologies with a share of 63%, 23%, 8%, 3%, and 3%, respectively [23].

Currently, multi-stage flash distillation (MSF) and multi-effect distillation (MED), reverse osmosis, and a combination of these technologies (hybrid desalination) are the dominant technologies for seawater desalination.

2.1.1. Phase-Change Desalination

The primary energy required for phase change technologies is thermal energy. MSF, MED, and vapor compression (VC), which could be mechanical (MVC) or thermal (TVC), are the most commercially available technologies in this category [24]. In the MSF process, vapor is generated by a sudden pressure reduction of seawater or brine when saline water enters an evacuated chamber stage by stage. MED is based on vapor generation using the absorption of thermal energy by saline water. In the VC process, after the generation of vapor from saline water, this vapor is converted into freshwater by thermal or mechanical compression.

The separation process in the desalination of water by freezing follows the solid–liquid phase-change phenomenon. In this process, the temperature of saline water is reduced to the freezing point, which ice crystals of pure water are formed within the salt solution. Refrigeration systems are used in this process to reduce the temperature. In the next step, these crystals can be separated and washed. A humidification/dehumidification (H/DH) process captures the water vapor, which is mixed with air. In this method, brine is used to increase the humidity in an air stream. In the next stage, freshwater is collected by condensing this humid air on the surface of cool coils. H/DH technologies have not matured industrially due to technical barriers [24].

2.1.2. Single-Phase Desalination

The primary types of energy needed for membrane-based desalination are electricity and hydraulic pressure. Reverse osmosis, electrodialysis and membrane distillation (MD) technologies are the most commonly utilized in this category. In the RO processes, electricity or shaft power is needed to drive high-pressure pumps. For the RO process, mechanical pressure is applied to overcome osmotic pressure and separate salt of saline water. In the ED process, electricity is used for the ionization of salts contained in the seawater. The membrane distillation process includes two streams: one hot saline stream and a cool freshwater stream. Water vapor is transported between these two streams because of a temperature difference of streams. With a 80 percent water recovery rate, ED technology has better performance compared to RO technology, which has about a 40 to 50 percent recovery rate. Capacitive deionization (CDI) is a novel desalination technology suitable for brackish water treatment with a low level of salinity [25]. CDI technology operates by applying a relatively low potential (around 1–1.5 V) to drive ions from brackish water to a charged electrode with a porous structure using static electrical force, thereby separating the salt from brackish water [26].

2.1.3. Hybrid Desalination

Hybrid desalination plants are typically co-located with power plants so as to use waste heat for a thermal desalination facility (MSF or MED) and a combination of a RO desalination plant. Combined thermal and RO plants are usually suitable for situations with wide diurnal or seasonal variation in power or water demand. In such countries, peak power demand during summer is 30 to 40 percent higher than the maximum power demand in winter. In the Middle East, this difference reaches up to 50 percent, while the demand for desalinated water is almost constant. Switching between the RO and thermal plants allows benefiting from cheap available energy, thus leading to the cheapest desalination process.

2.2. Renewable Energy and Desalination

Among renewable resources, hydropower and biomass sources are not suitable to combine with desalination technologies due to the requirement for water resources, which is limited in regions facing water scarcity. The authors in study [27] considered biomass resources to reduce the CO_2 footprint of desalination plants in Saudi Arabia but did not mentioned the source or type of biomass. In areas with abundant solar irradiance, the main focus has been on integrating the desalination process and solar energy since water scarcity is more likely to occur in these regions [28]. Solar energy, with 51 percent of worldwide renewable desalination capacity, has the highest share, following wind energy, which accounts for 30 percent [29]. Figure 2a,b shows the integration of desalination technologies with renewable energy resources and the share of each renewable technology in desalination worldwide. Geothermal and wave and tidal resources are the other options to couple with renewable resources, which are still in the research phase and are not yet economically feasible [10].

Previous studies [30] investigated co-locating pumped hydro storage systems with reverse osmosis desalination plants based on geographical and economic benchmarks in several cities in the USA, Iran, China, and Chile. The results indicated that pumped hydro systems can compensate for the intermittent nature of power production from photovoltaic panels and wind turbines and decrease the energy intensity needed for reverse osmosis plants. Average daily historical data were used for calculating the renewable energy (RE) production for a whole year, which did not describe the renewable production with sufficient accuracy to calculate the fluctuations resulting due to the intermittent nature of renewable power production. The authors in another study [31] proposed a spatial model to assess potential technical and economical viable site locations for desalination facilities powered by renewables (wind and solar). Depth of water resource, distance to current water facilities, salinity degree, the magnitude of local RE resources, and local water price were considered as criteria in this model. Among 1445 site locations, 193 site locations were recognized as economically viable for

RO desalination facilities, 145 of which were wind-powered desalination units. Solar-powered units were preferable at the remaining 48 sites.

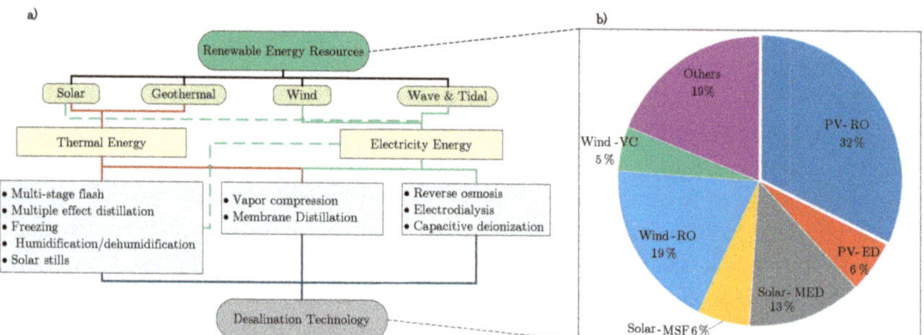

Figure 2. (**a**) The integration of desalination technologies with RE. (**b**) The share of renewable energy technologies in worldwide renewable-powered desalination processes in 2017.

There are three categories of technologies to harness ocean energy: thermal, mechanical, and chemical or salt gradient. The ocean mechanical (tidal and wave energy) and thermal energy technologies are more advanced than ocean chemical energy technology. Integrating the thermal energy technologies with phase-change desalination processes and using direct ocean mechanical energy (tidal, wave, and current energy) without transforming to electricity in desalination methods (which need hydraulic pressure) could improve the efficiency and economic feasibility of the integrated systems. Ocean salt gradient technology is still far away from being a reality. However, in the future, ocean salinity gradient energy is a promising ocean energy source, since the forward osmosis, pressure retarded osmosis, and reverse electrodialysis devices can be readily integrated into current desalination technologies as a recovery energy system without major reconstruction in plants. There are several limitations to developing ocean-based power generation, including technological and economic limitations of energy harvesting and transport and device maintenance underwater. Having said that, using ocean energy in desalination applications could solve the ocean energy technological defects relating to economic limitations by co-location in the future [23].

Solar thermal and geothermal resources are water-consuming resources. Water availability is an essential factor that must be considered to assess the potential of these resources in each region, which has not been considered in the majority of studies like [32,33]. A study examined the extent to which physical water scarcity can limit the deployment of geothermal and solar thermal energy resources to produce electricity in California [34]. This study first calculated the sustainable amount of extraction from the water resources and then determined the supportable capacity of these power plants from the available water supply based on technology and cooling type by 2050. For several areas in California, the estimated capacity of geothermal and solar thermal resources was found to be limited due to insufficient water availability, and without considering water limitations the assessment would not be realistic.

Table 1 indicates the renewable energy resources used for desalination purposes . As an example, thermal collectors produced thermal energy for sanitary hot water in study [35], which is not mentioned in this table; only solar and wind electricity powering the desalination system are indicated as energy resources for this study in Table 1. This table shows that solar and wind electricity are the most common sources of renewable energy for desalination among studies. RO desalination technology is the dominant technology that has been studied the most (56 studies). MSF desalination technology, which requires high temperatures for the process, is not popular among studies (two studies in the Middle East and another one in the American region) compared to MED technology (nine studies), which operates at low temperatures.

Table 1. Renewable energy resources used for desalination purpose. RO: reverse osmosis; MED: multiple-effect distillation; MVC: mechanical vapor compression; MSF: multi-stage flash; ED: electrodialysis.

Model Type	Energy Resource								Desalination Technology	Scale	Ref.
	Solar Electricity	Solar Thermal	Wind Turbine	Geothermal	Ocean Energy	Hydro Power	Diesel Generator	Hydrogen			
On-grid	✓	-	-	-	-	-	-	-	RO	Full plant	[36]
	✓	-	-	-	-	-	-	-	RO	Pilot scale	[37]
	✓	-	-	-	-	-	-	-	RO	Full plant	[38]
	✓	-	✓	-	-	-	-	-	RO and MED	Full plant	[39]
	✓	-	-	-	-	-	-	-	RO	Full plant	[40]
	✓	-	✓	-	-	-	✓	✓	RO	Pilot scale	[41]
	✓	-	-	-	-	-	-	-	RO	Pilot scale	[42]
	✓	-	✓	-	-	-	-	-	RO	Pilot scale	[43]
	✓	-	✓	-	-	-	-	-	RO	Full plant	[44]
	✓	-	✓	-	-	-	-	-	RO, MVC	Pilot scale	[45]
	✓	✓	-	-	-	-	-	-	RO	Full plant	[46]
	-	-	-	-	✓	-	-	-	RO	Pilot scale	[47]
	✓	-	✓	-	-	-	-	-	RO	Pilot scale	[48]
	✓	-	✓	-	-	-	-	-	RO	Pilot scale	[49]
	✓	-	✓	-	-	-	-	-	RO	Full plant	[50]
	✓	-	✓	✓	-	✓	-	-	RO	Pilot scale	[51]
	✓	-	-	-	-	-	-	-	RO	Full plant	[52]
	✓	-	✓	-	-	-	-	-	RO	Full plant	[53]
	✓	-	✓	-	-	-	-	-	RO	Pilot scale	[54]
	✓	-	-	-	-	-	-	-	RO	Full plant	[55]
	✓	-	-	-	-	-	-	-	RO	Lab scale	[56]
	✓	✓	-	-	-	-	-	-	RO	Full plant	[18]
	✓	✓	-	-	-	-	-	✓	MSF	lab scale	[57]
	✓	-	✓	-	-	-	-	✓	RO	Full plant	[31]
	✓	✓(Water preheating)	✓	-	-	-	-	-	RO	Lab scale	[58]
	✓	-	✓	-	-	-	-	-	RO	Pilot scale	[59]
	✓	-	✓	-	-	✓	-	-	RO	Full plant	[60]
	-	✓	-	-	-	-	-	-	MED	Full plant	[61]
	✓	-	✓	✓	-	✓	-	-	MED	Full plant	[62]
	✓	✓	✓	-	-	-	-	-	RO and MED	Full plant	[5]
Total number	28	6	17	2	1	3	1	3			31

Table 1. *Cont.*

Model Type	Energy Resource								Desalination Technology	Scale	Ref.
	Solar electricity	Solar thermal	Wind turbine	Geothermal	Ocean energy	Hydro power	Diesel generator	Hydrogen			
Off-grid	✓	-	-	-	-	-	-	-	RO, Solar-still	Lab scale	[63]
	-	-	-	-	✓	-	-	-	MED	Lab scale	[64]
	✓	✓	✓	-	-	-	-	-	RO, MSF	Lab scale	[65]
	✓	-	✓	-	-	-	-	✓	RO	Lab scale	[66]
	✓	-	✓	-	-	-	-	-	RO	Lab scale	[67]
	-	✓	-	-	-	-	-	-	MSF	Lab scale	[68]
	✓	-	-	-	-	-	-	-	RO	Pilot scale	[69]
	✓	-	-	-	-	-	-	✓	RO	Lab scale	[70]
	✓	✓	-	-	-	-	✓	-	RO, MED	Lab scale	[71]
	✓	-	-	-	-	-	-	-	ED	Full plant	[72]
	-	-	-	-	✓	-	-	-	RO	Full plant	[73]
	✓	-	-	-	-	-	-	-	RO	Pilot scale	[74]
	-	✓	-	✓	-	-	-	-	MED	Pilot scale	[75]
	✓	-	-	-	-	-	-	-	RO	Lab scale	[76]
	✓	-	✓	-	-	-	-	-	RO	Pilot scale	[77]
	✓	-	-	-	-	-	✓	✓	RO	Lab scale	[78]
	✓	-	✓	-	-	-	-	-	RO	Pilot scale	[79]
	✓	-	✓	-	-	-	-	-	RO	Pilot scale	[80]
	✓	-	✓	-	-	-	-	-	RO	Pilot scale	[81]
	✓	-	✓	-	-	-	-	-	RO, MD	Pilot scale	[35]
	✓	-	✓	-	✓	-	-	-	RO	Pilot scale	[82]
	-	-	-	-	-	-	-	-	Solar-still	Full plant	[83]
	✓	-	✓	-	-	-	-	-	RO	Full plant	[84]
	✓	-	-	-	-	-	-	-	RO	Lab scale	[85]
	-	✓	-	-	-	-	✓	-	MED, Solar-still	Lab scale	[86]
	✓	-	-	-	-	-	-	-	RO	Full plant	[87]
	✓	-	✓	-	-	-	✓	-	RO	Pilot scale	[88]
	✓	-	✓	-	-	-	-	-	RO	Pilot scale	[89]
	✓	-	✓	-	-	-	-	-	RO	Pilot scale	[90]
	-	✓	-	-	-	-	-	-	MED	Full plant	[91]
Total number	23	6	13	1	3	0	4	3			30

	On-grid	Off-grid	RO	MED	MSF	Solar still	lab scale	Pilot scale	Full plant
Number of studies/ Total studies	30/60	30/60	56/60	9/60	3/60	3/60	15 /60	23/60	22/60

2.3. System Configuration

Conventional thermal desalination technologies are now well-proven and mature. Therefore, a further improvement in these technologies is relatively limited. Continuous innovation in RO desalination technology in the last twenty years has reduced the energy consumption per unit of product water to 1.8 KWh/m^3 compared to the historic energy consumption range of 3 to 5.5 KWh/m^3, which is close to the theoretical minimum required energy for seawater desalination [92]. This means that a further significant reduction in energy consumption is not expected for RO technology. However, further significant advances in membrane technology (increasing in water productivity per area) are predicted, which could cause up to a 20 percent cost reduction in the next five years [4].

In Table 2, the production capacity of desalination units is divided into three levels by scale: small (less than 20,000 m^3/day), large (between 20,000 m^3/day to 200,000 m^3/day), and mega (more than 200,000 m^3/day).

Table 2. Brief configuration of water-energy systems.

Energy Sector Description		Water Sector Description		Type of Analysis	Ref.
Model Type	Energy Type	Application	Scale		
Centralized	Electricity	Potable	Small	Theoretical	[49]
	Electricity	Potable	Small	Theoretical	[43]
	Electricity	Potable and Agriculture	Small	Experimental	[56]
	Electricity	Potable	Large	Theoretical	[36]
	Electricity	Potable	Large	Experimental	[50]
	Electricity	Potable	Large	Theoretical	[38]
	Electricity	Potable	Small	Theoretical	[45]
	Electricity	Potable	Large	Theoretical	[46]
	Electricity	Potable	Large	Theoretical	[18]
	Electricity	Potable	Large	Theoretical	[39]
	Electricity and thermal	Potable	Large	Theoretical	[57]
	Electricity	Potable	Large	Theoretical	[84]
	Electricity	Potable	Large	Theoretical	[31]
	Electricity	Potable	Small	Theoretical	[74]
	Electricity	Potable	Small	Theoretical	[47]
	Electricity and thermal	Potable	Large	Theoretical	[58]
	Electricity	Potable	Small	Experimental	[85]
	Electricity	Overall	Large	Theoretical	[93]
	Electricity	Potable	Large	Theoretical	[59]
	Electricity	Potable	Small to Large	Theoretical	[48]
	Thermal	Potable	Small	Experimental	[48]
	Thermal	Agriculture	Large	Pilot-project	[91]
	Electricity	Potable	Large	Theoretical	[40]
	Electricity	Potable	Large	Theoretical	[60]
	Thermal	Potable	Large	Theoretical	[61]
	Electricity	Potable	Large	Theoretical	[41]
	Electricity and thermal	Potable	Laboratory scale	Experimental	[65]
	Electricity	Potable	Large	Theoretical	[52]
	Electricity	Potable	Small	Theoretical	[79]
	Electricity	Potable	Small	Theoretical	[42]
	Electricity	Potable	Small	Theoretical	[82]
	Electricity and thermal	Overall	Mega	Theoretical	[94]
Decentralized	Electricity	Potable	Small	Theoretical	[63]
	Electricity	Agriculture and Potable	Small	Theoretical	[73]
	Solar-stills	Agriculture	Small	Theoretical	[83]
	Electricity	Potable	Small	Theoretical	[87]
	Electricity	Potable	Small	Theoretical	[67]
	Electricity	Potable	Small	Theoretical	[89]
	Electricity	Potable	Small	Theoretical	[80]
	Electricity	Potable	Small	Theoretical	[69]
	Electricity	Potable	Small	Theoretical	[81]
	Electricity	Agriculture	Small	Theoretical	[70]
	Electricity	Potable	Small	Experimental	[35]
	Electricity	Potable	Small to large	Theoretical	[55]
Decentralized and Centralized	Electricity and thermal	Potable	Small to large	Theoretical	[5]
	Thermal	**Agriculture**	**Small**	**Experimental**	
Number of studies/ Total studies	8/45	7/45	25/45	7/45	

A centralized water system refers to systems in which desalinated water is produced in one unit and distributed among all target users, while a distributed water system includes more than one desalination unit that is providing water demands. These decentralized desalination plants (mostly small-scale) have a great potential to solve the intermittent power generation problem of variable renewable resources (namely, wind and solar). These desalination plants can effectively operate without energy storage systems (mostly batteries), as water can be desalinated based on energy availability and stored as the final product [95]. This direct consumption of renewable energy increases the efficiency of the whole system because storage systems such as battery systems have a typical charge-cycle efficiency of 75% to 98% [5]. Furthermore, high ambiance temperatures, which are common in regions facing water scarcity, increase the self-discharge rate and performance of batteries. A small-scale RO desalination unit coupled with a PV system with battery storage in Malaysia was tested for six months [85]. This experiment aimed to examine the system performance and find the optimal condition to operate an RO desalination unit. The results indicated that climatic conditions (such as high ambiance temperatures) significantly reduced the performance of the battery and PV system.

Although desalination units with larger capacity face technical limitations to operate as variable units, it is still possible to integrate them with renewable resources to a certain extent [94]. For instance, one of the main problems with an intermittent RO desalination unit is the biological fouling when the unit is not in operation. The pretreatment of intake feed-water can significantly decrease the fouling [29]. This membrane fouling needs to be considered in the desalination system design in order to avoid an under-sized system and unmet water demand [96].

Flexible load resources can support ancillary services in electric grids. In this regard, the key operating features to determine the adequacy of the flexible loads are as follows [59]:

- Initial response time: the response duration time to a change in a power set-point.
- Ramp rate: the change rate of power consumption.
- Settling time: the settle time duration after an operating power set-point change.
- Duration: the period after settling time required to maintain the settled changed power set-point.
- Power capacity: the rated power points for operating the flexible load resources, which vary from kilowatts to megawatts.
- Minimum turn-down: the lowest operating point of the flexible load, below which the flexible load resource must be turned off.

According to the above criteria, desalination is perceived as an attractive option to integrate with renewable energy resources as a flexibleelectric load. As an example, the minimum turn-down for RO desalination plants is unlimited. The authors in [59] tested the capability of an RO desalination plant to manage the variability of renewable energy production while supporting (1) the above mentioned ancillary services in an electric grid and (2) meeting water demand with the desired quality level. The results of this study showed that RO desalination plants are capable of responding quickly, settling fast and operating long enough at required power points while maintaining the desired water demand quantity and quality. The authors in one study [97] showed the potential of an MED plant for the peak shaving of a coal-fired power plant.

The levelized cost of energy in a 100 percent renewable energy scenario for Saudi Arabia decreased by 3 percent by integrating the desalination sector with the energy system in study [39]. The required battery storage was reduced because of the flexibility provided by RO desalination plants. The authors in another study [57] introduced an integrated process of MSF desalination, hydrogen production, and solar power production. This study considered solar hydrogen production as an energy storage and a hydrogen-fired power plant to overcome issues related to the intermittency of solar power generation. Furthermore, a centralized water system typically requires more energy for the water transfer and distribution than a decentralized water system. Water distribution pumps, with 70–80% of the energy consumed in a surface-water-based supply, are the highest energy-intensive components of conventional water supply systems [28].

As can be seen in Table 2, the number of studies that investigated the centralized water system and the decentralized water system are very few. Study [5] developed a novel methodology for the sustainable planning of energy and water supplies with a share of renewables and desalination. This study conducted a comprehensive comparison between centralized and decentralized desalination systems and their technical, economic and environmental impacts on the energy and water sectors. This study found that a decentralized water sector and renewable-powered MED disalination technology experienced synergistic benefits and avoided conflicts between the water and energy sectors in the Middle East. Researches have studied [55] an alternative water system to meet water demand in a region of Australia. This study first calculated the capacity and surplus of rooftop PV systems in this region through a spatial model. In the next step, based on solar power production and surplus, a distributed water system was optimized, producing water through the RO process. This distributed water desalination system leads to a decrease of around 10 percent in the levelized cost of water and about 20 percent decrease in the levelized cost of solar energy compared to a centralized water desalination system for this region.

3. Economic Aspect

Removing the salts from saline water is a high-cost and high energy-consuming process compared to other freshwater supply and treatment alternatives. This section discusses the prospects of the current desalination costs and expected future costs of desalination for different technologies. Desalination technology, plant size, feed-water (salinity, temperature, and biofouling elements), energy use, intake–outlet system (environmental regulations), and target water quality (municipal, industry, or agriculture use) are the main factors affecting the desalinated water costs. Table 3 summarizes the cost of desalination for current commercially viable desalination technologies. The authors in [93] developed a model to estimate the cost of providing municipal, industrial, and agricultural water demand using RO desalination plants powered by a combination of PV, wind energy, battery and power-to-gas plants for regions facing water scarcity (regions where more than 40 percent of the renewable water resources are being withdrawn) in 2030. The levelized cost of water (LCOW) for the described system was found to be 0.65 to 3.10 USD/m^3. Table 4 gives an overview of the current situation and the pros and cons of desalination technologies.

Table 3. Cost components of the dominated desalination technologies [4,9].

Technology	Total Cost USD/m^3	Amortised Capital (%)	Electrical Energy (%)	Thermal Energy (%)	Membranes (%)	Labor (%)	Chemicals (%)	Miscellaneous (%)
RO	0.6–2.86	38.2	31.6	-	3.9	13.2	9.2	3.9
MED	1.12–1.5	34.9	7.2	37.3	-	9.6	9.6	1.2
MSF	1.02–1.74	39.3	18.7	29	-	7.5	4.7	0.9

Table 4. Energy consumption of desalination technologies [4,9,23,94,98–102]. H/DH: humidification/dehumidification; VC: vapor compression; ED: electrodialysis; MD: membrane distillation.

Technology	H/DH	MSF	MED	VC	RO	FD	ED	MD
Thermal energy KWh/m^3	45–100	7.5–11	4–7	0 (MVC) 51.9–63 (TVC)	-	8-24	-	30–240
Electricity KWh/m^3	-	2.5–3.5	1.5–2.5	7–15 (MVC) 1.6–1.80 (TVC)	1.8–6	-	2.46–5.5	0.6–1.8

3.1. Desalination Technology and Plant Size

It has been found that the capital costs and operating costs of desalination technologies are nonlinear functions of their capacity [103]. By reviewing and analyzing data, a study [103] found different degrees of non-linearity and types of correlations for each desalination technology. The capital cost of thermal desalination plants is about 1.50 to 2.00 million USD/MLD (where MLD refers to million

Liter/day), which is higher than the capital cost for RO plants with about 1.3 million USD/MLD. Construction cost accounts for around 75 percent of the capital cost of thermal desalination units, while it is about 50 percent for RO plants. By contrast, RO desalination plants are more design-intensive (cost of engineering services, skilled labor, regulatory and etc.) and thermal desalination units require heavier physical investment due to higher technology maturity.

According to studies, the total cost of producing freshwater with MSF plants is between 1.02 and 1.74 USD/m^3. The total cost of water production, including capital cost and O&M costs, is reducing proportional to the plant size. For medium-size (between 20,000 to 100,000 m^3/day) MSF plants, the cost of water production is about 1.50 to 1.74 USD/m^3, while for large plants this cost is about 1 USD/m^3. Even though the total energy cost for the MSF process is higher than the energy cost for the RO process, the total cost of MSF water desalination is competitive to medium- and large-sized RO plants. Co-locating MSF plants with power plants lowers the steam cost, which is a requirement for this technology, and keeps the cost down for this technology. Besides, unlike RO product water, the water product of MSF technology is of immediate potable quality. It is noteworthy to mention that innovations in this technology have been limited recently, impacting costs slightly.

The total cost of freshwater production for MED technology is between 1.12 and 1.5 USD/m^3. Medium-scale MED facilities (between 20,000 to 100,000 m^3/day) are producing water at a lower cost than MSF plants with the same size due to lower capital and O&M costs. Although for small-scale (less than 20,000 m^3/day) and medium-sized plants, MED units are more energy efficient with cost advantages, MSF plants are easier to operate, which sometimes makes them the preferred option for investors due to lower risk. At larger facility sizes, the water produced by MSF technology is cheaper than MED water production.

For RO desalination facilities, the cost of water production varies between 0.6 and 2.86 USD/m^3. RO desalination plants have a significant cost advantage for small-scale (below 20,000 m^3/day) and medium-scale facilities compared to other desalination technologies. Moreover, further significant advances in membrane technology (increasing in water productivity per area) are predicted to cause an up to 20 percent cost reduction in the next five years [100] (see Table 5).

Table 5. Forecast of RO membrane productivity and costs [103].

Parameters	2016	Within 5 Years	Within 20 Years
Cost of product water (USD/m^3)	0.8–3	0.6–1.0	0.3–0.5
Electricity requirement (KWh/m^3)	3–4	1.8–3.2	1.1–2.4
Membrane productivity (m^3/membrane)	28–47	35–55	95–120

3.2. Feed-Water

Salinity, temperature and biofouling elements of feed-water are important factors for RO desalination efficiency, performance, and costs, while the thermal desalination process is mostly insensitive to these factors. For low salinity, RO units desalinate water with lower cost comparing to the thermal desalination technologies, mainly because of energy-saving and lower energy requirements. As an example, a Red Sea RO seawater desalination plant with an average salinity of TDS (total dissolved solids) 44 ppt requires 30 percent more energy comparing to plants desalinating Pacific Ocean or Atlantic Ocean seawater, which have a salinity of TDS 46 ppt, with all other conditions being the same [4].

Changes in feed-water quality and temperature affect the efficiency of RO plants because the membrane performance is sensitive to these changes during the thermal desalination process. Study [100] found that RO plants in the Persian Gulf require 16 percent more capital costs and 14 percent more recurrent costs compared to ones in the Mediterranean region because of the different source of water. This is mostly because of the higher salinity and biofouling potential of seawater in the Persian Gulf, which requires costly pretreatment and intake systems and a more frequent need to change or clean the membranes.

3.3. Target Product Water Quality

Target water demands in the current study are categorized into three groups (see Table 2), namely potable demand, industrial demand, and agriculture demand. Table 2 illustrates that studies in this field have mostly targeted potable water, and there are few studies which studied the agricultural or industrial sectors as the target water user.

The product water of the thermal desalination process has lower salt (TDS of 50 milligrams per liter), boron, and bromide levels compared to water desalinated through the RO process. The high level of salt and boron of desalinated RO water needs to pass through an additional RO stage to achieve good-quality water. The second-pass stage can increase by 10 to 25 percent in the total cost of the first-pass desalination process [4]. Furthermore, calcium-based compounds and chlorine (for disinfection) are added to desalinated water (which is typically soft) before distribution. For most of the industrial and agricultural applications, it is not necessary to design this second-pass stage. The current costs of RO product water of desalination include the second-pass stage, which can reach up to 25 percent of the total RO water desalination costs and there is a lack of study in evaluating RO units without this second-pass stage.

3.4. Energy

Energy is the most important factor affecting the extent and feasibility of desalination; from 30% to over 50% of the cost of water produced by desalination processes is related to energy (see Table 3). Energy affects not only the cost of water produced by desalination but also the technology of the desalination process. For example, to use thermal processes, the largest desalination plants are located in the Middle East regions, which have rich fossil fuel reserves [23]. Total recurrent costs for each unit product of RO desalination plants are about twice those of MSF desalination plants and three times more than the MED per unit of water production. The main share of this recurrent costs goes to energy.

Study [49] designed an energy supply for an RO plant to desalinate brackish water in a rural area in Australia. The RO desalination unit with variable capacity operation has been assumed to treat brackish water of a river as a water supply. The optimization results for a 25-year period showed that the hybrid system with a combination of wind turbines, rooftop solar photovoltaic and electricity grid had the minimum levelized cost of energy (COE) and minimum overall net present cost (NPC). However, the operation and maintenance cost of the desalination plant and the required energy for the water distribution system were considered in this model.

Another study [104] investigated several scenarios to achieve a net zero-emissions electricity supply system by 2050. In two scenarios, the future electricity demand for seawater desalination was forecast and considered. Although desalination demand was approximately 3 percent of the total electricity demand in 2050, the comparison among scenarios showed that sector coupling between the desalination and electricity sector does not cause a significant change to the future electricity supply. The reason is that the desalination electricity demand was constant in this study, causing the electricity system to extract no benefit from using desalination plants as flexible electric loads to compensate for the fluctuation of variable renewable power production. The authors in [94], by considering desalination units as flexible loads, found that the levelized cost of energy is 30 percent lower than the cost with desalination as non-flexible loads due to less required battery storage to compensate for the fluctuation of wind and photovoltaic electricity production in Iran.

The authors in another study [105] evaluated the overall costs of RO water desalination by considering several scenarios for energy supplies in Chile. This study divided these costs into two categories, including internal costs and external costs. Internal costs refer to investment costs as well as operation and maintenance costs, while the external costs cover a carbon tax on energy resources. The evaluation in this study showed that by considering this carbon tax, the unit cost of desalinated water for scenarios with renewable energy resources is lower than the scenarios powered by conventional energy resources.

Water distribution pumps are energy-intensive components, which should be considered for site selection, size of desalination plants, and type of water system (distributed water system with several small-size desalination units or centralized water systems with a large-size desalination unit). Water transport costs from coastal desalination plants are estimated in other studies [9,106]. Table 6 [5] shows the cost of desalinated water transfer in several countries. The cost of water transfer in some cases in this table is considerable compared to desalinated water costs. Figure 3 depicts the current state of development, potential capacity and estimated cost of water production for the integration of renewable energy resources with desalination technologies. For details, see the list of the reference studies to outline the development state of renewable-powered desalination technologies in [5].

Table 6. Transport costs of desalinated water ([5]).

City, Country	Distance (km)	Elevation (m)	Transport Cost (USD/m^3)
Beijing, China	135	100	1.1
Mexico City, Mexico	225	2500	2.4
Yemen, Sana	135	2500	2.2
Mexico City, Mexico	280	320	2.4
Crateus, Brazil	240	350	1.3
Zaragoza, Spain	163	500	1.4
Riyadh, Saudi Arabia	350	750	1.6
New Delhi, India	1050	500	1.9

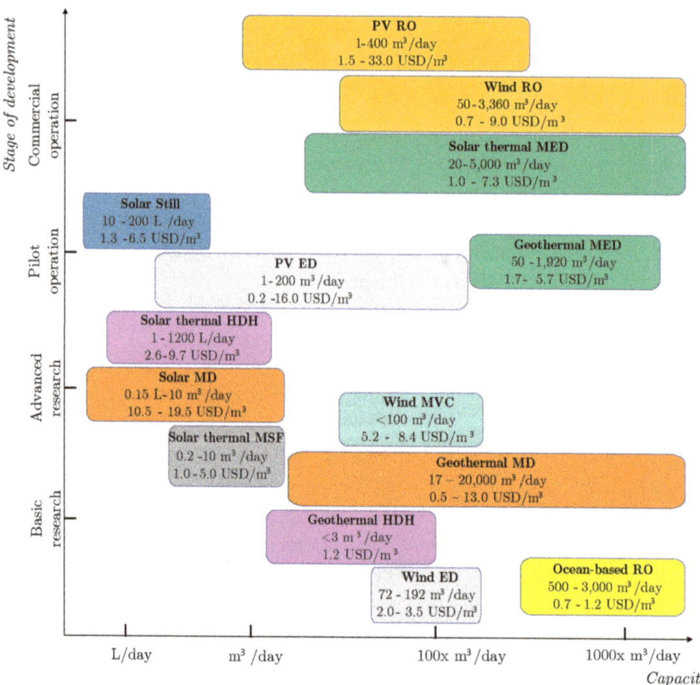

Figure 3. The status of the renewable energy operated desalination technologies (adapted from [5]).

The current share of renewable desalination is less than 1 percent of global desalination capacity [107]. The cost of renewable desalination is still higher than the cost of conventional desalination powered by fossil fuels. However, renewable technologies are experiencing a rapid cost reduction, making the renewable desalination already cost-competitive with the conventional desalination in remote regions (where the cost of electricity transmission and distribution is higher than the cost of decentralized electricity production). With this rapid cost reduction of renewable technologies, technical advances, and an improvement in the knowledge and experience by increasing the number of installations, the costs of renewable-powered desalination are likely to reduce significantly in the near future. It is expected that a major portion of fossil-fuel-powered desalination plants will be replaced by renewable-powered desalination with an average cost of 0.9 USD/m^3 by 2050 [108]. The renewable desalination was forecast to be sufficient only for domestic water supply in 2030, but to expand for further domestic and industrial water supply needs by 2050 [108,109].

As discussed, each desalination technology has its advantages and disadvantages. As shown in Table 1, RO is the most utilized desalination technology among previous studies indicating the potential of this technology for integration with renewables as a future sustainable solution for water-scarce regions. As an advantage, RO plants are scalable, typically consisting of several dozen units, and thereby its size can be expanded to meet the growing demand by adding more units as needed. The costs of RO desalination significantly decrease in treating lower salinity or brackish water due to lower energy requirements. By contrast, thermal distillation processes, namely MED and MSF technologies, need the same amount of energy regardless of salinity. As a result, thermal distillation processes are more competitive than RO desalination technology for high salinity waters when there is also high biofouling potential. MED technology is more competitive at a smaller scale compared to MSF technology, making this technology a better option for integration with renewables. Furthermore, MED operates at lower temperatures than MSF; as a result, its process is more compatible for integrating with renewable thermal power generation.

Solar and wind resources have been widely used for powering desalination among previous studies (see Table 1). Power generation from wind and photovoltaic requires zero or little water use [107]. Wind power is likely to be cheaper than photovoltaic power wherever it is available. Coastal areas usually benefit from a high availability of wind power resources. On the other hand, water-scarce regions or drylands are characterized by abundant solar radiation, which increases the capacity factor of solar resources and makes them more competitive compared to wind resources [110].

To sum up, the most promising combination of technologies is RO desalination technology with photovoltaic and MED desalination technology with solar thermal collectors. For large-scale units, wind power is more attractive wherever it is available, as it does not need a large area for installation, such as islands where often limited flat ground is available.

4. Environmental Aspect

Studies repeatedly addressed the two major environmental impacts of the desalination process, which are high carbon footprint due to energy consumption (indirect impacts) and intake seawater and effluent-associated pollution (direct impact). Thermal, chemical and saline pollution caused by the disposal of concentrate from the desalination process, which is most commonly discharged to the ocean, are the main environmental impacts of desalination. Evidently, the magnitude of these environmental impacts depends on the choice of desalination technology and site.

One study [63] compared three systems to provide potable water for a rural area in the UAE based on environmental impacts. These water systems included a solar still unit, a PV system coupled with an RO unit, and water delivery by truck from a central RO plant. The environmental impacts in this study were categorized as eco-toxicity, minerals, fossil fuels, carcinogens, respiratory inorganics, radiation, ozone layer depletion, acidification/eutrophication, respiratory organics, climate change, and land use. The result of a life cycle assessment indicated that the PV system, coupled with the RO unit, had the smallest environmental impact among these three proposed options.

4.1. Intake-Related Environmental Impacts

The main impact of intake facilities is on aquatic organisms. Subsurface intake wells (instead of direct saline or brackish water intake from the surface water) and constructed wetlands are commonly used to mitigate the environmental impacts of seawater intake for desalination. Adding an intake impact mitigation stage to desalination units can increase the capital cost by 5.3 percent and the annual O&M costs by 4.5 percent [111]. The authors in another study [36] assessed the environmental impacts of the open-intake pretreatment and subsurface intake pretreatment of two RO desalination plants in the Persian Gulf. A life cycle assessment highlighted that the subsurface intake method had fewer environmental impacts compared to open intake pretreatment.

As discussed, renewable-powered desalination is suitable for the distributed or decentralized water sector because of its size compatibility. The decentralized system increases the options of site locations for desalination units. Such multiple options enable the policymakers to choose site locations with a lower density of aquatic organisms.

4.2. Effluent-Related Environmental Impacts

Brine is a sub-product of the desalination process causing two environmental impacts including the effects of highly concentrated saline solution and metal components (copper, nickel, iron, chromium, zinc, etc., which are discharged with the brine during pre- and post-treatment processes [112]) that may heavily affect marine ecosystems. Besides, the outlet brine of the thermal desalination process could have a higher temperature than the ambient seawater and the amount of brine volume is much greater than the volume from RO process (thermal desalination technologies use almost twice as much saline water comparing to RO technology to produce the same amount of freshwater). On the other hand, the outlet from the RO process is more concentrated, which requires treatment prior to discharge.

To decrease the impacts of brine on marine ecosystems, first, the discharging of brine into sensitive ecosystems must be avoided through process design (technology, unit size, etc.) and the assessment of site location (site ecological value, hydro-geological and hydrodynamic conditions, etc.); second, the salinity must be reduced through active dilution processes, such as artificial diffusers or through natural local hydrodynamic conditions. Another study [113] introduced a geographic information systems multi-criteria framework to identify regions that where suitable for the deployment of RO units coupled with solar energy supplies. This study chose solar irradiance, ocean salinity, ocean temperature water stress, prevailing water prices, and population as factors for evaluation. Another study [73] examined the potential of wave energy resources to provide the power demand of current desalination plants for an island in Spain. This study investigated the selected location based on the environmental lens to assure that there is no vulnerable or sensitive area (such as reserves or marine and land habitats, with some environmental protection), which are leading to restrictions for the deployment of wave energy converters. The technical aspect and the compatibility of the desalination technologies with the proposed alternative energy resource were not considered in this research.

According to [23], adding ocean salinity gradient energy devices (e.g., PRO, FO, and RED) to existing desalination plants can decrease these environmental and social risks without a major change to current desalination infrastructures. Another study [30] introduced a hydro storage system with a reverse osmosis desalination plant that can help the problems related to brine disposal of reverse osmosis plants by diluting with seawater before releasing to sea. However, this study did not consider environmental issues due to an artificial lake with brine water, which has vast consequences for each ecology. For instance, the area proposed for usage as a lake for pumped hydro storage around the capital of Iran, Tehran, is located in important agricultural lands.

Another study [114] investigated the current status of regulations on discharge from desalination plants in Saudi Arabia. The results indicated that studies regarding the discharge from desalination plants are facing deficient statistics and a lack of supporting data. This study showed the necessity to impose more strict regulations on effluent water quality monitoring systems in desalination plants.

4.3. GHG Emissions of the Desalination Process

Powering desalination processes with renewable energy resources is considered the main solution to decrease the GHG emissions of desalination plants. It is noteworthy to mention that different renewable resources also have different levels of GHG emissions that need to be considered in studies. The authors in one study [115] provided an estimation about the potential reduction of life cycle GHG emissions of RO, MED, and MSF technologies when renewable energy supplies were utilized instead of conventional energy systems. This study showed that hydro-power has the best performance in decreasing GHG emissions compared to wind and solar supplies. In this study, desalination units were considered as fixed loads and the fluctuation of renewable power was neglected, which is not realistic.

Cities are aiming to increase the share of local water resources for enhancing water security, but the GHG footprint of these local sources should be compared to imported sources to reach solutions that are more effective. Including upstream (non-combustion) emissions of energy resources is another aspect that should be considered in models to achieve more holistic results. By excluding upstream emissions in a study [116], the carbon footprint of water supplies in Southern California were shown to be underestimated by up to 30 percent. The results of this study indicated that with current energy supplies in Southern California, importing water caused a lower carbon footprint than expanding local recycled water.

Another study [117] proposed an urban water system to apply several scenarios for the water sector for a city in Canada. The difference between scenarios included considering wastewater as a water resource and demand-side water management. This study calculated the amount of energy consumption related to the water sector, the water consumption of the energy sector and the CO_2 emissions related to the water sector due to energy usage for each scenario. This study concluded that designing these policies, scenarios, and plans (such as carbon mitigation) should not discourage economic or population growth. For reaching this aim, the most important step is the baseline measurements, which should be calculated based on alternative metrics, for example, emissions per GDP or per jobs.

5. Social Aspects

Palestine and its neighbor counties have access to the Mediterranean Sea, facing land limitations for developing large-scale renewable energy projects. Meanwhile, Jordan (a neighbor country) has an abundance of available land suitable for solar electricity production but has no access to the Mediterranean Sea, which is close to population areas facing water scarcity in Jordan. The authors of [46] examined the technical and economic feasibility of exchanging desalinated water with renewable electricity among countries. This study considered the exchange of desalinated water and RE as a potential solution to improve the relationship among these countries, which are facing critical political issues. Social acceptance has a vital role in the failure or success of such projects. For instance, according to a study [49], an RO desalination project was opposed and halted in Australia as a result of the lack of community engagement during the planning and the common perception of the RO as an environmentally unsustainable and energy-intensive solution. The study showed that increasing renewable energy resources in this region could improve the social acceptance of the desalination project. Deploying desalination units could have significant social impacts on society during the construction and operational stages. Social impacts of desalination units include lack of public trust and confidence in water supply, issues related to dust, noise and visual facilities, disturbance of the beach and sea-based recreational activities, pipelines, and the future residential and industrial development in the local area.

The authors in [45] studied two approaches to decide about desalination technology (RO and MVC) and source of energy (wind turbine or PV) for a desalination unit in an island in Italy. In the first approach, just the technical and economic feasibility were investigated, as a result of which the combination of RO technology with wind energy had the best performance. The list of considered criteria in the second approach included economic, environmental, social, legislation, policies, and technical aspects, as well as the specific location characteristics. Through a multi-criteria analysis with the participation of stakeholders, the combination of RO technology and PV ranked highest using the second approach. A survey with 333 respondents was conducted in Australia [9], with around 55 percent of respondents having very low to low levels of trust in the desalination water corporation, while 23 percent of them neither trusted nor distrusted the company. Social issues of desalination also depend on the size of the community. Desalination technologies might face fewer issues with large communities. A study [118] focused on the social acceptance of small-scale RO plants with solar electricity supply in remote areas in Central Australia. This study evaluated the unit's capacity to meet water demands (potable, non-potable domestic, and agriculture demand), the human resources availability to operate and maintain these units, and the attitudes of community members to prototype RO units. The results of interviews, questionnaires, and site visits indicated that these units were generally perceived positively, showing great potential for acceptance by communities. One of the best examples to depict the role of public acceptance is the case of Ngare Nanyuki in Northern Tanzania. Volcanoes in nearby mountains are the potable water supply for this area, which was considered as a symbol of pure water by their society. These natural water resources contain high levels of fluoride, which is toxic. Although a membrane-based desalination unit is producing high quality and healthy water for this region, people are still drinking directly from the toxic water sources [29].

5.1. Role of Culture

The massive deployment of desalination technologies in a society can trigger a cultural change in water consumption. Although water scarcity has accelerated recently, it is not a new phenomenon and in arid parts of the world watering techniques have been adapted to this resource-scarce condition for centuries. Another study [119] refers to the fact that the rise of water consumption in Lanzarote, Morocco, after the introduction of desalination technologies was a result of falling costs of supply and water availability rather than an increase in actual water demand. Hence, countries considering desalination units as their future water supply need to focus on the demand side to conserve traditional, water-saving consciousness developed by the local population and increase the awareness of society regarding the amount of energy and materials that are consumed in the desalination processes. As another example, a community in Greece destroyed solar stills, which were donated [29]. This example reveals that society needs to pay for a share of service costs, even if partially, to increase society's responsibility towards that service.

Another study [120] discussed the sustainable solutions to meeting future water demand in the UAE, including increasing the efficiency of the water sector and diversifying the water supplies (applying wastewater treatment and solar RO plants). Based on this article, treated wastewater is not a viable source for the domestic sector in Islamic countries due to religious beliefs.

5.2. Policy-Making

Decision-makers should consider both the demand side and the supply side to reach more effective roadmaps in the water sector. A study [121] compared the results of different water trajectories for two regions in Australia including South East Queensland and Perth between 2002 and 2014. Both areas encountered a water crisis for some time during this period. The decisions in Perth have been focused on the supply-side, while in South East Queensland, in addition to the supply side, demand-side management was considered in decisions. Another study [122] proposed a framework to evaluate the policies in the water and energy sectors to develop concentrated solar thermal technologies coupled with desalination technologies in South Africa. This study aimed to assist policy-makers in truly understanding the nexus and complexity of the water and energy systems they are attempting to influence. The size of the community also affects the social issues of desalination technologies. The authors in [9] discussed small communities in remote locations as an example. These locations usually rely on transported water and water supply infrastructure does not exist; in consequence, desalination projects face lesser social issues compared to locations with large communities, and in addition, these projects might seem attractive for them. The authors in another study [123] studied policies and legislation within the European Union in the Mediterranean region for developing autonomous desalination units powered by renewables. Policy barriers to decentralized desalination implementation generally stemmed from ignorance regarding these systems when desalination regulations were enacted. These decentralized systems were viewed as large fossil-fuel-based desalination systems with outright opposition in the regulations. As an example, small-scale water abstractions in the region were exempt from both the Water Framework Directive and the Drinking Water Standards Directive, while the decentralized autonomous desalination systems had to comply with strict regulations that apply to all drinking water supplies. Another study done in Turkey [124] indicated that the framework conditions in this country do not pose unnecessary barriers for the implementation of autonomous desalination units.

The authors of [125] first investigated the current and future status of the Jordanian water and energy sectors and then focused on linkages between these sectors. In the next step, this study addressed the stakeholders of both sectors and the solutions to enforce the links between these two sectors in Jordan's future policy-making. The study classified the stakeholders based on power, legitimacy, and urgency and provided solutions to increase the inter-linkages between key actors and groups for achieving sustainable integrated water and energy policy-making. According to Jordan's future energy plan, oil shale and nuclear energy are expected to increase from zero to 21% of total energy supply by 2022. The required water intensity of oil shale extraction is significantly larger than that for conventional oil extraction. Due to insufficient groundwater and surface water resources in the country, seawater desalination and reuse of municipal sewage are the only alternatives.

Water pricing policies need to consider the social and environmental elements above the technical aspect. Heavily subsidized prices can induce changes in user behavior and cause an unexpected increase in irresponsible utilization and inefficient resource use by ignoring the demand-side behavior in policy-making [9].

Table 7. Summary of focal factors in desalination systems powered by renewable resources.

Horizon	Description of the System	Nexus	Approach	Analysis	Uncertainty Level	Geographical Scale	Ref.
Review/ Understanding	• Applying ocean-based energy generation as an energy resource for desalination plant	Technical and Environmental	Optimization	Qualitative	-	-	[23]
	• Sustainable solutions to meet future water demand in the UAE	Social	Discussion	Qualitative	-	National	[120]
	• Investigation the policies regarding to autonomous desalination in the EU	Social	Discussion	Qualitative	-	Regional	[123]
	• Investigation the policies regarding to autonomous desalination in Turkey	Social	Discussion	Qualitative	-	National	[124]
	• Evaluating the social acceptance of RO units powering by solar electricity using surveys	Social	Discussion	Qualitative	-	Rural	[118]
	• Proposing a framework to evaluate water-energy polices	Policy	Discussion	Qualitative	-	Regional	[122]
	• Studying decentralized solar-powered desalination systems in remote regions	Sustainability	Review	Qualitative	-	-	[29]
	• Integrating wave energy converters with desalination technologies for commercialization of wave energy	Technical and Economic	Review	Qualitative	-	Island (remote)	[126]
	• Investigating the potential and development of ocean-based power generation for desalination systems	Sustainability	Review	Qualitative	-	-	[23]
Short-Term/ Operation	• Integration of MSF desalination, solar thermal power, and hydrogen production processes to achieve synergy	Technical	Process simulation	Quantitative	Deterministic	City	[57]
	• Co-locating pumped hydro storage with reverse osmosis desalination plant	Technical and Environmental	Optimization	Quantitative	Deterministic	City	[30]
	• Operating an MED desalination process by ocean energy (thermal energy which is harnessed from seawater temperature gradient)	Technical	Process simulation	Quantitative	Deterministic	Stand-alone	[64]
	• Proposing a tool for operating a reverse electrodialysis system to produce power (salinity gradient power)	Technical	Process simulation	Quantitative	Deterministic	Laboratory scale	[51]
	• Studying optimal climate conditions for operating small-scaled RO desalination units coupled with PV systems	Technical	Experimental	Quantitative	Deterministic	Laboratory scale	[85]
	• Studying the capability of an RO desalination plant to manage the variability of renewable energy production	Technical	Process simulation	Quantitative	Deterministic	City	[59]
	• Studying the performance of a combination of the MED process with solar still desalination powered by solar thermal and waste heat	Technical	Experimental	Quantitative	Deterministic	Laboratory scale	[86]
	• Operating an MSF desalination unit powered with a hybrid energy system including solar, geothermal, and ocean thermal energy	Technical	Process design	Quantitative	Deterministic	-	[127]
	• Modeling the integration of MED unit with solar and geothermal resources	Technical	Process simulation	Quantitative	Deterministic	Island	[40]
	• Introducing an energy management and control system for an RO desalination connected to a DC micro-grid (PV-Battery)	Technical and Economic	Fuzzy optimization	Quantitative	Deterministic	Island	[76]
	• Using concentrating solar power for a MED process and electricity production as a hybrid system	Technical and Economic	Process design	Quantitative	Deterministic	City	[61]
	• Evaluating the optimal operation of an MSF desalination system powered by solar thermal energy	Technical and Economic	Experimental	Quantitative	Deterministic	Laboratory scale	[68]
	• Considering membrane fouling during intermittent operation in designing PV powered RO installations	Technical and Economic	Process simulation	Quantitative	Deterministic	Laboratory scale	[96]
	• Shifting load using desalination demand as a flexible load for increasing the share of renewable resources in the energy system	Technical and Economic	Optimization	Quantitative	Deterministic	Island	[54]
	• Evaluating the potential of water desalination and distribution for load shifting in an off-grid remote energy system	Technical	Linear optimization	Quantitative	Monte-Carlo	Island	[80]

Table 7. *Cont.*

Horizon	Description of the System	Nexus	Approach	Analysis	Uncertainty Level	Geographical Scale	Ref.
Long-Term/ Planning	• Design a cost-effective energy system for small desalination plant	Economic	Optimization	Quantitative	Deterministic	Rural	[49]
	• Coupling PV and CSP with RO and MED plants to minimize the cost and to maximize the RE penetration in an island	Economic	Optimization	Quantitative	Deterministic	Island	[71]
	• Investigating the potential of RO plants to meet future water demand	Economic-Environmental	System dynamics	Quantitative	Deterministic	State	[128]
	• Proposing scenarios to achieve an electricity system with net-zero emission	Economic	LP Optimization	Quantitative	Deterministic	National	[104]
	• Minimizing the total cost and GHG emissions of a hybrid energy system coupled with an RO plant	Economic and Environmental	Multi-object Optimization	Quantitative	Stochastic	-	[43]
	• Evaluating life cycle GHG emissions of different desalination technologies coupling with renewables	Environmental	LCA	Quantitative	Deterministic	-	[115]
	• Evaluating the environmental impacts of different desalination technologies coupling with solar resources	Environmental	LCA	Quantitative	Deterministic	Rural	[63]
	• Considering carbon tax as an external cost of desalination process	Economic	-	Quantitative	Deterministic	National	[105]
	• Identifying regions that are suitable for deployment of RO units coupled with solar energy supplies	Economic and Technical	GIS	Quantitative	Deterministic	Global	[113]
	• Evaluating the potential of wave energy resources to provide the power demand of desalination plants	Environmental and Technical	-	Quantitative	Deterministic	Island	[73]
	• Evaluating the environmental impacts for open-intake pretreatment and subsurface intake pretreatment of RO desalination plants	Environmental	LCA	Quantitative	Deterministic	City	[36]
	• Evaluating the environmental impacts RO desalination plants powered by hybrid renewable energy resources and the electricity grid	Environmental	LCA	Quantitative	Deterministic	City	[50]
	• Studying the scenarios to achieve 100% RE in Iran by considering electricity demand of RO desalination by 2050	Economic	LP Optimization	Quantitative	Deterministic	National	[38]
	• Designing a sustainable desalination system powered with renewable energy resources	Sustainability	AHP	Quantitative	Deterministic	Island	[45]
	• Evaluating the feasibility of exchanging desalinated water with renewable electricity	Technical and Economic	Optimization	Quantitative	Deterministic	Multi-national	[46]
	• Evaluating the GHG emissions of different water sources	Environmental	LCA	Quantitative	Deterministic	City	[18]
	• Investigation on the economic impacts and CO_2 footprint of desalination units	Environmental and Economic	Triple-I light	Quantitative	Deterministic	City	[27]
	• Investigating the role of the desalination sector to achieve a 100 percent renewable energy system in Saudi Arabia	Technical and Economic	linear optimization	Quantitative	Deterministic	National	[39,40]
	• Achieving 100 percent renewable energy in India by considering desalination demand	Technical and Economic	linear optimization	Quantitative	Deterministic	National	[84]
	• Proposing a spatial model to assess potential technical and economically viable site locations for RO desalination facilities powered by renewables	Technical and Economic	Multi-criteria	Quantitative	Deterministic	Regional	[31]
	• Finding the optimal size and configuration of a small-scaled RO desalination unit coupled with a PV system (including battery storage and water storage)	Technical and Economic	Fuzzy Optimization	Quantitative	Deterministic	Island	[74]
	• Technical feasibility of using RO desalination units powered by wave energy as an alternative for imported water	Technical	Optimization	Quantitative	Deterministic	Island	[47]

Table 7. *Cont.*

Horizon	Description of the System	Nexus	Approach	Analysis	Uncertainty Level	Geographical Scale	Ref.
	• Calculating the optimal size of renewable energy supply (wind turbine and PV) for RO desalination units with a solar preheating water system	Technical and Economic	Optimization	Quantitative	Deterministic	Regional	[58]
	• Estimating the cost of providing water demand using renewable-powered RO desalination plants for regions facing water scarcity in 2030	Economic	Linear Optimization	Quantitative	Deterministic	Global	[93]
	• Developing a tool for sizing RO desalination plants powered by renewables units	Economic	Optimization	Quantitative	Deterministic	Island	[48]
	• Investigating the economic feasibility of desalinating agriculture drainage water using the MED process powered by solar thermal resources	Economic	Optimization	Quantitative	Deterministic	Region	[91]
	• Evaluating the technical and economic feasibility of RO desalination units powered by distributed PV-battery systems in Myanmar	Technical and Economic	Optimization	Quantitative	Deterministic	National	[87]
	• Considering desalination energy demand in the transition to a 100 percent renewable system in South and Central America	Technical and Economic	Optimization	Quantitative	Deterministic	Multi-National	[60]
	• Studying the benefits of the integration of RO desalination energy demand in the transition to a 100 percent renewable energy system for India and the South Asian Association for Regional Cooperation	Technical and Economic	Optimization	Quantitative	Deterministic	Multi-National	[129]
	• Minimizing the cost and CO_2 emissions of an energy system including PV, wind turbine, hydrogen electrolyzer, battery, and hydrogen storage coupled with an RO desalination unit	Economic and Environmental	Heuristic optimization	Quantitative	Deterministic	City	[41]
	• Investigating the role of RO desalination demand in the transition to a 100 percent solar electricity system in Pakistan by 2050	Economic	Linear optimization	Quantitative	Deterministic	National	[52]
	• Proposing a dynamic approach to consider the operation of an RO plant in sizing the PV and wind turbine energy system	Technical and Economic	Multi-objective optimization	Quantitative	Deterministic	Island	[79]
	• Forecasting CO_2 emissions from different energy systems providing desalination power demand for an Island by 2020	Environmental	Scenario-based	Quantitative	Deterministic	Island	[44]
	• Estimating the potential amount of desalination water powering with solar and wind electricity in Iran	Technical and Economic	Scenario-based	Quantitative	Deterministic	National	[130]
	• Investigating the technical and economic feasibility of RO units powered by off-grid PV systems in remote case studies in Iran	Technical and Economic	Fuzzy optimization	Quantitative	Deterministic	Rural	[69]
	• Evaluating on-grid decentralized or distributed renewable-powered desalination systems for sustainable water and energy supply planning	Sustainability	Hybrid approach	Quantitative	Deterministic	National	[5]
	• Comparing a centralized water desalination system to a distributed desalination system powered by solar electricity resources in Australia	Technical and Economic	Optimization	Quantitative	Deterministic	National	[55]
		Technical	Environmental	Economic	Social	Qualitative	Quantitative
	Number of studies/ Total studies	38/61	10/61	37/61	10/61	9/61	52/61

6. Results and Discussion

Table 7 depicts the aspects of sustainability that each study investigated. Technical, economic, and environmental aspects of desalination technologies receive considerable attention among studies [23,29,45] compared with the social aspect (only 10 studies).

Integrating desalination plants with renewable resources reduces GHG emissions from desalination facilities, which are currently one of the main environmental impacts of these systems. It also improves the social acceptance of desalination systems. How to consider uncertainty is a critical part of modeling and it gets more important in integrating variable renewable resources with desalination facilities. Table 7 shows that more studies are needed to cover uncertainties in the integrated renewable desalination systems (both supply-side and demand-side).

Concerning the technical aspect, there is a contradiction among studies. Figure 3 reveals that renewable desalination systems are suitable for a small capacity of water production, while many studies are focusing on large desalination capacities in this field of study. It is possible to operate large membrane-based desalination facilities (mainly RO) using the excess electricity production from variable renewables, but studies are limited and the techno-economic performance is still uncertain. Studies have investigated the potential of large RO desalination plants to compensate for the fluctuation of the variable renewable electricity production, which is relatively low compared to small-scaled desalination systems. The compatibility of other desalination technologies (small- to large-scale) needs further investigation.

There is a gap among studies in the investigation of distributed water systems, which are compatible with renewable resources and can operate as flexible loads to integrate with variable renewable electricity production. Only two studies investigated distributed water systems [5,55]. Instead of one centralized desalination unit, these distributed water systems include several small-scale desalination units. Distributed water systems can moderate the effluent-associated (concentrated salinity and thermal pollution) environmental impact of the desalination process by providing multiple options for site locations to distribute the brine discharge (in order to reduce the salinity) and avoid discharging brine into sensitive ecosystems.

Public education and participation in developing desalination projects are essential [9]. Labor costs account for more than 10 percent of the operating costs of desalination processes and this reaches 15 percent for RO plants. A study [131] estimated the urban–rural gap in 101 developed and developing countries using surveys and national labor statistics. The results indicated that average urban wage advantages were 38 percent, half of which cannot be explained by differences in average skills across the markets. Due to size compatibility, the decentralized desalination systems are suitable for deployment in rural areas, thereby mitigating the labor cost of desalination projects. Energy savings from water distribution and transfer are also considerable in the distributed water systems, which cause further cost reductions. Moreover, energy storage systems are responsible for a significant share of renewable energy cost, which is avoidable with the application of desalination units as flexible loads in the regions facing water scarcity and having access to water sources to be desalinated. Besides, the freedom to choose site locations for distributed desalination systems (multiple locations) is larger than that for centralized systems (only one location). Selecting locations in which the feed-water suits the desalination technology better (causing cost reduction, as discussed in previous sections) could mitigate environmental impacts, namely intake-related impacts and effluent-related impacts. Furthermore, presenting such multiple location alternatives allows the policy-makers to allocate desalination to other target users apart from potable water. It is notable that potable water has been the main target product for studies and there is no estimation for the cost of water produced for other purposes. As an example, water containing high levels of salt and boron, even after first-pass RO water production, needs to pass through an additional RO stage to achieve good-quality water that is suitable for drinking. The second-pass stage can increase the total cost of the first-pass desalination process by 10 to 25 percent. For most industrial and agricultural applications, it is unnecessary to design this second stage pass. The current cost of desalination product water includes the second-pass stage, which is responsible for up to 25 percent of total RO water desalination costs.

Conventional thermal desalination technologies are now well-proven and mature. Thus, further improvement in these technologies is relatively limited. Continuous innovation in RO desalination technology in the last twenty years has reduced the energy consumption per unit of product water to 1.8 KWh/m^3, which is close to the theoretical minimum required energy for seawater desalination. It means a further large reduction in energy consumption is not expected for desalination technologies. However, further significant advances in renewable energy technologies in the future are highly likely to decrease the water cost of renewable desalination. Besides, future advances in membrane technology (increasing water productivity per area) are predicted, which could cause up to a 20 percent cost reduction in the next five years. Experimental studies [132,133] showed that CDI technology is effective and reliable for brackish water treatment in remote areas in Australia. Advances in electrode materials are crucial to developing CDI technology in the future for decentralized brackish water treatment [134]; in the meantime, further investigation is required for the integration of this technology with renewables.

Considering all of these cost reductions and developments, desalination is an expensive water supply compared to other conventional fresh-water resources and needs heavy national investments and subsidies from governments for the next few decades. Policy-makers and governments should consider social equity in designing plans and budgeting as a social aspect of desalination projects. Above technical and environmental aspects, different potential target users and system configurations (centralized or decentralized) should be investigated based on equity, wealth distribution, and social acceptance. Furthermore, desalination is economically and politically important to achieving self-reliance for specific regions, such as the Middle East and Singapore. When achieving this goal, it is important to examine the resilience of desalination systems further. For instance, in the case of technical failure or war, the centralized desalination systems and potable desalination units are more vulnerable, while the distributed desalination systems and agriculture users are more resilient compared to domestic users. Table 8 compares the distributed and centralized desalination systems powered by renewable energy resources.

Table 8. Comparison between distributed and centralized desalination systems powered by renewable energy resources.

Sustainability Aspect	Decentralized Desalination System	Centralized Desalination System
Technical	• The technology is under development • High potential to compensate the fluctuation due to VRE • High reliability	• The technology is mature • Needs energy storage to operate with VRE
Economic	• High capital cost • Low maintenance cost • Saving cost from water distribution and transfer	• Low capital cost • High maintenance cost
Environmental	• Less thermal pollution • No need for diffuser to decrease the salinity of effluent • Multiple site locations to decrease the environmental impacts • Difficult to monitor the effluent (regulations)	• Impacts on marine ecosystem due to thermal pollution • Need for diffuser to decrease the salinity of effluent • Producing water in one location • It is practical to monitor effluent (regulations)
Social	• High reliability and security • Wealth distribution and equity	• Has less of a financial burden on society

7. Conclusions

The share of desalination in the fresh-water supply is increasing around the world. Currently, desalination facilities are mainly fossil-fuel-based, which has environmental impacts due to emissions. This review study sheds light on the potential of renewable energy resources to integrate with desalination technologies for making the whole system more sustainable. Renewable desalination systems are more suitable to produce water on a small scale; in consequence these systems need to be studied in the same scope. Decentralized or distributed renewable desalination systems (mainly membrane-based desalination technologies that are consuming electricity) are promising future systems. These systems have the potential to compensate for the fluctuating power production of variable renewables, to reduce GHG emissions, and to mitigate effluent-associated environmental issues, and are more reliable and secure. Ongoing progress in renewable desalination technologies will decrease the cost of renewable energy generation and water production. Further cost reduction is expected from targeting other sectors such as agriculture, and from distributed water systems, which need less energy for water distribution and transfer and labor costs, and benefit from desalination units as flexible loads in energy systems. Moreover, integrating renewables with desalination systems has positive impacts on the social acceptance of these systems. Considering the social aspect and demand-side in the decision-making process makes the transition models more realistic and effective towards sustainability.

Author Contributions: E.A. designed, conceptualized, and conducted the review, analyzed the results, visualized, and wrote the manuscript. B.M. reviewed, edited, and visualized. B.M., B.M.-I., and T.T. gave guidance, provided the materials, and helped to improve the quality of the work. All authors have read and agreed to the published version of the manuscript.

Funding: This research received no external funding.

Conflicts of Interest: The authors declare no conflict of interest.

References

1.   Mekonnen, M.M.; Hoekstra, A.Y. Four billion people facing severe water scarcity. *Sci. Adv.* **2016**, *2*, e1500323. [CrossRef] [PubMed]
2.   United Nations Environment Programme (UNEP). Options for Decoupling Economic Growth from Water Use and Water Pollution. 2017. Available online: https://www.resourcepanel.org/reports/options-decoupling-economic-growth-water-use-and-water-pollution (accessed on 7 May 2020).
3.   Katz, D. Book Review: "Let There Be Water: Israel's Solution for a Water-Starved World". *Water Econ. Policy* **2017**, *3*, 4. [CrossRef]
4.   World Bank. *The Role of Desalination in an Increasingly Water-Scarce World*; Water Papers; World Bank: Washington, DC, USA, 2019. [CrossRef]
5.   Ahmadi, E.; McLellan, B.; Ogata, S.; Mohammadi-Ivatloo, B.; Tezuka, T. An Integrated Planning Framework for Sustainable Water and Energy Supply. *Sustainability* **2020**, *12*, 4295. [CrossRef]
6.   Aliewi, A.; El-Sayed, E.; Akbar, A.; Hadi, K.; Al-Rashed, M. Evaluation of desalination and other strategic management options using multi-criteria decision analysis in Kuwait. *Desalination* **2017**, *413*, 40–51. [CrossRef]
7.   ALMAR Water Solution. Desalination Technologies and Economics: CAPEX, OPEX & Technological Game Changers to Come. 2017. Available online: https://www.cmimarseille.org/knowledge-library/desalination-technologies-and-economics-capex-opex-technological-game-changers-0 (accessed on 15 June 2020).
8.   Khan, J.; Arsalan, M.H. Solar power technologies for sustainable electricity generation: A review. *Renew. Sustain. Energy Rev.* **2016**, *55*, 414–425. [CrossRef]
9.   Gude, V. Desalination and sustainability—An appraisal and current perspective. *Water Res.* **2016**, *89*, 87–106. [CrossRef]
10. Alkaisi, A.; Mossad, R.; Sharifian-Barforoush, A. A Review of the Water Desalination Systems Integrated with Renewable Energy. *Energy Procedia* **2017**, *110*, 268–274. [CrossRef]
11. International Energy Agency. *World Energy Outlook 2016*; Organisation for Economic Co-Operation and Development OECD: Paris, France, 2016; p. 684, [CrossRef]
12. Pakdel, M.J.V.; Sohrabi, F.; Mohammadi-Ivatloo, B. Multi-objective optimization of energy and water management in networked hubs considering transactive energy. *J. Clean. Prod.* **2020**, *2020*, 121936. [CrossRef]
13. Kılkış, Ş. Sustainable development of energy, water and environment systems index for Southeast European cities. *J. Clean. Prod.* **2016**, *130*, 222–234. [CrossRef]
14. Cai, Y.; Cai, J.; Xu, L.; Tan, Q.; Xu, Q. Integrated risk analysis of water–energy nexus systems based on systems dynamics, orthogonal design and copula analysis. *Renew. Sustain. Energy Rev.* **2019**, *99*, 125–137. [CrossRef]

15. Xie, X.; Jia, B.; Han, G.; Wu, S.; Dai, J.; Weinberg, J. A historical data analysis of water–energy nexus in the past 30 years urbanization of Wuxi city, China. *Environ. Prog. Sustain. Energy* **2017**, *37*, 46–55. [CrossRef]

16. Stokes-Draut, J.; Taptich, M.; Kavvada, O.; Horvath, A. Evaluating the electricity intensity of evolving water supply mixes: The case of California's water network. *Environ. Res. Lett.* **2017**, *12*, 11400. [CrossRef]

17. Sperling, J.B.; Ramaswami, A. Cities and budget-based management of the energy-water-climate nexus: Case studies in transportation policy, infrastructure systems, and urban utility risk management. *Environ. Prog. Sustain. Energy* **2015**, *37*, 91–107. [CrossRef]

18. Stokes, J.; Horvath, A. Energy and air emission effects of water supply. *Environ. Sci. Technol.* **2009**, *43*, 2680–2687. [CrossRef]

19. Valek, A.M.; Sušnik, J.; Grafakos, S. Quantification of the urban water–energy nexus in Mexico City, Mexico, with an assessment of water-system related carbon emissions. *Sci. Total Environ.* **2017**, *590–591*, 258–268. [CrossRef]

20. Bolwig, S.; Bazbauers, G.; Klitkou, A.; Lund, P.D.; Blumberga, A.; Gravelsins, A.; Blumberga, D. Review of modelling energy transitions pathways with application to energy system flexibility. *Renew. Sustain. Energy Rev.* **2019**, *101*, 440–452. [CrossRef]

21. Cherp, A.; Vinichenko, V.; Jewell, J.; Brutschin, E.; Sovacool, B. Integrating techno-economic, socio-technical and political perspectives on national energy transitions: A meta-theoretical framework. *Energy Res. Soc. Sci.* **2018**, *37*, 175–190. [CrossRef]

22. Stercke, S.D.; Mijic, A.; Buytaert, W.; Chaturvedi, V. Modelling the dynamic interactions between London's water and energy systems from an end-use perspective. *Appl. Energy* **2018**, *230*, 615–626. [CrossRef]

23. Li, Z.; Siddiqi, A.; Anadon, L.D.; Narayanamurti, V. Towards sustainability in water–energy nexus: Ocean energy for seawater desalination. *Renew. Sustain. Energy Rev.* **2018**, *82*, 3833–3847. [CrossRef]

24. Kalogirou, S. Seawater desalination using renewable energy sources. *Prog. Energy Combust. Sci.* **2005**, *31*, 242–281. [CrossRef]

25. Zhang, W.; Mossad, M.; Yazdi, J.S.; Zou, L. A statistical experimental investigation on arsenic removal using capacitive deionization. *Desalin. Water Treat.* **2016**, *57*, 3254–3260. [CrossRef]

26. Zhang, W.; Jia, B. Toward anti-fouling capacitive deionization by using visible-light reduced TiO_2/graphene nanocomposites. *MRS Commun.* **2015**, *5*, 613–617. [CrossRef]

27. Tokui, Y.; Moriguchi, H.; Nishi, Y. Comprehensive environmental assessment of seawater desalination plants: Multistage flash distillation and reverse osmosis membrane types in Saudi Arabia. *Desalination* **2014**, *351*, 145–150. [CrossRef]

28. Vakilifard, N.; Anda, M.; Bahri, P.A.; Ho, G. The role of water–energy nexus in optimising water supply systems: Review of techniques and approaches. *Renew. Sustain. Energy Rev.* **2018**, *82*, 1424–1432. [CrossRef]

29. Kharraz, J.; Richards, B.; Schafer, A. Autonomous Solar-Powered Desalination Systems for Remote Communities. In *Desalination Sustainability: A Technical, Socioeconomic, and Environmental Approach*; Elsevier: Amsterdam, The Netherlands, 2017; pp. 75–125. [CrossRef]

30. Slocum, A.H.; Haji, M.N.; Trimble, A.Z.; Ferrara, M.; Ghaemsaidi, S.J. Integrated Pumped Hydro Reverse Osmosis systems. *Sustain. Energy Technol. Assess.* **2016**, *18*, 80–99. [CrossRef]

31. Aminfard, S.; Davidson, F.; Webber, M. Multi-layered spatial methodology for assessing the technical and economic viability of using renewable energy to power brackish groundwater desalination. *Desalination* **2019**, *450*, 12–20. [CrossRef]

32. Ramos, A.; Chatzopoulou, M.A.; Guarracino, I.; Freeman, J.; Markides, C.N. Hybrid photovoltaic-thermal solar systems for combined heating, cooling and power provision in the urban environment. *Energy Convers. Manag.* **2017**, *150*, 838–850. [CrossRef]

33. Kang, C.-N.; Cho, S.-H. Thermal and Electrical Energy Mix Optimization(EMO) Method for Real Large-scaled Residential Town Plan. *J. Electr. Eng. Technol.* **2018**, *13*, 513–520.

34. Tarroja, B.; Chiang, F.; AghaKouchak, A.; Samuelsen, S. Assessing future water resource constraints on thermally based renewable energy resources in California. *Appl. Energy* **2018**, *226*, 49–60. [CrossRef]

35. Uche, J.; Acevedo, L.; Cirez, F.; Uson, S.; Martinez-Gracia, A.; Bayod-Rujula, A. Analysis of a domestic trigeneration scheme with hybrid renewable energy sources and desalting techniques. *J. Clean. Prod.* **2019**, *212*, 1409–1422. [CrossRef]

36. Al-Kaabi, A.; Mackey, H. Environmental assessment of intake alternatives for seawater reverse osmosis in the Arabian Gulf. *J. Environ. Manag.* **2019**, *242*, 22–30. [CrossRef] [PubMed]

37. Birge, D.; Berger, A. Transitioning to low-carbon suburbs in hot-arid regions: A case-study of Emirati villas in Abu Dhabi. *Build. Environ.* **2019**, *147*, 77–96. [CrossRef]

38. Ghorbani, N.; Aghahosseini, A.; Breyer, C. Transition towards a 100% Renewable Energy System and the Role of Storage Technologies: A Case Study of Iran. *Energy Procedia* **2017**, *135*, 23–36. [CrossRef]

39. Caldera, U.; Bogdanov, D.; Afanasyeva, S.; Breyer, C. Role of Seawater Desalination in the Management of an Integrated Water and 100% Renewable Energy Based Power Sector in Saudi Arabia. *Water* **2018**, *10*, 3. [CrossRef]

40. Caldera, U.; Breyer, C. Impact of Battery and Water Storage on the Transition to an Integrated 100% Renewable Energy Power System for Saudi Arabia. *Energy Procedia* **2017**, *135*, 126–142. [CrossRef]

41. Abdelshafy, A.M.; Hassan, H.; Jurasz, J. Optimal design of a grid-connected desalination plant powered by renewable energy resources using a hybrid PSO-GWO approach. *Energy Convers. Manag.* **2018**, *173*, 331–347. [CrossRef]

42. Salama, L.; Abdalla, K. Design and analysis of a solar photovoltaic powered seawater reverse osmosis plant in the southern region of the gaza strip. *Desalin. Water Treat.* **2019**, *143*, 96–101. [CrossRef]

43. Li, Q.; Loy-Benitez, J.; Nam, K.; Hwangbo, S.; Rashidi, J.; Yoo, C. Sustainable and reliable design of reverse osmosis desalination with hybrid renewable energy systems through supply chain forecasting using recurrent neural networks. *Energy* **2019**, *178*, 277–292. [CrossRef]

44. Jaime Sadhwani, J.; Sagaseta de Ilurdoz, M. Primary energy consumption in desalination: The case of Gran Canaria. *Desalination* **2019**, *452*, 219–229. [CrossRef]

45. Marini, M.; Palomba, C.; Rizzi, P.; Casti, E.; Marcia, A.; Paderi, M. A multicriteria analysis method as decision-making tool for sustainable desalination: The asinara island case study. *Desalin. Water Treat.* **2017**, *61*, 274–283. [CrossRef]

46. Katz, D.; Shafran, A. Transboundary exchanges of renewable energy and desalinatedwater in the Middle East. *Energies* **2019**, *12*, 1455. [CrossRef]

47. Corsini, A.; Tortora, E.; Cima, E. Preliminary assessment of wave energy use in an off-grid minor island desalination plant. *Energy Procedia* **2015**, *82*, 789–796. [CrossRef]

48. Mentis, D.; Karalis, G.; Zervos, A.; Howells, M.; Taliotis, C.; Bazilian, M.; Rogner, H. Desalination using renewable energy sources on the arid islands of South Aegean Sea. *Energy* **2016**, *94*, 262–272. [CrossRef]

49. Fornarelli, R.; Shahnia, F.; Anda, M.; Bahri, P.A.; Ho, G. Selecting an economically suitable and sustainable solution for a renewable energy-powered water desalination system: A rural Australian case study. *Desalination* **2018**, *435*, 128–139. [CrossRef]

50. Shahabi, M.; McHugh, A.; Anda, M.; Ho, G. Environmental life cycle assessment of seawater reverse osmosis desalination plant powered by renewable energy. *Renew. Energy* **2014**, *67*, 53–58. [CrossRef]

51. Nagaraj, R.; Murthy, D.; Rajput, M. Modeling Renewables Based Hybrid Power System with Desalination Plant Load Using Neural Networks. *Distrib. Gener. Altern. Energy J.* **2019**, *34*, 32–46. [CrossRef]

52. Sadiqa, A.; Gulagi, A.; Breyer, C. Energy transition roadmap towards 100% renewable energy and role of storage technologies for Pakistan by 2050. *Energy* **2018**, *147*, 518–533. [CrossRef]

53. Khan, M.; Rehman, S.; Al-Sulaiman, F. A hybrid renewable energy system as a potential energy source for water desalination using reverse osmosis: A review. *Renew. Sustain. Energy Rev.* **2018**, *97*, 456–477. [CrossRef]

54. Hamilton, J.; Negnevitsky, M.; Wang, X.; Lyden, S. High penetration renewable generation within Australian isolated and remote power systems. *Energy* **2019**, *168*, 684–692. [CrossRef]

55. Vakilifard, N.; Bahri, P.A.; Anda, M.; Ho, G. An interactive planning model for sustainable urban water and energy supply. *Appl. Energy* **2019**, *235*, 332–345. [CrossRef]

56. Cavalcante, R., Jr.; Freitas, M.; da Silva, N.; de Azevedo Filho, F. Sustainable groundwater exploitation aiming at the reduction of water vulnerability in the Brazilian semi-arid region. *Energies* **2019**, *12*, 904. [CrossRef]

57. Gencer, E.; Agrawal, R. Toward supplying food, energy, and water demand: Integrated solar desalination process synthesis with power and hydrogen coproduction. *Resour. Conserv. Recycl.* **2018**, *133*, 331–342. [CrossRef]

58. Gold, G.; Webber, M. The energy-water nexus: An analysis and comparison of various configurations integrating desalination with renewable power. *Resources* **2015**, *4*, 227–276. [CrossRef]

59. Kim, J.; Chen, J.; Garcia, H. Modeling, control, and dynamic performance analysis of a reverse osmosis desalination plant integrated within hybrid energy systems. *Energy* **2016**, *112*, 52–66. [CrossRef]

60. De Barbosa, L.; Bogdanov, D.; Vainikka, P.; Breyer, C. Hydro, wind and solar power as a base for a 100% renewable energy supply for South and Central America. *PLoS ONE* **2017**, *12*, e0173820. [CrossRef] [PubMed]

61. Mata-Torres, C.; Escobar, R.; Cardemil, J.; Simsek, Y.; Matute, J. Solar polygeneration for electricity production and desalination: Case studies in Venezuela and northern Chile. *Renew. Energy* **2017**, *101*, 387–398. [CrossRef]

62. Aghahosseini, A.; Bogdanov, D.; Barbosa, L.; Breyer, C. Analysing the feasibility of powering the Americas with renewable energy and inter-regional grid interconnections by 2030. *Renew. Sustain. Energy Rev.* **2019**, *105*, 187–205. [CrossRef]

63. Jijakli, K.; Arafat, H.; Kennedy, S.; Mande, P.; Theeyattuparampil, V. How green solar desalination really is? Environmental assessment using life-cycle analysis (LCA) approach. *Desalination* **2012**, *287*, 123–131. [CrossRef]

64. Ng, K.C.; Shahzad, M.W. Sustainable desalination using ocean thermocline energy. *Renew. Sustain. Energy Rev.* **2018**, *82*, 240–246. [CrossRef]

65. Heidary, B.; Hashjin, T.; Ghobadian, B.; Roshandel, R. Optimal integration of small scale hybrid solar wind RO-MSF desalination system. *Renew. Energy Focus* **2018**, *27*, 120–134. [CrossRef]

66. Maleki, A. Design and optimization of autonomous solar-wind-reverse osmosis desalination systems coupling battery and hydrogen energy storage by an improved bee algorithm. *Desalination* **2018**, *435*, 221–234. [CrossRef]

67. Peng, W.; Maleki, A.; Rosen, M.A.; Azarikhah, P. Optimization of a hybrid system for solar-wind-based water desalination by reverse osmosis: Comparison of approaches. *Desalination* **2018**, *442*, 16–31. [CrossRef]

68. Darawsheh, I.; Islam, M.; Banat, F. Experimental characterization of a solar powered MSF desalination process performance. *Therm. Sci. Eng. Prog.* **2019**, *10*, 154–162. [CrossRef]

69. Mostafaeipour, A.; Qolipour, M.; Rezaei, M.; Babaee-Tirkolaee, E. Investigation of off-grid photovoltaic systems for a reverse osmosis desalination system: A case study. *Desalination* **2019**, *454*, 91–103. [CrossRef]

70. Rezk, H.; Sayed, E.; Al-Dhaifallah, M.; Obaid, M.; El-Sayed, A.; Abdelkareem, M.; Olabi, A. Fuel cell as an effective energy storage in reverse osmosis desalination plant powered by photovoltaic system. *Energy* **2019**, *175*, 423–433. [CrossRef]

71. Astolfi, M.; Mazzola, S.; Silva, P.; Macchi, E. A synergic integration of desalination and solar energy systems in stand-alone microgrids. *Desalination* **2017**, *419*, 169–180. [CrossRef]

72. Fernandez-Gonzalez, C.; Dominguez-Ramos, A.; Ibanez, R.; Irabien, A. Sustainability assessment of electrodialysis powered by photovoltaic solar energy for freshwater production. *Renew. Sustain. Energy Rev.* **2015**, *47*, 604–615. [CrossRef]

73. Fernandez Prieto, L.; Rodriguez Rodriguez, G.; Schallenberg Rodriguez, J. Wave energy to power a desalination plant in the north of Gran Canaria Island: Wave resource, socioeconomic and environmental assessment. *J. Environ. Manag.* **2019**, *231*, 546–551. [CrossRef]

74. Karavas, C.S.; Arvanitis, K.; Papadakis, G. Optimal technical and economic configuration of photovoltaic powered reverse osmosis desalination systems operating in autonomous mode. *Desalination* **2019**, *466*, 97–106.

75. Calise, F.; Macaluso, A.; Piacentino, A.; Vanoli, L. A novel hybrid polygeneration system supplying energy and desalinated water by renewable sources in Pantelleria Island. *Energy* **2017**, *137*, 1086–1106. [CrossRef]

76. Kyriakarakos, G.; Dounis, A.; Arvanitis, K.; Papadakis, G. Design of a Fuzzy Cognitive Maps variable-load energy management system for autonomous PV-reverse osmosis desalination systems: A simulation survey. *Appl. Energy* **2017**, *187*, 575–584. [CrossRef]

77. Li, Q.; Moya, W.; Janghorban Esfahani, I.; Rashidi, J.; Yoo, C. Integration of reverse osmosis desalination with hybrid renewable energy sources and battery storage using electricity supply and demand-driven power pinch analysis. *Process. Saf. Environ. Prot.* **2017**, *111*, 795–809. [CrossRef]

78. Kofinas, P.; Dounis, A.I.; Vouros, G.A. Fuzzy Q-Learning for multi-agent decentralized energy management in microgrids. *Appl. Energy* **2018**, *219*, 53–67. [CrossRef]

79. Giudici, F.; Castelletti, A.; Garofalo, E.; Giuliani, M.; Maier, H. Dynamic, multi-objective optimal design and operation of water–energy systems for small, off-grid islands. *Appl. Energy* **2019**, *250*, 605–616. [CrossRef]

80. Meschede, H. Increased utilisation of renewable energies through demand response in the water supply sector—A case study. *Energy* **2019**, *175*, 810–817. [CrossRef]

81. Padron, I.; Avila, D.; Marichal, G.; Rodriguez, J. Assessment of Hybrid Renewable Energy Systems to supplied energy to Autonomous Desalination Systems in two islands of the Canary Archipelago. *Renew. Sustain. Energy Rev.* **2019**, *101*, 221–230. [CrossRef]

82. Trapanese, M.; Frazitta, V. Desalination in small islands: The case study of Lampedusa (Italy). In Proceedings of the OCEANS 2018 MTS/IEEE Charleston, Charleston, SC, USA, 22–25 October 2019. [CrossRef]

83. El-Kady, M.; El-Shibini, F. Desalination in Egypt and the future application in supplementary irrigation. *Desalination* **2001**, *136*, 63–72. [CrossRef]

84. Gulagi, A.; Bogdanov, D.; Breyer, C. The role of storage technologies in energy transition pathways towards achieving a fully sustainable energy system for India. *J. Energy Storage* **2018**, *17*, 525–539. [CrossRef]

85. Alghoul, M.; Poovanaesvaran, P.; Mohammed, M.; Fadhil, A.; Muftah, A.; Alkilani, M.; Sopian, K. Design and experimental performance of brackish water reverse osmosis desalination unit powered by 2 kW photovoltaic system. *Renew. Energy* **2016**, *93*, 101–114. [CrossRef]

86. Park, C.D.; Lim, B.J.; Chung, K.Y.; Lee, S.S.; Kim, Y.M. Experimental evaluation of hybrid solar still using waste heat. *Desalination* **2016**, *379*, 1–9. [CrossRef]

87. Thompson, M.; Baker, R.; Yong, N. Technical and economic evaluation of an off-grid solar desalination system in Myanmar. *J. Water Supply: Res. Technol.-AQUA* **2016**, *65*, 354–360. [CrossRef]

88. Gokcek, M. Integration of hybrid power (wind-photovoltaic-diesel-battery) and seawater reverse osmosis systems for small-scale desalination applications. *Desalination* **2018**, *435*, 210–220. [CrossRef]

89. Liu, W.; Wang, D.; Yu, X.; Wang, W.; Lan, Y.; Wang, X.; Yu, J. Multi-objective planning research on micro energy network considering desalination. *Energy Procedia* **2019**, *158*, 6502–6507. [CrossRef]

90. Ye, B.; Jiang, J.; Cang, Y. Technical and economic feasibility analysis of an energy and fresh water coupling model for an isolated island. *Energy Procedia* **2019**, *158*, 6373–6377. [CrossRef]

91. Stuber, M. Optimal design of fossil-solar hybrid thermal desalination for saline agricultural drainage water reuse. *Renew. Energy* **2016**, *89*, 552–563. [CrossRef]

92. Elimelech, M.; Phillip, W.A. The Future of Seawater Desalination: Energy, Technology, and the Environment. *Science* **2011**, *333*, 712–717. [CrossRef]

93. Caldera, U.; Bogdanov, D.; Breyer, C. Local cost of seawater RO desalination based on solar PV and wind energy: A global estimate. *Desalination* **2016**, *385*, 207–216. [CrossRef]

94. Ahmadi, E.; McLellan, B.; Ogata, S.; Tezuka, T. Modelling the water–energy-nexus to assist the design of economic and regulatory support instruments towards sustainability. In Proceedings of the Chemeca 2019: Chemical Engineering Megatrends and Elements, Sydney, NSW, Australia, 29 September–2 October 2019; p. 550.

95. Freire-Gormaly, M.; Bilton, A.M. Experimental quantification of the effect of intermittent operation on membrane performance of solar powered reverse osmosis desalination systems. *Desalination* **2018**, *435*, 188–197. [CrossRef]

96. Freire-Gormaly, M.; Bilton, A. Design of photovoltaic powered reverse osmosis desalination systems considering membrane fouling caused by intermittent operation. *Renew. Energy* **2019**, *135*, 108–121. [CrossRef]

97. Xue, Y.; Ge, Z.; Yang, L.; Du, X. Peak shaving performance of coal-fired power generating unit integrated with multi-effect distillation seawater desalination. *Appl. Energy* **2019**, *250*, 175–184. [CrossRef]

98. Gude, V. Geothermal source potential for water desalination—Current status and future perspective. *Renew. Sustain. Energy Rev.* **2016**, *57*, 1038–1065. [CrossRef]

99. Giwa, A.; Akther, N.; Housani, A.; Haris, S.; Hasan, S. Recent advances in humidification dehumidification (HDH) desalination processes: Improved designs and productivity. *Renew. Sustain. Energy Rev.* **2016**, *57*, 929–944. [CrossRef]

100. Voutchkov, N. *Desalination Project Cost Estimating and Management*; CRC Press: Boca Raton, FL, USA, 2018.

101. Gopi, G.; Arthanareeswaran, G.; Ismail, A.F. Perspective of renewable desalination by using membrane distillation. *Chem. Eng. Res. Des.* **2019**, *144*, 520–537. [CrossRef]

102. Gude, V. Energy storage for desalination processes powered by renewable energy and waste heat sources. *Appl. Energy* **2015**, *137*, 877–898. [CrossRef]

103. Bhojwani, S.; Topolski, K.; Mukherjee, R.; Sengupta, D.; El-Halwagi, M. Technology review and data analysis for cost assessment of water treatment systems. *Sci. Total Environ.* **2019**, *651*, 2749–2761. [CrossRef]

104. Solomon, A.; Bogdanov, D.; Breyer, C. Solar driven net zero emission electricity supply with negligible carbon cost: Israel as a case study for Sun Belt countries. *Energy* **2018**, *155*, 87–104. [CrossRef]

105. Molinos-Senante, M.; González, D. Evaluation of the economics of desalination by integrating greenhouse gas emission costs: An empirical application for Chile. *Renew. Energy* **2019**, *133*, 1327–1337. [CrossRef]

106. Zhou, Y.; Tol, R.S.J. Evaluating the costs of desalination and water transport. *Water Resour. Res.* **2005**, *41*, 10. [CrossRef]

107. Gude, G. *Renewable Energy Powered Desalination Handbook: Application and Thermodynamics*; Butterworth-Heinemann: Oxford, UK, 2018.

108. World Bank. *Beyond Scarcity: Water Security in the Middle East and North Africa*; World Bank Group: Washington, DC, USA, 2017.

109. Negewo, B.D. *Renewable Energy Desalination: An Emerging Solution to Close the Water Gap in the Middle East and North Africa*; World Bank Publications: Washington, DC, USA, 2012.

110. Parrillo, V.N. *Encyclopedia of Social Problems*; Sage publications: Southend Oaks, CA, USA, 2008.

111. Rodriquez, M. California Environmental Protection Agency, Water Quality Control Plan Ocean Waters of California. 2015. Available online: https://www.waterboards.ca.gov/water_issues/programs/ocean/docs/cop2015.pdf (accessed on 7 May 2020).

112. Meerganz von Medeazza, G. "Direct" and socially-induced environmental impacts of desalination. *Desalination* **2005**, *185*, 57–70. [CrossRef]

113. Grubert, E.; Stillwell, A.; Webber, M. Where does solar-aided seawater desalination make sense? A method for identifying sustainable sites. *Desalination* **2014**, *339*, 10–17. [CrossRef]

114. Van der Merwe, R.; Lattemann, S.; Amy, G. A review of environmental governance and its effects on concentrate discharge from desalination plants in the Kingdom of Saudi Arabia. *Desalin. Water Treat.* **2013**, *51*, 262–272. [CrossRef]

115. Raluy, R.; Serra, L.; Uche, J. Life cycle assessment of desalination technologies integrated with renewable energies. *Desalination* **2005**, *183*, 81–93. [CrossRef]

116. Fang, A.J.; Newell, J.P.; Cousins, J.J. The energy and emissions footprint of water supply for Southern California. *Environ. Res. Lett.* **2015**, *10*, 114002. [CrossRef]

117. Chhipi-Shrestha, G.; Hewage, K.; Sadiq, R. water–energy-Carbon Nexus Modeling for Urban Water Systems: System Dynamics Approach. *J. Water Resour. Plan. Manag.* **2017**, *143*, 04017016. [CrossRef]

118. Werner, M.; Schafer, A. Social aspects of a solar-powered desalination unit for remote Australian communities. *Desalination* **2007**, *203*, 375–393. [CrossRef]

119. Von Medeazza, G. Water desalination as a long-term sustainable solution to alleviate global freshwater scarcity? A North-South approach. *Desalination* **2004**, *169*, 287–301. [CrossRef]

120. Giwa, A.; Dindi, A. An investigation of the feasibility of proposed solutions for water sustainability and security in water-stressed environment. *J. Clean. Prod.* **2017**, *165*, 721–733. [CrossRef]

121. Lam, K.L.; Lant, P.A.; O'Brien, K.R.; Kenway, S.J. Comparison of water–energy trajectories of two major regions experiencing water shortage. *J. Environ. Manag.* **2016**, *181*, 403–412. [CrossRef]

122. Brent, A.; Mokheseng, M.; Amigun, B.; Tazvinga, H.; Musango, J. Systems dynamics modelling to assess the sustainability of renewable energy technologies in developing countries. *WIT Trans. Ecol. Environ.* **2011**, *143*, 13–24. [CrossRef]

123. Gibbons, J.; Papapetrou, M.; Epp, C. Assessment of EU policy: Implications for the implementation of autonomous desalination units powered by renewable resources in the Mediterranean region. *Desalination* **2008**, *220*, 422–430. [CrossRef]

124. Sozen, S.; Teksoy, S.; Papapetrou, M. Assessment of institutional and policy conditions in Turkey: Implications for the implementation of autonomous desalination systems. *Desalination* **2008**, *220*, 441–454. [CrossRef]

125. Siddiqi, A.; Kajenthira, A.; Anadon, L.D. Bridging decision networks for integrated water and energy planning. *Energy Strategy Rev.* **2013**, *2*, 46–58. [CrossRef]

126. Foteinis, S.; Tsoutsos, T. Strategies to improve sustainability and offset the initial high capital expenditure of wave energy converters (WECs). *Renew. Sustain. Energy Rev.* **2017**, *70*, 775–785. [CrossRef]

127. Azhar, M.; Rizvi, G.; Dincer, I. Integration of renewable energy based multigeneration system with desalination. *Desalination* **2017**, *404*, 72–78. [CrossRef]

128. Sahin, O.; Siems, R.; Richards, R.; Helfer, F.; Stewart, R. Examining the potential for energy-positive bulk-water infrastructure to provide long-term urban water security: A systems approach. *J. Clean. Prod.* **2017**, *143*, 557–566. [CrossRef]

129. Gulagi, A.; Choudhary, P.; Bogdanov, D.; Breyer, C. Electricity system based on 100% renewable energy for India and SAARC. *PLoS ONE* **2017**, *12*, e0180611. [CrossRef]

130. Mollahosseini, A.; Abdelrasoul, A.; Sheibany, S.; Amini, M.; Salestan, S. Renewable energy-driven desalination opportunities—A case study. *J. Environ. Manag.* **2019**, *239*, 187–197. [CrossRef]
131. Artz, G.M.; Hoque, M.; Orazem, P.F.; Shah, U. Urban-Rural Wage Gaps, Inefficient Labor Allocations, and GDP per Capita, Iowa State University Digital Repository. 2016. Available online: https://lib.dr.iastate.edu/cgi/viewcontent.cgi?article=1006&context=econ_workingpapers (accessed on 24 June 2020).
132. Mossad, M.; Zhang, W.; Zou, L. Using capacitive deionisation for inland brackish groundwater desalination in a remote location. *Desalination* **2013**, *308*, 154–160. [CrossRef]
133. Zhang, W.; Mossad, M.; Zou, L. A study of the long-term operation of capacitive deionisation in inland brackish water desalination. *Desalination* **2013**, *320*, 80–85. [CrossRef]
134. Jia, B.; Zhang, W. Preparation and application of electrodes in capacitive deionization (CDI): A state-of-art review. *Nanoscale Res. Lett.* **2016**, *11*, 64. [CrossRef]

© 2020 by the authors. Licensee MDPI, Basel, Switzerland. This article is an open access article distributed under the terms and conditions of the Creative Commons Attribution (CC BY) license (http://creativecommons.org/licenses/by/4.0/).

Article

The Impact of Government Subsidy on Renewable Microgrid Investment Considering Double Externalities

Deng Xu and Yong Long *

School of Economics and Business Administration, Chongqing University, Chongqing 400030, China;
xudeng@cqu.edu.cn
* Correspondence: Dr_ylong@yeah.net

Received: 13 May 2019; Accepted: 3 June 2019; Published: 5 June 2019

Abstract: Since microgrids require public support to make economic sense, governments regularly subsidize renewable microgrids to increase their renewable energy market penetration. In this study, we investigated the optimal subsidy level for governments to correct the market failure of microgrids and analyzed the impacts of regulation on the interaction between a microgrid and a distribution network operator (DNO). Specifically, we proposed economic rationales for government subsidies for microgrids regarding public interest benefits in relation to double externalities (learning spillover effect and environmental externality). We incorporated the double externalities into a three-echelon game model in an electricity supply chain with one regulator, one microgrid, and one DNO, in which the regulator decides the subsidy level to achieve maximal social welfare. We found that the double externalities and double marginalization caused underinvestment in microgrid capacity in the scenario without government intervention. The government could choose the appropriate subsidy level to achieve the system optimum, which led to a triple win for the microgrid, the DNO, and the social planner. Our analytical results also showed that the microgrid gained more benefits from regulation than the DNO. The microgrid may offer a negative wholesale price to the DNO in exchange for more opportunities to import electricity into the grid, especially when the investment cost is sufficiently low. Our study suggests that supporting microgrids requires a subsidy phase-out mechanism and alternative market-oriented policies with the development of the microgrid industry.

Keywords: microgrid; distribution network operator; double externalities; subsidy

---

1. Introduction

Alongside issues such as increasingly serious environmental pollution, energy supply security, greenhouse gas emissions, and climate change, there is a global consensus that all nations should strive together to reduce carbon emissions. Many countries seek approaches to gradually adjust their energy supply structure. As one of the contracting states of the Paris Climate Agreement, China promised to contribute to limiting the temperature increase of 1.5 °C through low-carbon development in the energy field. Renewable energy plays a key role in the process of energy structure transition, for which China faces the dual challenges of energy demand growth and carbon emission reduction. Microgrids are a prospective and accessible technology that can use distributed renewable energy resources to generate electricity and are also important components of smart grids. Some key benefits of microgrids can be summarized as follows:

Microgrids make the most use of local distributed renewable energy and reduce transmission losses by satisfying local consumers.

The smart energy management system and storage devices enable microgrids to manage intermittent renewable energy and realize real-time power balancing within the system.

Microgrids can provide cooling, heating, and electricity at the same time (e.g., the combined cooling, heating, and power, CCHP, system), which satisfies multidimensional service requirements [1].

Since microgrids can operate either grid-connected or islanded, they are more flexible, controllable, and reliable than distributed generation systems [2]. This helps to reduce the effects of major power disruption events and buffer the impact of intermittent renewable energy generation on power grids, which increases the opportunities for microgrids to access the power network.

Microgrids can provide differentiated services for sensitive users who may require higher power quality and a more resilient energy system (e.g., hospitals and data centers) [3].

The market practice of combining battery swap stations for electric vehicles and microgrids provides a possible channel to efficiently shift the peak electricity load [4].

China realized that its energy structure heavily relied on thermal power for its rapid economic growth, which limited high-quality development and green innovation in Chinese industry. China is devoted to supporting the efficient utilization of renewable energy and making efforts to deepen power system reform. The installed generation capacity share of thermal power, including coal and gas, decreased from 73.44% in 2010 to 60% in 2018 (Figure 1). Meanwhile, the installed generation capacity of renewable energy has maintained rapid development. The cost of renewable energy generation has declined gradually with technical progress, which has somewhat enhanced the competitive power of renewable energy.

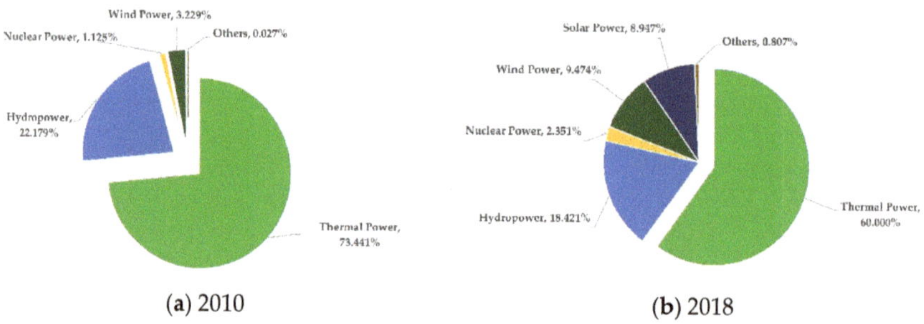

(a) 2010   (b) 2018

Figure 1. Gross installed generation capacity in China in 2010 and 2018 (source: China Electricity Council).

In July 2015, the China National Energy Administration elaborated on the significance of developing renewable microgrids and put forward the first batch of microgrid demonstration projects. Subsequently, another 24 grid-connected and 4 isolated microgrids were afforded renewable microgrid demonstration projects in May 2017. China learned international microgrid experience from both government-sponsored pilot projects and commercial projects [5], and a supporting document, "Proposed regulation for promoting the construction of grid-connected microgrids", was also announced by the National Energy Administration in July 2017.

Though microgrids have attracted much attention as an innovative technology, and the government spares no effort in advancing the microgrids industry, this industry is still in the early stages of development, and most of the projects are built for demonstration and research purposes. The low growing rate and the lack of investment are mainly due to the high initial investment cost and unfavorable economic benefits. The total proposed, under development, and operational power capacity of global microgrids reached 24,981 MW in 2018 [6]. There is no doubt that microgrid investment and operation costs will decrease as the microgrid industry matures and thus realizes large-scale commercial applications. Some incentive schemes, such as renewable energy investment subsidies, tax reductions, and feed-in tariffs, have been widely recognized [7]. However, the Chinese government does not publish detailed subsidy schemes for renewable microgrids. It is notable that renewable microgrids can receive renewable energy subsidies, which have been established for distributed generators using renewable sources. Since there are differences of technology systems, market features, and stages of development between microgrids and large-scale distributed generation, the current incentive mechanism for distributed generators may not be appropriate for microgrids [8].

Therefore, in this study, we first attempted to answer the following question:

What are the economic rationales of government subsidies for renewable microgrids?

Although some researchers have determined that the deployment of renewable microgrids can boost employment, ensure energy security, adjust the energy supply structure, share investment risks, and so forth, these are insufficient reasons to implement government subsidies from the perspective of the free market economy. In this paper, we argue that the economic rationales for government subsidies for renewable microgrids derive from market failures along with double externalities. One rationale for such subsidies is the learning spillover effect. Key microgrid technologies, such as distributed energy storage, controls and supervisory systems, and protection and automation, still require an immense amount of research. The R&D of an individual company is beneficial for the whole microgrid industry through the spillover effect and significantly contributes to microgrid cost reduction. Therefore, it is necessary to offer subsidies for microgrids due to the weak protection of intellectual property rights since China is still in a transition stage. Another rationale is the environmental externality of renewable energy. While microgrids can reduce carbon emissions by substituting for electricity generated by fossil fuels, the societal benefits cannot be included in the profit function of an individual microgrid. Consequently, the market price fails to signal the real cost of a microgrid when competing with conventional power. Simply put, we maintain that the economic rationales of renewable microgrid subsidies derive from their public interest benefits, in terms of both technology and the environment, and government intervention helps to internalize the double externalities.

Based on the above-stated reasoning, we built a three-echelon Stackelberg model consisting of a social planner, a renewable microgrid, and a distribution network operator (DNO) to address the following questions:

How should the government determine the optimal subsidy level in order to maximize social welfare? What are the effects of government subsidies on the interaction between a microgrid, a DNO, and social welfare?

Our model incorporated the double externalities to investigate the optimal subsidy level and the loss of social welfare without government intervention. The results showed that government subsidies lead to a lower wholesale price, which may be negative, and the DNO also benefits from the microgrid incentive scheme. Incremental social welfare is mainly determined by the spillover effect rate, environmental externality level, and potential demand.

The first contribution of this paper is analyzing the economic rationales of government subsidies for renewable microgrids and exploring the optimal subsidy scheme from the perspective of social welfare by incorporating the learning spillover effect and the environmental externality. Further, we introduce an electricity supply chain model with a microgrid and DNO and analyze how the government can achieve system coordination by offering an appropriate subsidy. Comparisons of some of the decision variables between the decentralized scenario and the government intervention scenario are also offered to illustrate the impact of subsidies on the interaction between the key stakeholders.

The rest of this paper is organized as follows: Section 2 reviews the literature related to microgrid subsidies. Section 3 outlines the model formulation and basic results under the centralized and decentralized scenarios. In Section 4, we introduce government subsidies to demonstrate their effects on the interaction between a microgrid and a DNO. Section 5 provides some numerical examples. Section 6 summarizes the article with some policy implications.

2. Literature Review

Our paper is related to the stream of literature on regulating technology spillover in R&D. Currently, most studies have found that a single firm cannot internalize the social benefits of R&D due to the spillover effect and, therefore, a social planner should attempt to correct market failure by regular intervention, such as with subsidies. Howell [9] focused on the American Small Business Innovation

Research (SBIR) grant program in the energy industry and demonstrated that early-stage grants help new ventures acquire more venture capital and patents because firms in emerging sectors face high technological uncertainty and financing constraints. Researchers have also analyzed the possible ways in which government subsidies positively affect a firm's innovation performance. First, from a resource-based perspective, government subsidies replenish the shortage of innovation resources, reduce output uncertainty, and share the risks for small firms [10–12]. Second, government subsidies are useful signals for small firms to attract private investors from the viewpoint of signal theory [13,14]. There is serious information asymmetry between firms and outside investors due to the uncertainty of innovation activities. The government plays a dominant role in a transition economy. Government subsidies can generate positive signals so that a firm's strategy can match government industry planning and the firm may maintain a good relationship with the government, which is beneficial for the firm to gain more innovative resources from outside and form alliances [15]. These papers addressed the impact of government subsidies on innovative enterprises and showed how government subsidies affect firms' innovation input and output, mainly from the industry level, and thus are quite different from our work.

This paper is also related to the stream of literature on technology and products with environmental externalities, particularly from the perspective of a social planner. For example, Drake et al. [16] considered a firm with both clean and dirty technologies and investigated its asset allocation reaction to different technologies under various governmental regulations, such as emissions tax, emissions cap-and-trade, and investment subsidies. Yu et al. [17] formulated a mathematical model to explore the effects of environmental awareness and green subsidies on manufacturers' production decisions regarding green products. The model was solved by a Euler algorithm, and a case involving automobiles with different green levels was proposed to validate the accuracy of the algorithm. Rayamajhee et al. [18] compared the impacts of exogenous and endogenous externality mitigation funds on hydroelectricity generation considering the negative environmental externality of crop damage. The results showed that an endogenous externality mitigation fund is better when weighing the income of energy production and externality damages. These papers focused on production and technology with external market effects which generate benefits or costs to society. The study scope included energy resources for heating in broiler production [19], electric vehicles [20,21], conventional and renewable energy investment [22], and so on. These papers, therefore, did not consider the electricity supply chain and learning spillover effect, which is our focus here.

This paper also builds on the operations management literature for microgrids. Along with the development of the microgrid industry, there has been a substantial amount of research on microgrid technical issues, such as microgrid control strategies [23], microgrid energy management [24], microgrid system resilience [25], and microgrid planning and design [26], in recent years. However, few studies have considered microgrid operations management, especially from the government's standpoint. Haghi et al. [27] analyzed the effects of a feed-in tariff (FIT) and capital grants on hydrogen production in a microgrid and found that it is more cost-efficient for a government to incentivize hydrogen production with grid electricity than with wind power, while hub operators prefer incentives for wind power. Chen et al. [28] investigated the impact of certified emission reductions on the deployment of solar photovoltaic community microgrids and verified that certified emission reductions can turn some unprofitable areas of deploying microgrids into profitable areas. Lo Prete et al. [29] used cooperative game theory to explore the optimal regulated electricity price for formulating microgrid cooperation among market players in a small electricity network. The intervention tool in their paper regulated the electricity retail price to achieve certain return requirements, which is different from our paper. Chen et al. [30] demonstrated the significant effect of price and cost subsidies on a microgrid energy storage system using a real options approach, which did not consider the electricity supply structure. Our paper is closely related to the works of Long and co-workers [8,31], who built a multistage incentive model for a microgrid project considering the microgrid industry chain and studied the impacts of government subsidies on various participants. However, since their work did not consider the economic rationales of government intervention, the regulator's decision relatively lacked a theoretical basis. Unlike their model, we have incorporated the technology spillover effect and environmental externality into a three-player model, and the optimal subsidy level was obtained to correct market failure.

Finally, to the best of our knowledge, this paper is the first to analyze the cause and correction of microgrid market failure in the context of the electricity supply chain. The contribution of this paper is that we have analyzed microgrid incentives to coordinate a microgrid and a DNO, which benefits not only the participants in the electricity supply chain but also the environment and society. Our study reveals the economic rationales of government subsidies for microgrids and thus jointly considers two types of externalities in the electricity supply chain of a microgrid, which helps to avoid new distortions due to government intervention. In addition, our paper completes the existing literature on microgrid subsidies by demonstrating that the government must account for the interaction between a microgrid and a DNO when determining the optimal subsidy level.

3. Model and Preliminary Results

3.1. Question Description and Model Assumption

The problem is defined in Figure 2. The electricity generated by a microgrid firstly satisfies its own user, and the remaining electricity is sold to the DNO at the wholesale price ω, which is a general operation mode of renewable microgrids. The DNO determines the transmission price l and sells electricity to end consumers. Thus, the final electricity price p can be represented as $p = \omega + l$. This kind of microgrid project is known as "self-generation, self-consumption, and the remainder is exported to the grid", which is the most encouraged pattern of renewable microgrids.

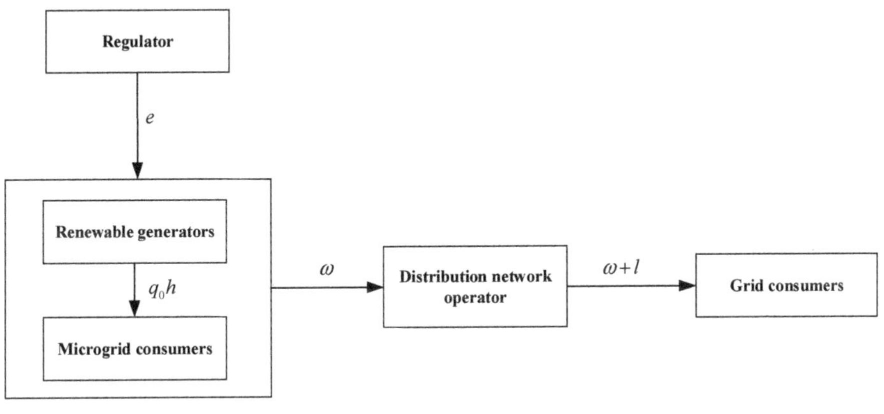

Figure 2. Model setting.

We considered the situation faced by a social planner who attempts to design an incentive policy to correct market failures of renewable microgrids with the spillover effect and environmental externality. The social planner offers the generation subsidy $S(q)$, which depends on the generated quantity q:

$$S(q) = e(q_0 + q_1) \tag{1}$$

where e denotes the marginal subsidy rate, and q_0 and q_1 are the electricity demands of the microgrid user and the end consumers, respectively. We also assumed that q_0 is an exogenous constant, and h is the microgrid benefits from per unit of electricity, which is also consistent with reality, since the investment level and microgrid user demand are usually determined before deployment.

We assumed that the market demand which the DNO faces is as follows:

$$D = a - bp \tag{2}$$

where D is the electricity demand quantity, $a(a > 0)$ is the market size, and $b(b > 0)$ is the sensitivity coefficient of the electricity price. It is worth noting that the demand means the total electricity demanded by end consumers in the whole period of the project rather than the instantaneous energy consumption.

For simplicity, we assumed that the microgrid incurs zero generation costs. This is a common assumption because renewable energy resources (e.g., solar and wind energy) are theoretically inexhaustible. Thus, we mainly took the investment cost into consideration. The investment cost I is a function of the installed capacity of the renewable microgrid m. Similar to the work of Kök (2018) [22], the investment cost function was assumed to be

$$I(m) = cm \tag{3}$$

where c denotes the cost of unit installed capacity. Specifically, the generation quantity of renewable energy cannot always reach the maximum value of installed capacity owing to the intermittency of solar and wind power. The effective generation time of solar power is in the daytime, while the generation pattern of wind power heavily depends on the richness of wind resources in different geographical locations. The function of electricity quantity can be expressed as

$$q = \delta m \tag{4}$$

where the capacity availability δ shows the phenomenon of renewable energy intermittency. We assumed $\delta(0 \leq \delta \leq 1)$ is a constant for ease of analysis.

Microgrids are still in the early stage of industrial development, and most microgrids are implemented as demonstration projects which are approved and supported by the energy sector. One of the guiding principles of promoting microgrids is maximizing the utilization of renewable energy, such as solar, wind, biomass, and so forth. Microgrids possess a unique feature in that they have two aspects of externalities: the technology spillover effect and environmental externality. The total externalities of the microgrid are denoted by E_T, which is a function of both investment level and generation quantity:

$$E_T = \lambda I(m) + \theta q, \ 0 < \lambda < 1 \tag{5}$$

where λ and θ are the spillover effect rate and the environmental externality level, respectively. Equation (5) shows that the spillover effect is related to both the investment level and the innovation environment in specific regions. Thus, $\lambda = 0$ means that the demonstration microgrid cannot contribute to industrial development, and $\lambda = 1$ implies complete spillover. The environmental externality of a microgrid derives from its substitution of the fossil fuels used in conventional generators.

In order to guarantee the intelligibility of this research, we assumed that the abovementioned parameters satisfy the following conditions:

$$c - c\lambda - \delta\theta > 0, \ a\delta > bc \tag{6}$$

$$(a + b\theta)\delta + bc(\lambda - 1) \geq 0 \tag{7}$$

In the following analysis, the subscript $i \in \{m, g\}$ refers to the microgrid and the DNO, respectively, and the notation $j \in \{C, D, I\}$ denotes to the centralized scenario, the decentralized scenario, and the scenario with government intervention, respectively.

3.2. Centralized Scenario: Maximizing Social Welfare

As one benchmark scenario, we first studied the centralized scenario, where a social planner directly determines the generation quantity of the microgrid. The objective of the social planner is to optimize social welfare, which includes the microgrid's profit, consumer surplus, externality benefits, and government cost. The consumer surplus can be written as

$$v^C = \int_{p^*}^{p^{\max}} D(p)dp \tag{8}$$

where p^* and $p^{\max}$ are the optimal electricity price and maximum price, respectively. The social welfare in the centralized scenario is given by

$$W_C = (a - bp)p + q_0 h - \frac{c}{\delta}(a - bp + q_0) + \frac{1}{2b}(a - bp)^2 + \frac{\lambda c + \delta\theta}{\delta}(a - bp + q_0) \tag{9}$$

The social planner directly chooses the optimal electricity price to maximize social welfare. We obtained the following equation by solving for "$\partial W_C / \partial p = 0$":

$$p_C^* = \frac{c - c\lambda - \delta\theta}{\delta} \tag{10}$$

which is equivalent to

$$q_C^* = \frac{(a + b\theta)\delta + bc(\lambda - 1)}{\delta} \tag{11}$$

Thus, the optimal social welfare can be written as

$$W_C^* = \frac{(c(\lambda - 1) + \delta\theta)^2 b^2 + 2[((a + x) + q_0 h)\delta + c(\lambda - 1)(a + q_0)]\delta b + a^2\delta^2}{2b\delta^2} \tag{12}$$

3.3. Decentralized Scenario: No Government Intervention

Under the scenario without government intervention, the problem was analyzed as a Stackelberg game, where the renewable microgrid acts as a leader, and the DNO is a follower. The microgrid investor maximizes its profit by deciding the optimal wholesale price. Then, the DNO chooses the transmission price to maximize the profit. The above problem was solved using backward induction. The DNO's profit function is given by

$$\pi_g^D = (a - b(l + \omega))l \tag{13}$$

The DNO's profit function π_g^D reaches optimization when the optimal transmission price is

$$l_D^* = \frac{a - b\omega}{2b} \tag{14}$$

The microgrid's profit function is

$$\pi_m^D = (a - b(\omega + l))\omega + q_0 h - \frac{c}{\delta}(a - b(\omega + l) + q_0) \tag{15}$$

We substituted Equation (14) into Equation (15) and took the derivative of Equation (15) with respect to ω in order to optimize the microgrid's profit. Thus, we obtained the optimal wholesale price of the microgrid:

$$\omega_D^* = \frac{a\delta + bc}{2\delta b} \tag{16}$$

Substituting Equation (16) into Equation (14) yielded the equilibrium transmission price:

$$l_D^* = \frac{a\delta - bc}{4\delta b} \tag{17}$$

The electricity demand quantity of end consumers is

$$q_1^D = \frac{a\delta - bc}{4\delta} \tag{18}$$

Therefore, the optimal profits of the renewable microgrid and DNO are

$$\pi_g^D = \frac{(a\delta - bc)^2}{16b\delta^2} \tag{19}$$

$$\pi_m^D = \frac{(a^2 + 8bq_0h)\delta^2 - 2bc(a + 4q_0)\delta + b^2c^2}{8b\delta^2} \tag{20}$$

Under the scenario without government intervention, social welfare is characterized by the following function:

Social Welfare = Microgrid's Profit + DNO's Profit + Consumer Surplus + Externality Benefit.

Therefore, we obtained the social welfare:

$$W_D^* = \frac{3(a\delta - bc)^2}{32b\delta^2} + \frac{(a^2 + 8bq_0h)\delta^2 - 2bc(a + 4q_0)\delta + b^2c^2}{8b\delta^2} + \frac{(c\lambda + \delta\theta)((a + 4q_0)\delta - cb)}{4\delta^2} \tag{21}$$

3.4. Comparison between Centralized and Decentralized Cases

The main difference between the centralized and the decentralized cases is that the microgrid cannot internalize the spillover effect and environmental externality without intervention, while the social planner can take them and consumer surplus into consideration.

Proposition 1. *The optimal electricity quantity under the centralized case is higher than that under the decentralized case, and the social welfare is also higher in the centralized scenario; that is, $q_C^* > q_D^*$, $W_C^* > W_D^*$.*

Proof. Proposition 1 can be simply proved by demonstrating that

$$q_C^* - q_D^* = \frac{(4b\theta + 3a)\delta + 4(\lambda - \frac{3}{4})bc}{4\delta} \tag{22}$$

Since there is $0 < \lambda < 1$, we have $q_C^* - q_D^* > [(4b\theta + 3a)\delta - 3bc]/4\delta$. From Equation (7), we obtained $q_C^* - q_D^* > 0$. The social welfare loss of the decentralized case compared with the centralized case is

$$W_C^* - W_D^* = \frac{(4bc\lambda + 4b\delta\theta + 3a\delta - 3bc)^2}{32b\delta^2} \tag{23}$$

so that $W_C^* - W_D^* > 0$. □

Proposition 1 states that the system tends to generate more electricity and reaches a higher level of social welfare in the centralized scenario. The social welfare loss is a joint result of double marginalization and double externalities. This suggests that the government has motivations to implement an incentive scheme to promote the microgrid owner to increase investment, since society receives externality benefits when the microgrid generates more electricity.

4. Analysis of Government Intervention

When the social planner attempts to overcome the market failure by providing certain subsidies corresponding to technology spillover and environmental externality, the timing in the three-echelon Stackelberg model is as follows. First, the government establishes its subsidy scheme to pursue social welfare maximization. Second, the microgrid investor determines the optimal electricity wholesale price. Third, the DNO chooses its optimal transmission price.

The DNO's profit can be written as

$$\pi_g^I = (a - b(l + \omega))l \tag{24}$$

We sought the values of transmission price l that maximize the DNO's profit. The optimal transmission price is

$$l_I^* = \frac{a - b\omega}{2b} \tag{25}$$

The microgrid's profit is given by

$$\pi_m^I = (a - b(\omega + l))\omega + q_0 h - \left(\frac{c}{\delta} - e\right)(a - b(\omega + l) + q_0) \tag{26}$$

We substituted Equation (25) into Equation (26) and obtained the optimal electricity wholesale price by solving for "$\partial \pi_m^I / \partial \omega = 0$":

$$\omega_I^* = \frac{a\delta + bc - b\delta e}{2\delta b} \tag{27}$$

Hence, the optimal transmission price can be rewritten as

$$l_I^* = \frac{(a + be)\delta - bc}{4b\delta} \tag{28}$$

Proposition 2. *A higher government subsidy level results in higher transmission price and lower wholesale price.*

Proof. By taking the derivative of Equations (27) and (28) with respect to e, we obtained $\partial \omega_I^* / \partial e < 0, \partial l_I^* / \partial e > 0$. □

Proposition 2 indicates that the government subsidy induces the microgrid to reduce the electricity wholesale price so that more electricity can be on sale to the power grid. Meanwhile, the DNO attempts to gain more profits from the subsidy by increasing the transmission price.

It is clear that social welfare can be concluded to be

Social Welfare = Microgrid's Profit + DNO's Profit + Consumer Surplus + Externality Benefit − Government Cost

$$W_I = (a - b(l + \omega))(l + \omega) + q_0 h - \frac{c}{\delta}(a - b(\omega + l) + q_0) + \frac{1}{2b}(a - b(\omega + l))^2 + \frac{\lambda c + \delta \theta}{\delta}(a - b(\omega + l) + q_0) \tag{29}$$

We substituted Equations (27) and (28) into Equation (29) and took the derivative of Equation (29) with respect to e. The optimal subsidy rate can be found at $\partial W_I / \partial e = 0$ since there is $\partial^2 W_I / \partial e^2 < 0$:

$$e_I^* = \frac{4bc\lambda + 4b\delta\theta + 3a\delta - 3bc}{\delta b} \tag{30}$$

Proposition 3. *The optimal subsidy level increases with the spillover effect rate and the environmental externality level.*

Proof. By taking the derivative of Equation (30) with respect to λ and θ, respectively, we obtained $\partial e_I^*/\partial \lambda > 0, \partial e_I^*/\partial \theta > 0$. $\quad\square$

According to Proposition 3, the economic rationales of government subsidies for microgrids derive from two aspects: (1) Since renewable microgrids are an emerging industry and some critical technology and business models are not mature in the early stage, microgrid demonstration projects play an import role in the spillover effect. (2) Renewable microgrids possess environmental externalities due to the substitution of conventional power generation. As the microgrid industry gradually matures, government subsidies should decrease with the lower spillover effect rate and environmental externality level.

Substituting the optimal subsidy rate into Equations (29) and (30) yields the optimal electricity wholesale price and transmission price:

$$\omega_I^* = \frac{(-2\theta\delta - 2c(\lambda-1))b - a\delta}{b\delta}, l_I^* = \frac{(\theta\delta + c(\lambda-1))b + a\delta}{b\delta} \tag{31}$$

Proposition 4. *There is a threshold value such that (i) if $c \geq \bar{c}$, then $\omega_I^* \geq 0$; (ii) on the other hand, if $c < \bar{c}$, then $\omega_I^* < 0$, where we defined the threshold cost $\bar{c}$ as $\bar{c} = \frac{\delta(2b\theta+a)}{2b(1-\lambda)}$.*

This threshold value suggests that, for the microgrid whose marginal investment cost is low enough (i.e., $c < \bar{c}$), the renewable microgrid prefers offering a negative wholesale price to the DNO so that more electricity can be imported into the electricity grid. Proposition 4 can partially give insights into the common phenomenon of negative electricity prices in some free electric markets. Negative electricity prices have been found in some electricity spot markets, such as the United States, the European Union, and Australia, and it will likely appear more frequently with the increasing installed capacity of renewable energy [32]. Microgrids face the trade-off between the negative price and government subsidy under the situation that the electricity demand is lower than the power supply. Proposition 4 demonstrates that the microgrid induces a negative price to stimulate the DNO to purchase its excess power as long as the gain from the government subsidy can offset the transfer payments to the DNO.

Next, we considered the DNO's and microgrid's profits with government intervention:

$$\pi_g^I = \frac{((\delta\theta + c(\lambda-1))b + a\delta)^2}{b\delta^2} \tag{32}$$

$$\pi_m^I = \frac{2(\delta\theta + c(\lambda-1))^2 b^2 + 4\delta((a+q_0)\theta + \frac{1}{4}q_0 h)\delta + c(\lambda-1)(a+q_0))b + 2a(a + \frac{3}{2}q_0)\delta^2}{b\delta^2} \tag{33}$$

Therefore, we obtained the social welfare:

$$W_I^* = \frac{(c(\lambda-1) + \delta\theta)^2 b^2 + 2[((a+x) + q_0 h)\delta + c(\lambda-1)(a+q_0)]\delta b + a^2\delta^2}{2b\delta^2} \tag{34}$$

Proposition 5. *The whole system can be coordinated if, and only if, $e = e_I^*$ where e_I^* is given in Equation (30), and the effects of government subsidies on the relative stakeholders are as follows:*

(i)   The government can choose the appropriate subsidy rate to achieve the system optimum of social welfare.

(ii)   The government subsidy scheme always benefits both the microgrid and the DNO.

(iii) The incremental social welfare under government intervention increases with both the spillover effect rate and the environmental externality level.

Proof. By Equations (12) and (34), we have $W_I^* = W_C^*$, which means the social welfare approaches its highest value in the centralized scenario when the regulator properly sets the subsidy level. Comparing the microgrid's profit and the DNO's profit between the decentralized scenario and the government intervention scenario, we obtained

$$\pi_g^I - \pi_g^D > 0, \pi_m^I - \pi_m^D > 0 \tag{35}$$

The incremental social welfare can be represented by

$$\Delta W = W_I^* - W_D^* = \frac{(4bc\lambda + b\delta\theta + 3a\delta - 3bc)^2}{32b\delta^2} \tag{36}$$

Differentiating, we obtained

$$\partial\Delta W/\partial\lambda > 0, \partial\Delta W/\partial\theta > 0 \tag{37}$$

□

Proposition 5 states that the incentive scheme makes the three parties better off, which demonstrates the economic necessity of government intervention for renewable microgrids. Proposition 5 also implies that government intervention not only internalizes the spillover effect and environment externality but also eliminates the double marginalization effect. With higher values of spillover effect rate and environmental externality level, the government is more willing to rectify the market failure. However, along with the developing and maturing microgrid industry, the government also needs to design a subsidy phase-out mechanism responding to declining spillover effect and environmental externality rates.

5. Numerical Examples

In this section, we present a numerical study to compare the equilibrium solutions between the decentralized scenario and the government intervention scenario and explore some sensitivity analyses with respect to the model parameters.

Our first numerical example focuses on the effects of government intervention on the optimal electricity quantity, stakeholders' profits, and social welfare. The following parameter values were used for the numerical example, which satisfied the conditions of (6) and (7). Although our choice of parameter values was somewhat arbitrary, they are easy for a regulator and manager to visualize:

$b = 1.2$, $q_0 = 1.3$, $h = 5$, $c = 2$, $\lambda = 0.4$, $\delta = 0.6$, $\theta = 1$, and $a_1 = 10$, $a_2 = 15$, $a_3 = 20$, $a_4 = 25$, and $a_5 = 30$.

Table 1 summarizes the optimal electricity quantity imported into the grid, the microgrid's profit, the DNO's profit, and social welfare under the decentralized scenario and the government intervention scenario for a varying from 10 to 30. This table shows that, regardless of the potential market demand a, the government subsidy fostered the microgrid to increase investment and improve social welfare. Though the subsidized object was a microgrid, the DNO also took a share of the benefits by setting a reasonable transmission price. However, the microgrid extracted more benefits from the government subsidy than the DNO because of the first-mover advantage. Another observation from Table 1 is that incremental social welfare by government subsidy increased with the value of potential market demand a. It is not surprising that a higher potential market demand leads to more incremental social welfare, which creates more incentives for the government to encourage consumption. Meanwhile, the government faces bigger challenges due to the increased intervention cost.

Table 1. The effect of government intervention on the results.

	Decentralized Scenario				Government Intervention				Incremental	
a	q_1^D	π_g^D	π_m^D	W_D^*	q_1^I	π_g^I	π_m^I	W_I^*	ΔW	$\frac{W_I^*-W_D^*}{W_D^*}$
10	1.50	1.88	5.92	15.26	8.80	64.53	162.87	37.47	22.21	1.46
15	2.75	6.30	14.77	33.67	13.80	158.70	367.45	84.55	50.88	1.51
20	4.00	13.33	28.83	61.20	18.80	294.53	655.37	152.47	91.27	1.49
25	5.25	22.97	48.10	97.84	23.80	472.03	1026.61	241.22	143.38	1.47
30	6.50	35.21	72.58	143.60	28.80	691.20	1481.20	350.80	207.20	1.44

Next, we considered a numerical example to address the sensitivity of government subsidy costs with respect to the spillover rate and environmental externality level. Let $b = 1.2$, $q_0 = 1.3$, $h = 5$, $c = 2$, and $\delta = 0.6$.

Figure 3 presents the impacts of the spillover rate λ and the environmental externality level θ on the government intervention cost when $a = 10$, $a = 15$, and $a = 20$. The overall policy cost to achieve social welfare maximization decreased with a lower value of λ and θ for three different market sizes. It can also be seen that the policy cost increased significantly as the market size a increased. This result provides insights into subsidy policy formulation in the long term. The government intervention cost is relatively low in the early stage of the microgrid industry, but it may hinder the government from achieving social welfare maximization as the market size becomes larger and larger. It is necessary and rational to design a subsidy phase-out mechanism in the long run for these reasons: (1) The learning spillover rate tends to decline with the microgrid industry gradually transforming into its mature stage. (2) The environmental externality decreases with the cleaner development trend of thermal power. Some microgrid support policies based on market mechanisms such as a renewable portfolio standard, tradable green certificates, and carbon cap-and-trade are some alternative options to motivate microgrid investment without relying on government financial support.

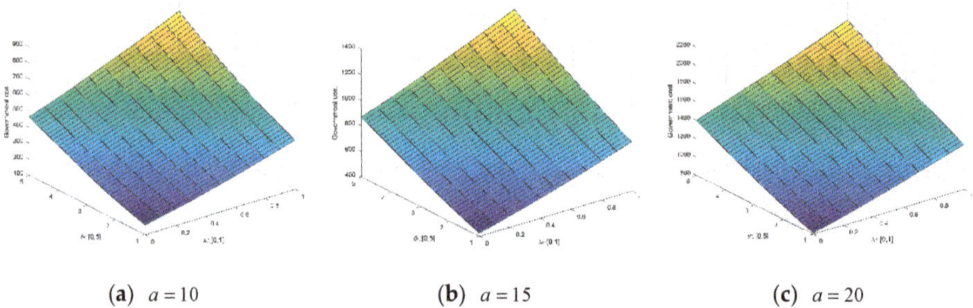

(**a**) $a = 10$ (**b**) $a = 15$ (**c**) $a = 20$

Figure 3. The effects of the learning spillover rate and environmental externality level on government subsidy cost.

In a third numerical experiment, we explored the effect of government subsidies on the profits of a microgrid, DNO, and social welfare. We considered a situation where $b = 1.2$, $q_0 = 1.3$, $h = 5$, $c = 2$, $\lambda = 0.4$, $\delta = 0.6$, $\theta = 1$, and $a_1 = 10$.

Figure 4 represents the social welfare, microgrid's profit, and DNO's profit as a function of the government subsidy coefficient e. As e increased, the microgrid's and DNO's profits increased significantly, especially when the subsidy nearly reached 55.94. However, the goal of the government was to maximize social welfare, and too high or too low subsidy levels were both harmful from the perspective of the social planner. The optimal subsidy coefficient in this case was 24.33, which was obtained by substituting the selected parameter values into Equation (30).

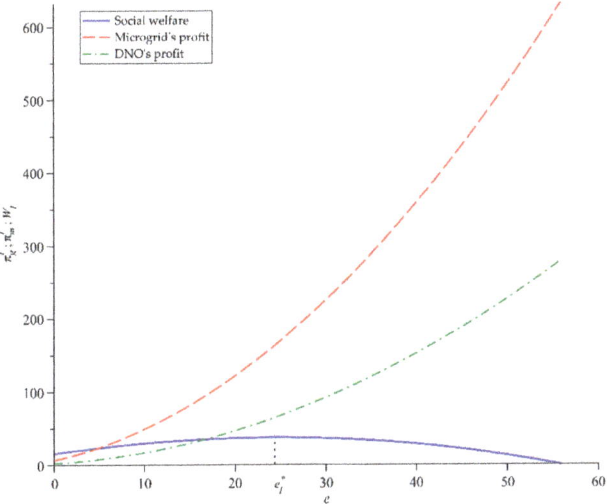

Figure 4. The effect of government subsidies on stakeholders' profits and social welfare.

6. Conclusions

The market practice of microgrid demonstration projects has attracted a considerable amount of interest because of the potential to boost renewable energy utilization. Making economic sense of microgrids with a premium cost requires profit-seeking investors to pursue public incentives in the early stage of the microgrid industry. Therefore, understanding the economic rationales of government intervention and exploring the effects of government subsidies on the interaction between a microgrid and a DNO is crucial.

What differentiates microgrids from other forms of power generation is that microgrids have learning spillover effects and environmental externalities at the same time. This paper examined the decision-making process of choosing the optimal government subsidy level to maximize social welfare and investigated the reactions of a microgrid and a DNO under three different scenarios. A framework was developed to address the economic rationales and impacts of government subsidies on renewable microgrids from the perspective of a social planner. To focus our attention on solving the market failure of renewable microgrids caused by the spillover effect and environmental externalities, we considered a three-stage Stackelberg model to present the necessity of government subsidies and their effects. Unlike the extant literature on government intervention for microgrids, we considered the double externalities and proposed a model of an electricity supply chain with a microgrid and a DNO to derive the optimal subsidy strategy to solve market failures by coordinating the electricity supply chain.

Our results suggest that the government can use subsidy instruments to achieve the system optimum compared with the scenario without government intervention, and the microgrid acquires more benefits than the DNO. In the presence of a government subsidy, the microgrid may offer a negative electricity price to the DNO to earn more opportunities to import electricity into the grid. We observed that, in addition to increasing social welfare and the microgrid's profit, coordinating the electricity supply chain also benefits the DNO and the consumers.

Our results demonstrate that we need to incorporate the spillover effect and environmental externality into the electricity supply chain for microgrids to correct market failure. There is an optimal subsidy level to coordinate the electricity supply chain, which is determined by the spillover effect rate, environmental externalities, and some other model parameters. Our study can provide useful guidelines for regulators to correct market failures. First, the regulator needs to design a subsidy phase-out mechanism, since the spillover effect and environmental externality gradually decline with the development of the microgrid industry. Second, the government needs to strengthen intellectual property protection so that the learning spillover effect can be corrected by market means (e.g., patents). Lastly, some market-oriented policies, such as tradable green certificates, carbon trading, and so forth, could be analyzed and implemented to substitute subsidies, which would help to alleviate the government's financial burden.

We encourage extending our model to analyzing the impact of government subsidies on the electricity supply chain under uncertain demand. It would also be of interest to study the effects of government intervention on microgrids when considering electricity storage and investor's risk preference. Moreover, it would be useful to study other forms of government intervention on microgrids, such as a renewable portfolio standard, tradable green certificates, and carbon cap and trade.

Author Contributions: D.X. wrote the paper; Y.L. contributed ideas and guidance.

Funding: This work was supported by the National Social Science Foundation of China (Grant No. 17ZDA065).

Conflicts of Interest: The authors declare no conflict of interest.

References

1. Varasteh, F.; Nazar, M.S.; Heidari, A.; Shafie-khah, M.; Catalão, J.P.S. Distributed energy resource and network expansion planning of a CCHP based active microgrid considering demand response programs. *Energy* **2019**, *172*, 79–105. [CrossRef]

2. Hussain, A.; Bui, V.-H.; Kim, H.-M. Microgrids as a resilience resource and strategies used by microgrids for enhancing resilience. *Appl. Energy* **2019**, *240*, 56–72. [CrossRef]

3. Van den Broeck, G.; Stuyts, J.; Driesen, J. A critical review of power quality standards and definitions applied to DC microgrids. *Appl. Energy* **2018**, *229*, 281–288. [CrossRef]

4. Long, Y.; Xu, D.; Li, X. Channel coordination of battery supplier and battery swap station of micro-grid with uncertain rental demand. *Soft Comput.* **2018**, 1–13, In press. [CrossRef]

5. Romankiewicz, J.; Marnay, C.; Zhou, N.; Qu, M. Lessons from international experience for China's microgrid demonstration program. *Energy Policy* **2014**, *67*, 198–208. [CrossRef]

6. Navigant Research. Microgrid Deployment Tracker 2018. Report Abstract. Available online: https://www.navigantresearch.com/reports/microgrid-deployment-tracker-2q18 (accessed on 12 May 2019).

7. Husein, M.; Chung, I.-Y. Optimal design and financial feasibility of a university campus microgrid considering renewable energy incentives. *Appl. Energy* **2018**, *225*, 273–289. [CrossRef]

8. Long, Y.; Pan, C.; Wang, Y. Research on a Microgrid Subsidy Strategy Based on Operational Efficiency of the Industry Chain. *Sustainability* **2018**, *10*, 1519. [CrossRef]

9. Howell, S.T. Financing Innovation: Evidence from R&D Grants. *Am. Econ. Rev.* **2017**, *107*, 1136–1164.

10. Wang, C.; Nie, P.-Y.; Peng, D.-H.; Li, Z.-H. Green insurance subsidy for promoting clean production innovation. *J. Clean. Prod.* **2017**, *148*, 111–117. [CrossRef]

11. Xie, X.; Huo, J.; Qi, G.; Zhu, K.X. Green Process Innovation and Financial Performance in Emerging Economies: Moderating Effects of Absorptive Capacity and Green Subsidies. *IEEE Trans. Eng. Manag.* **2016**, *63*, 101–112. [CrossRef]

12. Bronzini, R.; Piselli, P. The impact of R&D subsidies on firm innovation. *Res. Policy* **2016**, *45*, 442–457.

13. Yan, Z.; Li, Y. Signaling through government subsidy: Certification or endorsement. *Financ. Res. Lett.* **2018**, *25*, 90–95. [CrossRef]

14. Chen, J.; Heng, C.S.; Tan, B.C.Y.; Lin, Z. The distinct signaling effects of R&D subsidy and non-R&D subsidy on IPO performance of IT entrepreneurial firms in China. *Res. Policy* **2018**, *47*, 108–120.

15. Kleer, R. Government R&D subsidies as a signal for private investors. *Res. Policy* **2010**, *39*, 1361–1374.

16. Drake, D.F.; Kleindorfer, P.R.; Van Wassenhove, L.N. Technology Choice and Capacity Portfolios under Emissions Regulation. *Prod. Oper. Manag.* **2016**, *25*, 1006–1025. [CrossRef]

17. Yu, Y.; Han, X.; Hu, G. Optimal production for manufacturers considering consumer environmental awareness and green subsidies. *Int. J. Prod. Econ.* **2016**, *182*, 397–408. [CrossRef]

18. Rayamajhee, V.; Joshi, A. Economic trade-offs between hydroelectricity production and environmental externalities: A case for local externality mitigation fund. *Renew. Energy* **2018**, *129*, 237–244. [CrossRef]

19. Mansilha, R.B.; Collatto, D.C.; Lacerda, D.P.; Wolf Motta Morandi, M.I.; Piran, F.S. Environmental externalities in broiler production: An analysis based on system dynamics. *J. Clean. Prod.* **2019**, *209*, 190–199. [CrossRef]
20. Hao, H.; Ou, X.; Du, J.; Wang, H.; Ouyang, M. China's electric vehicle subsidy scheme: Rationale and impacts. *Energy Policy* **2014**, *73*, 722–732. [CrossRef]
21. Huang, J.; Leng, M.; Liang, L.; Liu, J. Promoting electric automobiles: Supply chain analysis under a government's subsidy incentive scheme. *IIE Trans.* **2013**, *45*, 826–844. [CrossRef]
22. Kök, A.G.; Shang, K.; Yücel, Ş. Impact of Electricity Pricing Policies on Renewable Energy Investments and Carbon Emissions. *Manag. Sci.* **2018**, *64*, 131–148. [CrossRef]
23. Han, H.; Hou, X.; Yang, J.; Wu, J.; Su, M.; Guerrero, J.M. Review of Power Sharing Control Strategies for Islanding Operation of AC Microgrids. *IEEE Trans. Smart Grid* **2016**, *7*, 200–215. [CrossRef]
24. Shi, W.; Li, N.; Chu, C.-C.; Gadh, R. Real-Time Energy Management in Microgrids. *IEEE Trans. Smart Grid* **2017**, *8*, 228–238. [CrossRef]
25. Li, Z.; Shahidehpour, M.; Aminifar, F.; Alabdulwahab, A.; Al-Turki, Y. Networked Microgrids for Enhancing the Power System Resilience. *Proc. IEEE* **2017**, *105*, 1289–1310. [CrossRef]
26. Jung, J.; Villaran, M. Optimal planning and design of hybrid renewable energy systems for microgrids. *Renew. Sustain. Energy Rev.* **2017**, *75*, 180–191. [CrossRef]
27. Haghi, E.; Raahemifar, K.; Fowler, M. Investigating the effect of renewable energy incentives and hydrogen storage on advantages of stakeholders in a microgrid. *Energy Policy* **2018**, *113*, 206–222. [CrossRef]
28. Chen, W.; Wei, P. Socially optimal deployment strategy and incentive policy for solar photovoltaic community microgrid: A case of China. *Energy Policy* **2018**, *116*, 86–94. [CrossRef]
29. Lo Prete, C.; Hobbs, B.F. A cooperative game theoretic analysis of incentives for microgrids in regulated electricity markets. *Appl. Energy* **2016**, *169*, 524–541. [CrossRef]
30. Chen, W.; Zeng, Y.; Xu, C. Energy storage subsidy estimation for microgrid: A real option game-theoretic approach. *Appl. Energy* **2019**, *239*, 373–382. [CrossRef]
31. Long, Y.; Wang, Y.; Pan, C. Incentive Mechanism of Micro-grid Project Development. *Sustainability* **2018**, *10*, 163. [CrossRef]
32. Zhou, Y.; Scheller-Wolf, A.; Secomandi, N.; Smith, S. Electricity Trading and Negative Prices: Storage vs. Disposal. *Manag. Sci.* **2016**, *62*, 880–898. [CrossRef]

 © 2019 by the authors. Licensee MDPI, Basel, Switzerland. This article is an open access article distributed under the terms and conditions of the Creative Commons Attribution (CC BY) license (http://creativecommons.org/licenses/by/4.0/).

Article

A Spatial-Based Integration Model for Regional Scale Solar Energy Technical Potential

Younes Noorollahi [1,2], Mohammad Mohammadi [1], Hossein Yousefi [1] and Amjad Anvari-Moghaddam [3,4,*]

[1]   Renewable Energy and Environmental Dep., Faculty of New Sciences and Technologies, University of Tehran, Tehran 15119-43943, Iran; noorollahi@ut.ac.ir (Y.N.); m.mohammady@alumni.ut.ac.ir (M.M.); Hosseinyousefi@ut.ac.ir (H.Y.)
[2]   Energy Modelling and Sustainable Energy Systems (METSAP) Research Lab., Faculty of New Sciences and Technologies, University of Tehran, Tehran, Iran
[3]   Department of Energy Technology, Aalborg University, 9220 Aalborg, Denmark
[4]   Faculty of Electrical and Computer Engineering, University of Tabriz, 5166616471 Tabriz, Iran
*   Correspondence: aam@et.aau.dk

Received: 25 January 2020; Accepted: 27 February 2020; Published: 2 March 2020

Abstract: One of the main objectives of human society in the present century is to achieve clean and sustainable energy through utilization of renewable energy sources (RESs). In this paper, the main purpose is to identify the locations that are suitable for solar energy in the Kurdistan province of Iran. Initially, solar-related data are collected, and suitable criterion and assessment methods are chosen according to the available data. Then, the theoretical potential of solar energy is assessed and the solar radiation map is prepared. Moreover, the technical potential of various solar technologies is evaluated in that study area. These technologies include concentrating solar power (CSP) and photovoltaic (PV) in power plant applications, and rooftop PV panels and solar water heaters in general applications. The results show that the Kurdistan province has the potential capacity for 691 MW of solar photovoltaic power plants and 645 MW of CSP plants. In the case of using solar water heaters, 283 million cubic meters of natural gas and 1.2 million liters of gasoline could be saved in fuel consumption. The savings in the application of domestic PV will be 10.2 MW in power generation.

Keywords: solar potential assessment; resource mapping; geographic information systems (GIS); site selection; Iran

---

1. Introduction

Human society has experienced great development in recent years, but this growing development has been associated with high energy consumption, which has been mainly met by fossil fuels [1]. According to statistics from the International Energy Agency (IEA), world energy consumption in 2014 was equal to 9425 million tons of oil equivalent (Mtoe). However, the share of fossil fuels in the production of this energy was 84.5%. In addition, 32,381 million tons (Mt) of CO_2 were produced in this year, which has been nearly doubled compared to the level reported in 1973 [2]. These statistics show the importance of preventing the emission of greenhouse gasses and, consequently, the importance of finding appropriate solutions, such as optimization, increasing the efficiency of current energy systems, and moving towards the use of renewable energy sources (RESs) [3]. One of the main challenges of RESs is their high dependence on time and location. Accordingly, it is necessary to fully investigate renewable energies geographically and their distribution in different spatial and temporal scales [4].

Solar energy is one of the renewable energies that has attracted considerable attention in recent years. So far, several studies have been conducted in different countries on the solar energy resource assessment and application of relevant technologies [5,6]. Zavylska and Brooks [7] calculated different components of solar radiation by using existing data in one of the urban areas of South Africa (Durban). On a larger scale, Noorollahi et al. [8] developed the solar resource map in south Iran and evaluated the theoretical and technical potential map. Viana et al. [9] have estimated the use of concentrating photovoltaic (PV) in Brazil. Journée et al. [10] employed meteorological data to build radiation data in some European countries. Fleuri [11] calculated the amount of solar energy potential in the concentrating plant in South Africa. In Chile, Ortega et al. [12] have tried to investigate the status of solar energy with radiation data that have been extracted using meteorological stations and validating them by using satellite data. Watson et al. [13] have used a geographic information systems (GIS)-based approach to identify suitable locations for the wind and solar farms in the south of England with an emphasis on environmental considerations. Polo et al. [14] have created the required map of solar radiation in Vietnam with the integration of remote sensing and ground measurement data and also estimated the solar energy potential by applying concentrated solar power (CSP) and PV technologies using GIS. Sun et al. [15] have investigated the potential of solar PV technologies (large-scale PV plant and roof-top PV) using solar radiation maps and the limiting factors in Fujian Province in China. He and Kammen [16] investigated the potential for solar energy production using a large data set of hourly solar radiation on a provincial level for the whole of China. Stökler et al. [17] created a solar atlas for Pakistan based on a combination of satellite data and the amount of solar radiation during two years in nine different sites. Gautam et al. [18] have studied the feasibility of generating electricity from PV systems on the rooftops of urban areas in Nepal. Moreover, in recent studies, there have been multi-criteria decision-making approaches using GIS algorithms to evaluate suitable locations for constructing the solar farms in different regions such as Iran [19], the southeast of Spain [20], Karapinar in Turkey [21], southern Morocco [21], and for the whole of Europe [22].

Yet, perhaps, complete data and research among different countries are related to the US, which have been published in the form of various reports. In this regard, reference books and guidelines published by the National Renewable Energy Laboratory (NREL), which is related to the generation and measurement of solar radiation data [23], and its potential survey [24] that can be used also in the rest of the world. In conjunction with the direct solar heat applications, the potential calculation method in the use of solar water heaters has been published by NREL [25]. Based on this, technical, economic, and environmental analysis of these systems were developed by this Institute [26].

However, most of the above researches have focused on calculating the solar radiation map or studying some particular solar technologies. This paper examines the solar energy potential by taking into account all the available solar technologies, including PV, CSP, and water-heating system in a particular land for integrated application. The Kurdistan province of Iran has been selected as a case study since there is no fossil resources in that region and the use of RESs could be an important step towards distributed generation and energy independence in the province. However, the proposed methods of research can be applied to other provinces of Iran. Iran is in a situation that includes different climate and therefore has the potential to exploit various renewable sources. So considering these resources for energy production could provide technical, economic, environmental and social advantages for the country [27]. Despite these benefits and high potential, the use of renewable energy is still very low and the production of energy from these sources is less than 0.2% [28]. Therefore, more attention should be paid to RESs and more research is needed to develop the use of such resources.

2. Study area

Kurdistan province is one of the 31 provinces of Iran, which is located in the west of the country, and its capital is the city of Sanandaj. Kurdistan province, with an area of 28,200 km^2, is located between 34°44′ N to 36°30′ N latitude and 45°31′ E to 48°16′ E longitude. This area is 1.7% of the country's total area and is ranked 16 in the country. According to the latest divisions in 2016 Kurdistan has 10 counties, 29 cities, 31 districts, 86 rural districts, and 1697 villages with inhabitants and 187 deserted villages. Counties of this province include Baneh, Bijar, Dehgolan, Diwandarreh, Sarvabad, Saqqez, Sanandaj, Qorveh, Kamyaran, and Marivan. Based on the 2016 population and housing census, the province has a population of about 1.6 million—71% urban and 29% are rural dwellers. The relative density of the population is 51.2 people per km². The location map of the study area shown in Figure 1.

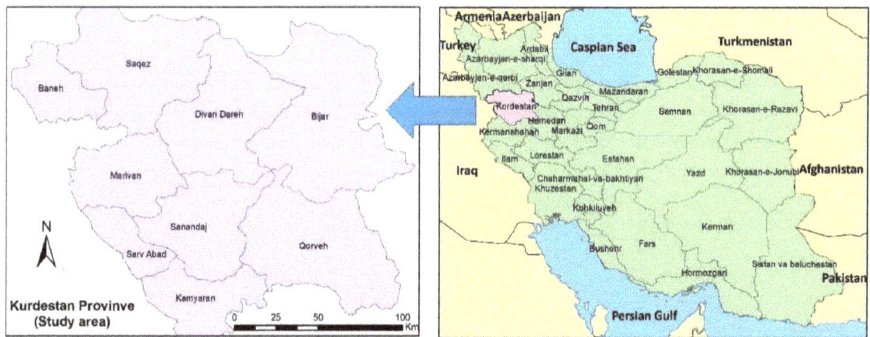

Figure 1. The location map of the study area.

In terms of climate, Kurdistan is a mountainous region, which has extensive plains and strewn valleys across the area. The lowest elevation is 900 meters above sea level (m.a.s.l) and highest is 3300 m.a.s.l, this height difference in elevation makes the creation of different climates. This province has been chosen as a case study and an example in the country and so the methods of research can be applied to other provinces of Iran too.

3. Methodology

In this paper, a GIS-based evaluation and integration method [29,30] is used for assessing the solar resource in Kurdistan. GIS is a powerful computer tool that enables the user to record, store, and organize databases, and analyze spatial and geographical data in different coordinate systems. Due to the nature of renewable energy, it is necessary to take several geographical analyses in their resources assessment [31]. In this paper, different data for the evaluation of solar resources in Kurdistan were obtained mostly from the Renewable Energy and Energy Efficiency Organization of Iran (Satba) [32] and also other national and International agencies and organizations. The solar irradiances were recorded in more than 150 stations in Iran during last 10 years by Satba and many other local power companies.

For solar resource assessment and in general RESs assessment, two types of potentials are usually considered—the geographical (theoretical) potential and technical potential. The technical potential is derived from the conversion potential, taking into account additional restrictions regarding the area that is realistically available for energy generation, while the theoretical potential identifies the physical upper limit of the energy available from a certain source [8].

The theoretical potential of solar energy is defined as the total amount of annual solar radiation in suitable areas for solar applications, which should take into account multiple constraints in the assessment phase to achieve the appropriate areas [8]. The technical potential is defined as the amount of the total theoretical potential that can be converted into electricity using existing technologies [8]. However, the technical potential can be limited more and more by economic assessments and evaluating competitiveness with non-renewable technologies [33]. In this paper, both the theoretical and technical potentials of the solar resources have been calculated.

In the solar power plant applications, solar radiation map in the province is prepared using data from meteorological stations, sunshine duration, and the Solar Radiation tool [34]. The technical, economic, environmental, and geographical constraints are used to eliminate unsuitable areas and identifying suitable sites for solar applications. According to the remaining locations and assessing the amount of producible power from CSP and PV power plants, the capacity for solar power plants in the study area can be calculated. For the general application of solar power, the available roof area for this application is calculated in residential buildings. According to this area, the production capacity of domestic PV can be assessed. The effect of using solar water heaters in fuel consumption reduction can be calculated by computing fuel consumption for hot water.

4. Solar Resource Assessment

In calculating the solar energy potential, according to the available technologies, it is necessary to determine which one of the technologies will be applicable. Therefore, the following technologies are selected among the existing technologies because they have been widely used worldwide and have been employed in various provinces of Iran.

- Power plant applications

    - CSP plants
    - PV farms

- General Applications

    - Rooftop PV panels
    - Solar water heaters

In the power plant applications, it is necessary to obtain the solar radiation map in the whole province. Therefore, the next section describes how to create the province's solar radiation map.

4.1. Solar Radiation Map

Radiation data can be obtained in several ways. Generally, these methods can be divided into two main categories:

- The satellite data
- Ground measured data

At present, various meteorological satellites are active in the atmosphere, but due to lack of access, unfortunately, in this study, it is not possible to use the satellite data. The ground data is commonly recorded by weather stations or radiation survey meters such as pyrometers. In all meteorological stations, which are located in the Kurdistan province, the solar radiation is not recorded and archived. The only data that are recorded for a long time, and are associated with radiation, are the number of sunny hours per day, available at all stations in the province—considering that data of the sunny hours per day are available in the stations. Recorded data are used to extract the radiation data in the stations. For this purpose, an efficient Angstrom–Prescott model [35] is used, which has long been used in solar radiation research, and its efficiency has been proven. This model is a statistical model with two constant factors. Data for Sanandaj (the most reliable data) in recent years is used to determine the constants of the model. In summary, it can be said the Angstrom model uses the Equation (1):

$$\frac{\overline{H}}{\overline{H}_0} = a + b\left(\frac{n}{N}\right) \tag{1}$$

where $\overline{H}$ is actual solar radiation on a horizontal surface that has reached to ground. $\overline{H}_0$ is solar radiation outside the atmosphere at the same point. Both are in W/m^2. n is the number of sunny hours that is measured in the station, and N is the total number of hours per day. The Angstrom model is applied to Sanandaj station (airport), and the obtained coefficients will be applied to the other stations' data. $\overline{H}/\overline{H}_0$ in the vertical axis and n/N in the horizontal axis (average per month as one point) is drawn. This diagram is shown in Figure 2.

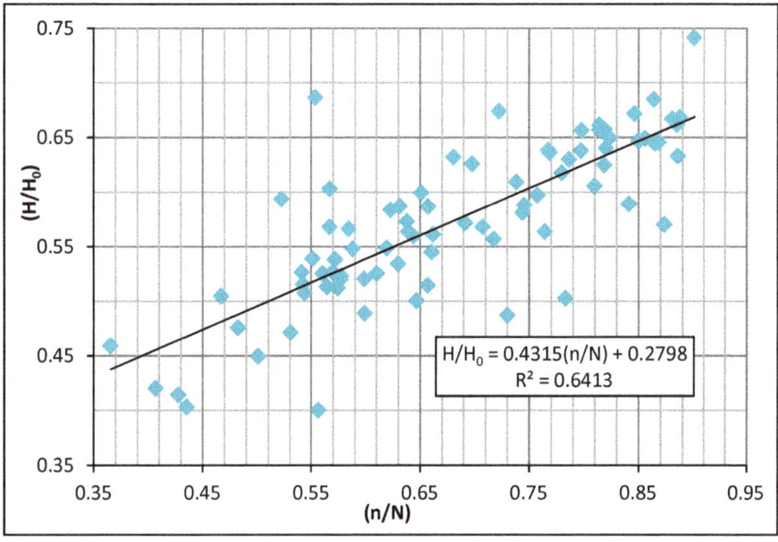

Figure 2. Calculated Angstrom model coefficients in Sanandaj+ station.

As seen in Figure 1, coefficients "a" and "b" are 0.2798 and 0.4315, with standard errors of 2.52% and 3.61%, respectively, in Sanandaj station, and the R^2 is 0.6413. These factors are applied to the other stations in the province to calculate radiation value from the sunshine duration. Average radiation in all stations in the province can be calculated accordingly, as reported in Table 1. Similar results obtained in other studies [32,36] indicate the effectiveness and accuracy of the results obtained from the Angstrom method in this paper [37].

Table 1. Annual radiation obtained from Angstrom model in the province stations.

Station	Computed Annual Radiation (kWh/m²)
Saqqez	1713.57
Marivan	1747.58
Sanandaj	1734.76
Qorveh	1770.55
Kamyaran	1791.93
Bijar	1745.71
Diwandarreh	1759.68
Baneh	1709.67

In the next step, it is necessary to consider the topographic conditions in providing a radiation map. By using the digital elevation model (DEM) [38] and Solar Radiation tool and by applying clearness and diffuse coefficients 0.5 and 0.3, respectively, using clearness index from Homer Energy [39] (in this map it is assumed that these coefficients are constant in the whole province) the radiation map can be obtained. Radiation at the stations is calculated from this map and radiation obtained from the Angstrom model is divided on it and the resulting number will be called the coefficient of each station. The value of this coefficient is interpolated in the whole area and the resulted map is called the factor map. By multiplying the factor map in the radiation map obtained from the Solar Radiation tool the radiation map of the province is achieved. DEM map and the final map of solar radiation for the whole province are shown Figures 3 and 4, respectively.

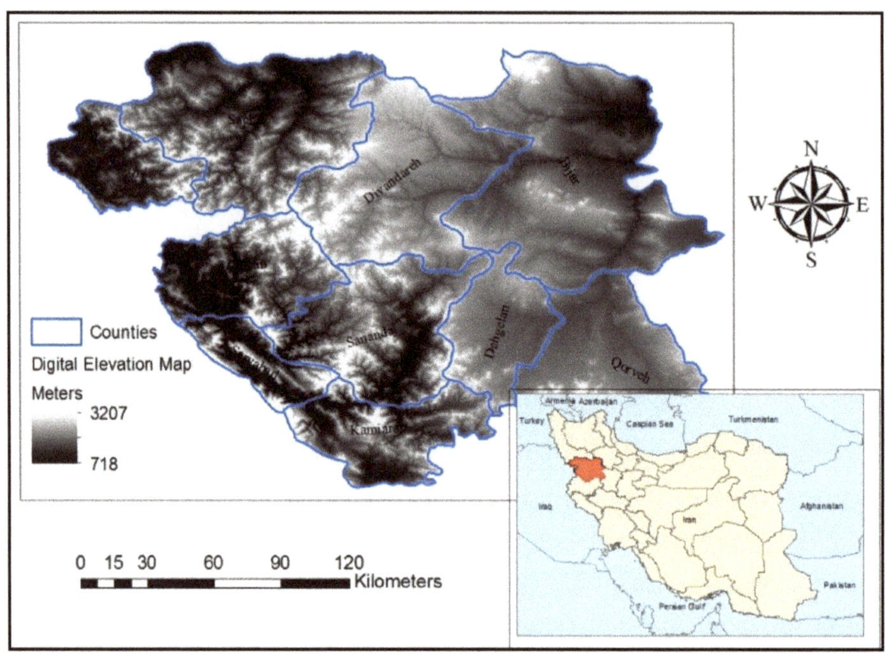

Figure 3. Digital elevation model (DEM) map in Kurdistan Province.

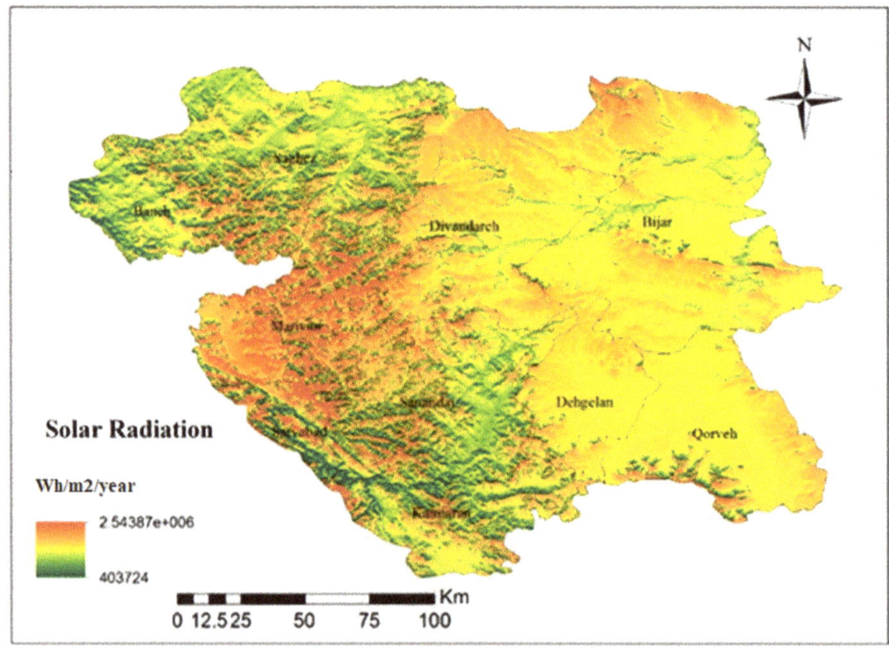

Figure 4. Final solar radiation map for Kurdistan province per year.

4.2. Site Selection

The main objective of the solar resource assessment is the evaluation of radiation values, as well as finding potential areas where solar power generation is possible with available technologies. In this case, according to the accuracy of the available data, narrowing the entire area to appropriate areas continues until the suitable areas are finally identified. The assessment process is based on a systematic survey process. At each step, less attractive areas are removed and the process continues on promising areas. Data and information management in the resource assessment process requires the integration and interpretation of results to identify suitable areas. A huge amount of technical and non-technical information is needed to identify suitable areas. The restrictive analytical method [40] is used to identify potentially suitable locations. In this regard, the study area can be divided into two parts based on the possibility and unfeasibility of applying solar technologies. The Boolean logic (restrictive) method uses pre-defined constraints (as shown in Figure 4) for identifying the suitable location for placing the solar technologies. Boolean logic uses a binary condition for input and checks a binary condition for the outputs. Logical math tools consider the value 0 for false conditions and the value of 1 for the true conditions. With respect to each criterion, the study area is divided into two discrete classes: 1 for areas with the possibility of constructing solar power plants, and the 0 for unfeasibility of the areas [41]. The main steps in the restrictive method are as follows:

- creating a conceptual model;
- determining and localization of the desired criterion for site selection;
- collecting the required data;
- assessment of the study area;
- identification of promising and probable areas based on desired criteria for each data layer;
- using a data integration method based on the conceptual model;
- determining the suitable areas for the construction of the solar power plant (site selection).

A flow diagram (conceptual model) of data intergration method using restrictive data layers to select potential solar power plants sites is shown in Figure 5. Based on this integration method 12, data layers are applied to evaluate the study area for finding and defining suitable areas. The restrictive method's data layers are divided into four main groups: technical, economic, environmental, and geographical constraints. Previous studies have used many layers to apply their criteria and constraints. These indicators have been assessed and localized for Iran, according to national and local laws. Criteria and constraints related to solar site selection are given in Table 2.

By considering the economic indicators and given the need to create a temporary road from project sites to transport links, the distance to these links should not exceed a specified limit because it will increase operational costs. In this research, this distance is intended for up to 10 km.

Given the need to avoid comfort disturbance during construction, as well as the need to keep the industrial areas away from population centers, distance from residential areas should be observed. A 2000 m buffer zone for the cities and 1000 m for the villages are considered in this study.

Consideration of 250 m buffer zone for the transport links is due to property rights that are considered in the road property rights laws. Moreover, according to the power lines property rights laws, there should be a required minimum safety corridor around power lines with consideration to safety clearances. High-voltage transmission lines at different voltages have different limits, where the highest degree is taken into account here, which corresponds to 750 kV lines.

For wetlands, coastlines, forests, and faults a 500 m buffer zone is considered, while this is around 2000 m for environmental protected areas and 700 m for historical and cultural sites.

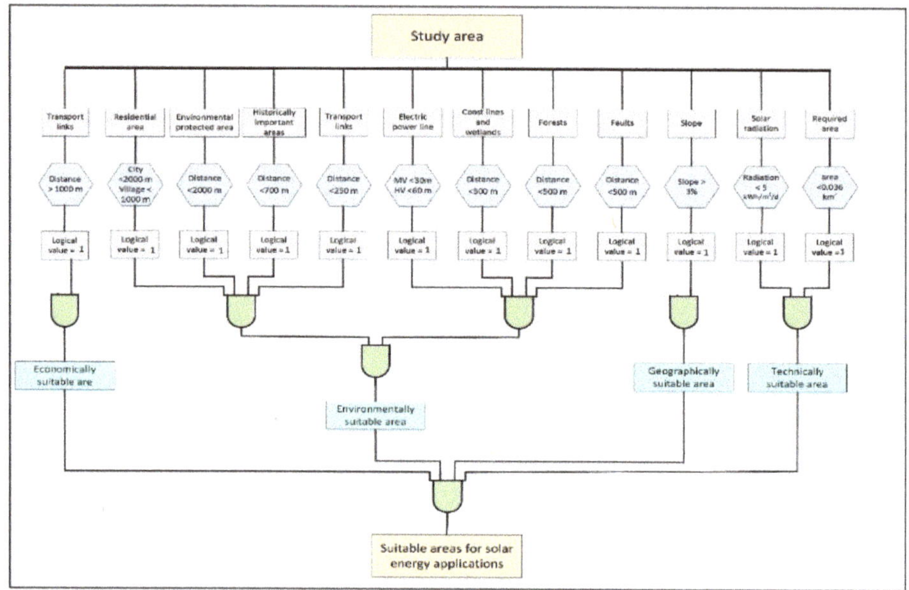

Figure 5. The conceptual model of the restrictive method.

Table 2. Criteria and constraints of solar site selection.

Category	Data	Criteria (Unsuitable Land)	References
Economic	Transport links	Area in distance > 10,000 m	[42]
Environmental	Residential area	Area in distance to cities < 2000 m and to village < 1000 m	[42,43]
	Environmental protected area	Area in distance < 2000 m	[44]
	Historically important areas	Area in distance < 700 m	[41]
	Transport links	Area in distance < 250 m	[42,45]
	Electric power line	The area in distance to medium-voltage lines < 30 m and to high voltage lines < 60 m	[46]
	Coastlines and wetlands	Area in distance < 500 m	[41]
	Forests	Area in distance < 500 m	[41]
	Faults	Area in distance < 500 m	[41]
Geographical	Slope	Area with slope of > 3%	[47]
Technical	Solar radiation	Area with radiation of < 5 kWh/m²/d	[47]
	Required area	Area with area < 0.036 km²	[47]

By applying these economic and environmental restrictions on radiation maps, unsuitable areas for the construction of solar power plants will be removed from the primary radiation map. A series of regions are also omitted due to technical reasons. According to the analysis carried out in [47], a minimum amount of radiation is required (here 5 kWh/m²/d) at a location to obtain power in an acceptable range from solar power plants. Accordingly, some parts are removed as well, and so radiation maps will be limited even more.

The photovoltaic panels with a nominal capacity of 1 kW, occupies an area of about 10 m². In order to install a power plant with 2 MW capacity, an area of over 20,000 m² is required [47]. This also applies in relation to CSP. From the perspective of geographical constraints in solar power plants due to the construction costs and also for receiving maximum sunlight, the slope should be less than 3%.

As appears from the conceptual model in Figure 4, due to technical, economic, environmental, and geographical constraints with addressing the entire area of study, logical value for the areas that are not suitable for the construction of solar plants is zero, and these areas will be removed. Finally, all the appropriate technical, economic, environmental, and geographical layers are integrated to obtain final proper areas. It can be seen that the final areas are selected as suitable sites that satisfy all constraints, and in all criteria have a logical value of 1.

By applying economic and environmental constraints, suitable areas map for solar power plant construction will be as Figure 6. Province slope map was created using DEM maps. Geographically permissible area map (areas with slopes less than 3%) is shown in Figure 7. Finally, by applying the technical constraints, and the integration of four layers of suitable areas, in terms of technical, economic, environmental, and geographical constraints. Figure 8 is achieved as the final map of suitable areas for solar power plant applications. According to the map of Figure 8, it can be seen that an area of about 62 km^2 is available in the province for the construction of solar power plants.

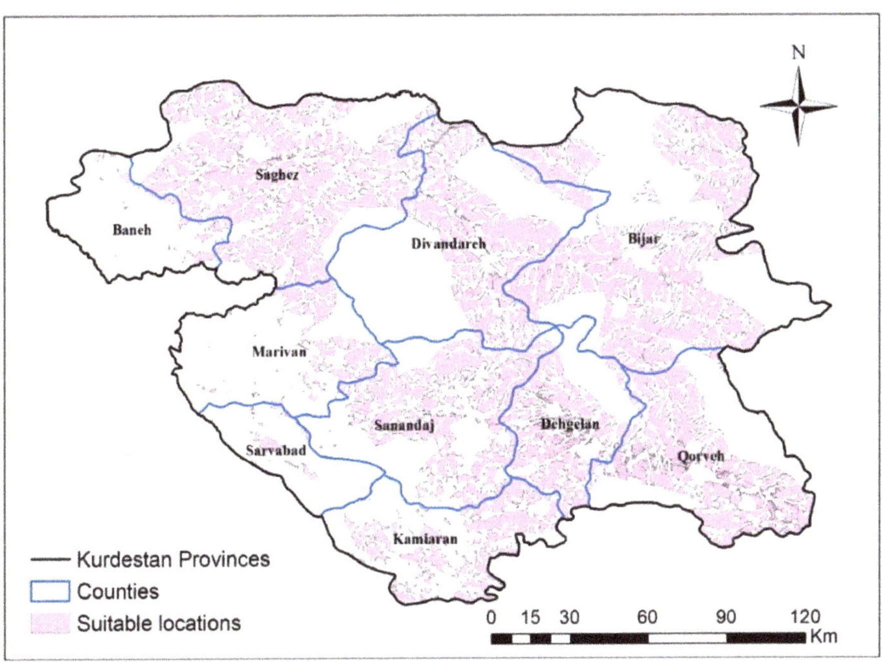

Figure 6. Suitable locations based on economic and environmental considerations.

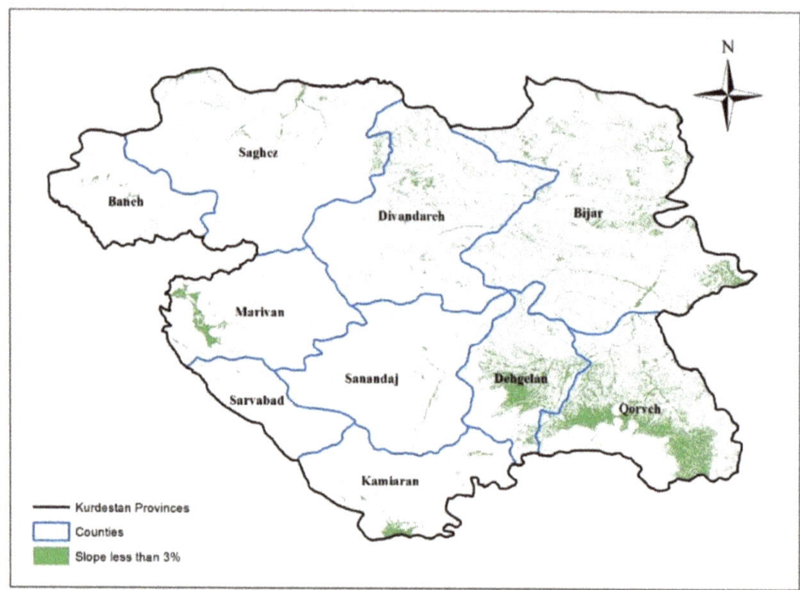

Figure 7. Selected area based on the geographical criterion.

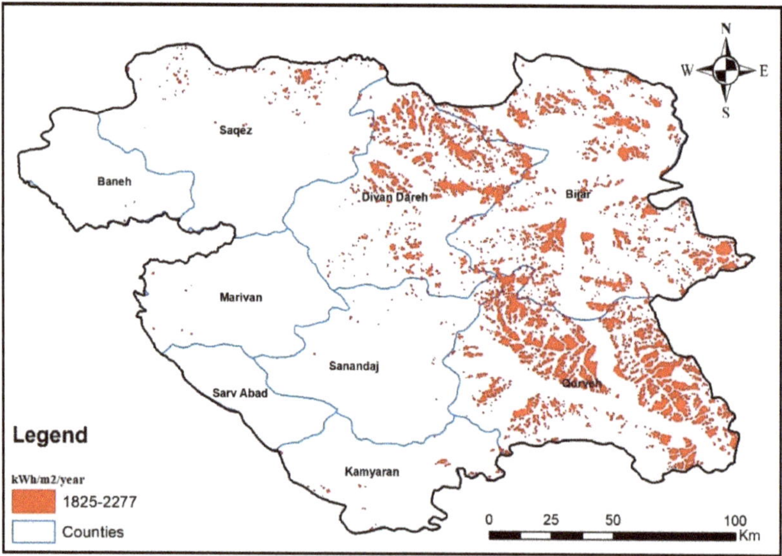

Figure 8. Final suitable area based on a restrictive method for solar power plant applications.

4.3. Solar Power Production Estimation

4.3.1. CSP Plants

As mentioned, in this study two application of solar power plant is intended. The first case relates to a CSP system with the following characteristics [47]:

- dry-cooled;

- six hours of storage;
- the amount of solar multiple equal to 2;
- the power density of 32.895 MW/km^2.

The performance of the system will be different for different environmental conditions. According to an analysis done with the System Advisor Model (SAM), various coefficients values (called capacity factors) are obtained for different radiation values [47]. By categorizing solar radiation to different classes, Table 3 is achieved.

Table 3. Capacity factors for CSP.

Class	kWh/m^2/d	Capacity Factor
1	5–6.25	0.32
2	6.25–7.25	0.39
3	7.25–7.5	0.43
4	7.5–7.75	0.43
5	> 7.75	0.45

Finally, the potential of solar energy (in MWh) in the CSP system will be obtained from the following Equation.

$$P = A\overline{P}_{max} \times h_{yr} \times z \tag{2}$$

where A is the available area for the construction of solar power in each class. $\overline{P}_{max}$ is the maximum power produced by this system, which is equal to 32.895 MW/km^2. h_{yr} is the number of hours in a year, which is equal to 8760, z is the capacity factor in each class and is obtained from Table 3. Therefore, from Equation (2), the potential of solar energy in each class will be achieved, and the total potential of different classes will be equal to the potential of this application. The amount of radiation at selected suitable locations for solar power plant applications (Figure 7) are placed in class 1 of Table 3. Therefore, the capacity factor (z) will be equal to 0.315. According to the map of Figure 7, the amount of available area (A) is equivalent to 62.289 km^2. By inserting these numbers in Equation (2), the evaluated power in the application of CSP plants will be equal to 645 MW.

4.3.2. PV Farms

PV power plants usually use a lot of same panels in a wide range of ground, so each panel reveals the function of the whole system. The considered system, which its analysis is available in [47] has the following characteristics:

- single-axis tracking, with a north-south axis of rotation;
- the slope angle of zero degrees;
- maximum power of 48 MW/km^2.

These systems will create a full-function only in designed conditions. According to the analysis conducted by NREL, the performance of these systems under various conditions is a coefficient of maximum power. These coefficients in different climatic conditions, and in various US states, have been presented in [47]. The capacity factor of 0.231 is considered for Kurdistan province 3030. Therefore, the power generated by the small PV power plant systems, in Kurdistan province is calculated from Equation (2), and with a capacity factor of (Z) 0.231, maximum power of 48 MW/km^2, and an available area of 62.289 km^2. By inserting these numbers in Equation (2), the evaluated power in the solar PV power plants application will be equal to 691 MW. This is the potential power of the solar PV power plants in Kurdistan province.

In the final evaluation of solar power plants in Kurdistan province from PV and CSP plants, it should be mentioned that solar power plant potential is not the sum of these two numbers, because if you use one of them, another cannot be used, and this amount depends on the amount of the use of these two applications based on desired scenarios.

4.4. Residential Applications

Another application in the use of solar energy is related to small-scale applications. In this regard, two applications are considered in this study: Rooftop PV panels and solar water heaters. In the residential areas, the installation location of this equipment is the flat and non-shadow surface of rooftops, which is obtained as fallows.

4.4.1. Available Area for Residential Applications

In accordance with the Master Plan maps of urban areas, and after analyzing in the ArcMap software, and separation of residential buildings, the surface areas of these sites will be achieved (Figure 9). Unfortunately, Master Plan maps in Marivan and Sarvabad cities, and urban areas that are not located in the center of the counties were not available. Thus, from the population in cities with Master Plan maps, we tried to estimate the other cities' residential areas. Thus, by calculating the per capita residential areas in cities with maps, and their average, and then by multiplying this average by the population of cities without Master Plan map, the residential area can be calculated in these cities.

Figure 9. Calculation of urban residential areas by using the ArcMap.

In rural areas, a roof area of 100 m^2 can be considered for each family [48]. According to the statistics of rural households in each city, we can also calculate the rural residential area. Population, occupied area by residents and residential area per capita in cities with Master Plan in Kurdistan province, is presented in Table 4.

The average residential area is 27.79 m^2/person. Using this average, the residential area is calculated in other urban areas (urban areas without a Master Plan map). By gathering the urban areas of each county together, the urban residential areas in each county will be achieved (the second column of Table 5). Furthermore, with regard to the consideration of 100 m^2 for each rural household, the approximated rural residential area can be obtained for each county (the third column of Table 5) and by summing these two values, the entire residential area in the counties can be achieved (fourth column of Table 5).

Table 4. Population, area of residential areas, and residential area per capita in cities with Master Plan.

City	Population	Residential Areas (m^2)	Residential Area Per Capita (m^2/pp)
Baneh	85,190	2,360,803	27.71
Bijar	47,926	1,657,237	34.58
Dehgholan	23,074	562,866	24.39
Diwandarreh	26,654	664,040	24.91
Saqqez	139,738	3,805,918	27.24
Sanandaj	373,987	870,4320	23.27
Qorveh	71,232	2,417,813	33.94
Kamyaran	52,907	1,391,334	26.30
Average			27.79

Table 5. Residential areas in the counties of Kurdistan province in the general applications of solar energy.

County	Urban Residential Area (m^2)	Rural Residential Area (m^2)	County Residential Area (m^2)	Available Area (m^2)	Technical PV Potential (kW)
Baneh	2,502,940	930,400	3,433,340	60,427	897.34
Bijar	1,778,334	1,112,600	2,890,934	50,880	755.57
Dehgolan	652,000	994,100	1,646,100	28,971	430.22
Diwandarreh	715,570	1,272,000	1,987,570	34,981	519.47
Sarvabad	138,301	1,221,300	1,359,601	23,929	355.35
Saqqez	3,869,677	1,618,500	5,488,177	96,592	1434.39
Sanandaj	8,740,257	2,068,600	10,808,857	190,236	2825.01
Qorveh	2,878,077	1,370,800	4,248,877	74,780	1110.48
Kamyaran	1,487,167	1,354,400	2,841,567	50,012	742.68
Marivan	3,392,582	1,137,100	4,529,682	79,722	1183.87
Total	26,154,905	13,079,800	39,234,705	690,530	10,254.37

According to the analysis conducted by NREL [49], it can be assumed that in lands with residential applications, 8% of the area is flat and on the roof; thus, can be considered as a potential for installing solar equipment. In addition, other factors, such as shading, roof structure, and so on will restrict this amount. According to this analysis, in the cold conditions (similar to Kurdistan province) 22% of the previous value is appropriate for using solar energy [49]. Therefore, available areas for solar energy in general applications in each county, obtained by multiplying the coefficients of 0.08 and 0.22 in the residential area of the counties, which is shown in the 5th column of Table 5. After identifying this area, we must determine the potential solar energy production from the two above-mentioned applications.

4.4.2. Rooftop PV Panels

In the use of solar PV panels, the density of power generation in the area is considered around 110 W/m2. On average, the efficiency of small-scale PV systems can also be considered as 13.5% [47]. So by multiplying these two amounts in the available area, the potential of the producible power of photovoltaic panels can be obtained. This amount is presented in the last column of Table 5, for each county. This amount is more than 10.2 MW in total for Kurdistan province.

4.4.3. Solar Water Heater

In Iran, natural gas and diesel are mainly used for water heating. Therefore, by having gas and diesel consumption in the residential sector in each county, the amount of fuel needed to generate hot water can be achieved. For this purpose, the following assumptions are made in this study.

- About 23% of fuel consumption in the residential sector is for hot water generation [25];
- solar water heater systems will reduce about 50% of fuel consumption for hot water [25];
- the efficiency of water heating systems based on gas and diesel is expected to be 50% [50];

- the energy content of natural gas and diesel used in residential sector is 37.2 MJ/m^3 and 36.6 MJ/l respectively.

According to the above cases, we can consider the following equations:

$$V_{11} = V_1 \times 37.2 \tag{3}$$

$$V_{21} = V_2 \times 36.6 \tag{4}$$

$$P = [(V_{11} + V_{21}) \times 0.5 \times 0.23]/0.5 \tag{5}$$

where, V_1 is the volume of natural gas consumed in the residential sector (m^3), V_2 is the volume of diesel consumed in the residential sector (L), and V_{11} is energy from the consumed natural gas (MJ), V_{21} is energy from consumed diesel (MJ) and P is the potential of solar energy for domestic hot water supply (MJ). Hence, the potential for energy saving will be achieved in each county. Statistics related to fuel consumption has been extracted from 3737, and the percentage of consumption in the residential section for 2011 has been extracted from 2727. Iran's energy balance report for 2011 indicates that 70.18% of natural gas and 1.23% of diesel consumption is in the residential sector (as well as gas and diesel consumption in each county of the province is shown in Table 6, see the second and third columns).

Table 6. Calculation of fuel consumption in the residential sector and its savings potentials by application of solar water heater in the Kurdistan.

County	Diesel Demand of Residential Sector (L)	Natural Gas Demand of Residential Sector (m^3)	Diesel Saving (L)	Natural Gas Saving (m^3)	Potential Solar Water Heater Capacity (MW)	Energy Saving (MJ)
Baneh	299,388	89,829,201	68,859	20,660,716	24.45	771,098,892
Bijar	552,666	112,286,501	127,113	25,825,895	30.61	965,375,647
Dehgolan	361,258	54,739,669	83,089	12,590,124	14.95	471,393,685
Diwandarreh	390,856	38,598,485	89,897	8,877,652	10.58	333,538,859
Sarvabad	146,734	12,632,231	33,748	2,905,413	3.46	109,319,549
Saqqez	679,391	132,638,430	156,260	30,506,839	36.17	1,140,573,520
Sanandaj	1,270,842	530,553,719	292,294	122,027,355	144.29	4,550,115,566
Qorveh	567,723	106,672,176	130,576	24,534,601	29.09	917,466,237
Kamyaran	407,011	52,634,298	93,612	12,105,888	14.39	453,765,266
Marivan	582,559	98,952,479	133,989	22,759,070	27.00	851,541,395
Total	5,111,694	1,229,537,190	1,175,690	282,793,554	334.97	10,562,950,439

Given that solar water heaters will reduce about 23% of fuel consumption in the residential sector, natural gas and gasoline savings potentials by application of solar water heater in each county have been computed and are shown in Table 6, (fourth and fifth columns). Finally, the amount of energy saving in fuel consumption, as a result of using solar water heaters in each county can also be calculated according to Equation (5). These results are shown in Table 6, in the sixth and seventh columns, separately for each county.

4.4.4. Evaluation of Total Solar Energy Potential

As mentioned, the available area will be for the use of both systems (not simultaneously); and so it cannot be represented total solar energy potential in general application. This will be possible by considering more scenarios, close to reality forecasts, and more detailed assessment. Therefore, solar potential map in these two general applications is presented separately for each application. The results of this section are shown in the map of Figure 10 as a potential use of solar water heater and Figure 11 shows domestic PV potential.

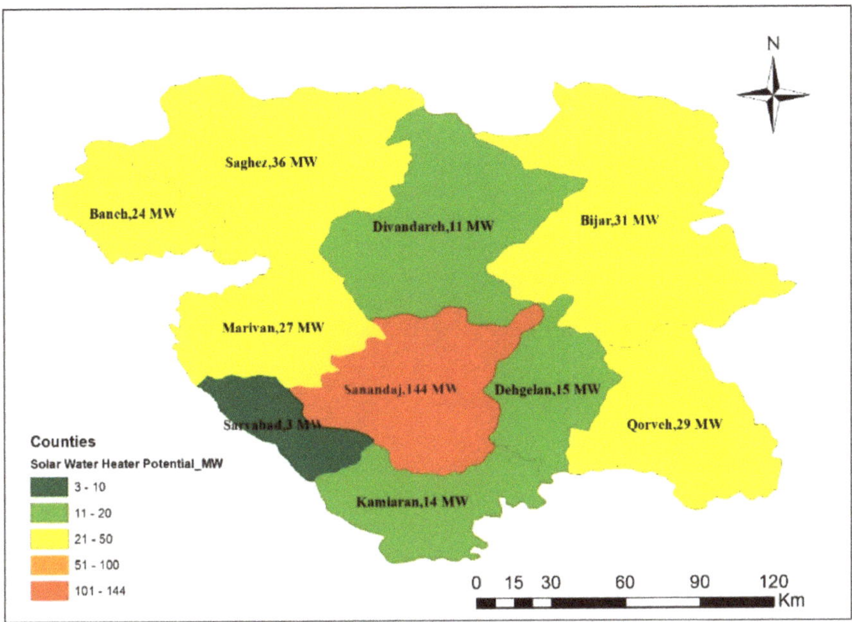

Figure 10. Potential of solar water heaters in Kurdistan.

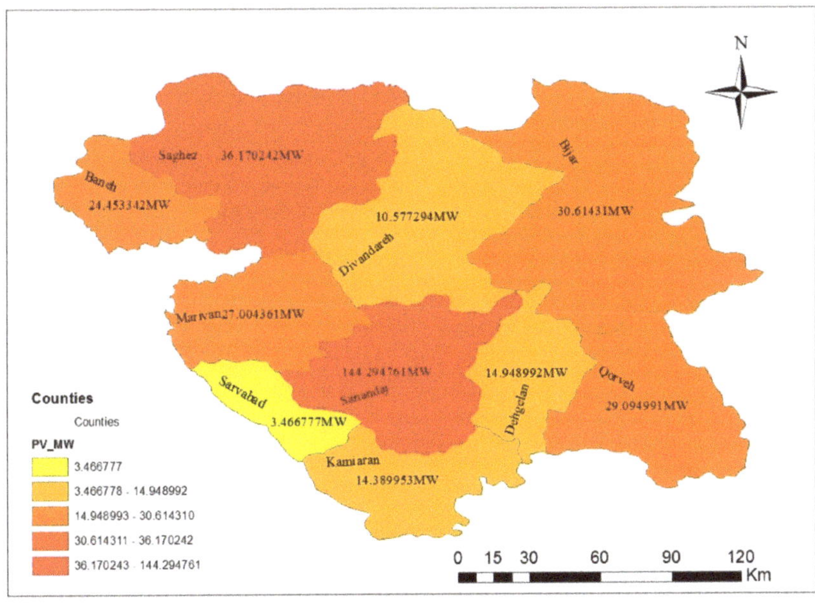

Figure 11. Potential of domestic photovoltaic in Kurdistan.

The results show that in terms of the potential use of solar water heater, Sanandaj is the highest and Sarvabad is the lowest priority. The same is true for domestic PV potential. In fact, in the general applications of solar potential, there is a direct proportion to the population. These results also show that fuel consumption saving by using solar water heaters is 1.2 million liters for diesel and

282.8 million cubic meters of natural gas. As well as by implementation of the domestic PV systems, power consumption is reduced more than 10 MW. The technical potential of solar energy in Kurdistan province, using various solar technologies are summarized in Table 7. As can be seen in the table, the results indicated that the Kurdistan province has capability to produce 691 MW of power from PV panels, and 645 MW from CSP plant, which are more than total power demand of the province. It means that by developing full capacity of solar potential in the Kudrestan total power demand can be supplied in the future. In the case of using solar energy for water heating, totally, 283 million cubic meters of natural gas and 1.2 million liters of gasoline could be saved. By developing solar energy, system energy in the area the energy supply security could be increased in wintertime when the high snowfall blocks the roads for fuel transportation.

Table 7. The technical potential of solar energy in Kurdistan province (PV: Photovoltaic, CSP: concentrating solar power)

Technology Type	Technical Potential	Unit	Consumption Sector
PV power plant	691	MW	Electricity
CSP plant	645	MW	Electricity
Rooftop PV panels	10.2	MW	Electricity
Solar water heater	282	Million m^3	Heating-natural gas
Solar water heater	1.2	Million liter	Heating-diesel

5. Conclusions

In this paper, solar energy resources in Iran's Kurdistan province were identified and evaluated. At first, solar resource-related data were collected in Kurdistan province and an appropriate assessment method was selected based on available data. Then, the theoretical potential of the solar energy in the province was evaluated and solar energy atlas was prepared. In the following, the technical potential of various solar technologies was studied. These technologies include CSP plant and PV farms in power plant applications, and rooftop PV panels and solar water heaters in general domestic applications.

In solar power plant applications, primarily using data from meteorological stations, sunlight duration, and Solar Radiation tool, radiation map was prepared in Kurdistan province. Then, using the technical, economic, environmental, and geographical constraints, some areas were discarded and finally, the suitable areas were identified. The available technical potential in the use of solar power plants was evaluated by considering the suitable areas and the recoverable amount of power in the application of PV and CSP plants. For general application, the available area in residential rooftops was calculated. Finally, with respect to this area, the production capacity of domestic PV was evaluated. The effect of solar water heater implementation in gas and diesel consumption value was calculated using fuel consumption for hot water and resulted in saving in the use of this application.

The results show that the Kurdistan province is capable of achieving, 691 MW of PV power generation, and 645 MW of CSP plant. In the case of using solar water heaters, 283 million cubic meters of natural gas and 1.2 million liters of gasoline could be saved. The savings in the application of domestic PV panels will be 10.2 MW in power generation.

Future works will be dedicated to the assessment of the potential of other renewable energy sources (RESs), comprehensive evaluation of the use of RESs, and finally, evaluation of different scenarios to increase the share of these sources in the consumption pattern of the Kurdistan province.

Author Contributions: Y.N. and M.M. designed the model and gathered the input data. H.Y. and Y.N. executed the GIS analysis and site selection, and accomplished writing of the paper. A.A.-M. revised the manuscript and checked the results. All authors have read and agreed to the published version of the manuscript.

Funding: This research received no external funding.

Conflicts of Interest: The authors declare no conflict of interest.

Sustainability **2020**, *12*, 1890

Nomenclature

$\overline{H}$	Actual solar radiation on a horizontal surface in Watts/m^2
$\overline{H}_0$	Solar radiation outside the atmosphere in Watts/m^2
n	Number of sunny hours
N	Total number of hours per day
SAM	System Advisor Model
$\overline{P}_{max}$	The maximum power produced
h_{yr}	The number of hours in a year
Z	Capacity factor
CSP	Concentrating solar power
RES	Renewable energy sources
PV	Photovoltaic
GIS	Geographic information systems
IEA	International energy agency
Mtoe	Million tons of oil equivalent
Mt	Million tons
NREL	National Renewable Energy Laboratory
DEM	the digital elevation model

References

1. Mohammadi, M.; Noorollahi, Y.; Mohammadi-ivatloo, B.; Yousefi, H. Energy hub: From a model to a concept—A review. *Renew. Sustain. Energy Rev.* **2017**, *80*, 1512–1527. [CrossRef]
2. OECD/IEA. *Key World Energy Statistics 2016*; International Energy Agency: Peris, France, 2016; pp. 1–77.
3. Mohammadi, M.; Ghasempour, R.; Razi Astaraei, F.; Ahmadi, E.; Aligholian, A.; Toopshekan, A. Optimal planning of renewable energy resource for a residential house considering economic and reliability criteria. *Int. J. Electr. Power Energy Syst.* **2018**, *96*, 261–273. [CrossRef]
4. Noorollahi, Y.; Itoi, R.; Yousefi, H.; Mohammadi, M.; Farhadi, A. Modeling for diversifying electricity supply by maximizing renewable energy use in Ebino city southern Japan. *Sustain. Cities Soc.* **2017**, *34*, 371–384. [CrossRef]
5. Østergaard, P.A.; Duic, N.; Noorollahi, Y.; Mikulcic, H.; Kalogirou, S. Sustainable development using renewable energy technology. *Renew. Energy* **2019**. [CrossRef]
6. Eslami, S.; Gholami, A.; Bakhtiari, A.; Zandi, M.; Noorollahi, Y. Experimental investigation of a multi-generation energy system for a nearly zero-energy park: A solution toward sustainable future. *Energy Convers. Manag.* **2019**, *200*, 112107. [CrossRef]
7. Zawilska, E.; Brooks, M.J. An assessment of the solar resource for Durban, South Africa. *Renew. Energy* **2011**, *36*, 3433–3438. [CrossRef]
8. Ghasemi, G.; Noorollahi, Y.; Alavi, H.; Marzband, M.; Shahbazi, M. Theoretical and technical potential evaluation of solar power generation in Iran. *Renew. Energy* **2019**, *138*, 1250–1261. [CrossRef]
9. Viana, T.S.; Rüther, R.; Martins, F.R.; Pereira, E.B. Assessing the potential of concentrating solar photovoltaic generation in Brazil with satellite-derived direct normal irradiation. *Sol. Energy* **2011**, *85*, 486–495. [CrossRef]
10. Journée, M.; Müller, R.; Bertrand, C. Solar resource assessment in the Benelux by merging Meteosat-derived climate data and ground measurements. *Sol. Energy* **2012**, *86*, 3561–3574. [CrossRef]
11. Fluri, T.P. The potential of concentrating solar power in South Africa. *Energy Policy* **2009**, *37*, 5075–5080. [CrossRef]
12. Ortega, A.; Escobar, R.; Colle, S.; de Abreu, S.L. The state of solar energy resource assessment in Chile. *Renew. Energy* **2010**, *35*, 2514–2524. [CrossRef]
13. Watson, J.J.W.; Hudson, M.D. Regional Scale wind farm and solar farm suitability assessment using GIS-assisted multi-criteria evaluation. *Landsc. Urban Plan.* **2015**, *138*, 20–31. [CrossRef]
14. Polo, J.; Bernardos, A.; Navarro, A.A.; Fernandez-Peruchena, C.M.; Ramírez, L.; Guisado, M.V.; Martínez, S. Solar resources and power potential mapping in Vietnam using satellite-derived and GIS-based information. *Energy Convers. Manag.* **2015**, *98*, 348–358. [CrossRef]

15. Sun, Y.; Hof, A.; Wang, R.; Liu, J.; Lin, Y.; Yang, D. GIS-based approach for potential analysis of solar PV generation at the regional scale: A case study of Fujian Province. *Energy Policy* **2013**, *58*, 248–259. [CrossRef]

16. He, G.; Kammen, D.M. Where, when and how much solar is available? A provincial-scale solar resource assessment for China. *Renew. Energy* **2016**, *85*, 74–82. [CrossRef]

17. Stökler, S.; Schillings, C.; Kraas, B. Solar resource assessment study for Pakistan. *Renew. Sustain. Energy Rev.* **2016**, *58*, 1184–1188. [CrossRef]

18. Gautam, B.R.; Li, F.; Ru, G. Assessment of urban roof top solar photovoltaic potential to solve power shortage problem in Nepal. *Energy Build.* **2015**, *86*, 735–744. [CrossRef]

19. Azizkhani, M.; Vakili, A.; Noorollahi, Y.; Naseri, F. Potential survey of photovoltaic power plants using Analytical Hierarchy Process (AHP) method in Iran. *Renew. Sustain. Energy Rev.* **2017**, *75*, 1198–1206. [CrossRef]

20. Sánchez-Lozano, J.M.; Teruel-Solano, J.; Soto-Elvira, P.L.; Socorro García-Cascales, M. Geographical Information Systems (GIS) and Multi-Criteria Decision Making (MCDM) methods for the evaluation of solar farms locations: Case study in south-eastern Spain. *Renew. Sustain. Energy Rev.* **2013**, *24*, 544–556. [CrossRef]

21. Uyan, M. GIS-based solar farms site selection using analytic hierarchy process (AHP) in Karapinar region, Konya/Turkey. *Renew. Sustain. Energy Rev.* **2013**, *28*, 11–17. [CrossRef]

22. Perpiña Castillo, C.; Batista e Silva, F.; Lavalle, C. An assessment of the regional potential for solar power generation in EU-28. *Energy Policy* **2016**, *88*, 86–99. [CrossRef]

23. Habte, A.; Stoffel, T.; Perez, R.; Myers, D.; Gueymard, C.; Blanc, P.; Wilbert, S. *Best Practices Handbook for the Collection and Use of Solar Resource Data for Solar Energy Applications*, 2nd ed.; National Renewable Energy Laboratory (NREL): Golden, CO, USA, 2017; p. 238.

24. Renné, D.; George, R.; Wilcox, S.; Stoffel, T.; Myers, D.; Heimiller, D. *Solar Resource Assessment*; National Renewable Energy Lab.(NREL): Golden, CO, USA, 2008.

25. Denholm, P. *The Technical Potential of Solar Water Heating to Reduce Fossil Fuel Use and Greenhouse Gas Emissions in the United States The Technical Potential of Solar Water Heating to Reduce Fossil Fuel Use and Greenhouse Gas Emissions in the United States*; National Renewable Energy Lab.(NREL): Golden, CO, USA, 2007.

26. Hudon, K.; Merrigan, T.; Burch, J.; Maguire, J. Low-Cost Solar Water Heating Research and Development Roadmap. *Contract* **2012**, *303*, 275–3000.

27. Noorollahi, Y.; Kheirrouz, M.; Asl, H.F.; Yousefi, H.; Hajinezhad, A. Biogas production potential from livestock manure in Iran. *Renew. Sustain. Energy Rev.* **2015**, *50*, 748–754. [CrossRef]

28. MOPEE. *Energy Balance Sheet*; MOPEE: Tehran, Iran, 2011.

29. Harder, C. *The ArcGIS Book: 10 Big Ideas about Applying The Science of Where*, 2nd ed.; Esri Press: New York, NY, USA, 2018.

30. Noorollahi, Y.; Itoi, R.; Fujii, H.; Tanaka, T. GIS model for geothermal resource exploration in Akita and Iwate prefectures, northern Japan. *Comput. Geosci.* **2007**, *33*, 1008–1021. [CrossRef]

31. Moghaddam, M.K.; Noorollahi, Y.; Samadzadegan, F.; Sharifi, M.A.; Itoi, R. Spatial data analysis for exploration of regional scale geothermal resources. *J. Volcanol. Geotherm. Res.* **2013**, *266*, 69–83. [CrossRef]

32. Satba Solar energy resource map of Iran. Available online: http://www.satba.gov.ir/en/home (accessed on 23 November 2018).

33. Sabziparvar, A.A.; Shetaee, H. Estimation of global solar radiation in arid and semi-arid climates of East and West Iran. *Energy* **2007**, *32*, 649–655. [CrossRef]

34. Esri. *Using ArcGIS TM 3D Analyst*, 1st ed.; Esri Press: New York, NY, USA, 2010.

35. Paulescu, M.; Stefu, N.; Calinoiu, D.; Paulescu, E.; Pop, N.; Boata, R.; Mares, O. Ångström-Prescott equation: Physical basis, empirical models and sensitivity analysis. *Renew. Sustain. Energy Rev.* **2016**, *62*, 495–506. [CrossRef]

36. SolarGIS Solar resources assessment maps. Available online: https://solargis.com/products/solar-resource-assessment-study/overview (accessed on 7 December 2019).

37. Besarati, S.M.; Padilla, R.V.; Goswami, D.Y.; Stefanakos, E. The potential of harnessing solar radiation in Iran: Generating solar maps and viability study of PV power plants. *Renew. Energy* **2013**, *53*, 193–199. [CrossRef]

38. NCC. *Iran Digital Elevtion Database*; NCC: Tehran, Iran, 2018.

39. HomerEenergy Calculates Clearness Index. Available online: https://www.homerenergy.com/products/pro/docs/latest/how_homer_calculates_clearness_index.html (accessed on 11 February 2020).

40. Noorollahi, Y.; Itoi, R.; Fujii, H.; Tanaka, T. GIS integration model for geothermal exploration and well siting. *Geothermics* **2008**, *37*, 107–131. [CrossRef]
41. Noorollahi, Y.; Yousefi, H.; Mohammadi, M. Multi-criteria decision support system for wind farm site selection using GIS. *Sustain. Energy Technol. Assess.* **2016**, *13*, 38–50. [CrossRef]
42. Satkin, M.; Noorollahi, Y.; Abbaspour, M.; Yousefi, H. Multi criteria site selection model for wind-compressed air energy storage power plants in Iran. *Renew. Sustain. Energy Rev.* **2014**, *32*, 579–590. [CrossRef]
43. Bennui, A.; Rattanamanee, P.; Puetpaiboon, U.; Phukpattaranont, P.; Chetpattananondh, K. Site selection for large wind turbine using GIS. In Proceedings of the PSU-UNS International Conference on Engineering and Environment, Phuket, Thailand, 10–11 May 2007; pp. 561–566.
44. Gass, V.; Schmidt, J.; Strauss, F.; Schmid, E. Assessing the economic wind power potential in Austria. *Energy Policy* **2013**, *53*, 323–330. [CrossRef]
45. Abbaspour, M.; Satkin, M.; Mohammadi-Ivatloo, B.; Hoseinzadeh Lotfi, F.; Noorollahi, Y. Optimal operation scheduling of wind power integrated with compressed air energy storage (CAES). *Renew. Energy* **2013**, *51*, 53–59. [CrossRef]
46. MOE. Office of Legal Affairs Transmission and distribution of electricity airlines privacy. Available online: http://law.moe.gov.ir/ (accessed on 13 September 2019).
47. Lopez, A.; Roberts, B.; Heimiller, D.; Blair, N.; Porro, G.U.S. Renewable Energy Technical Potentials: A GIS-Based Analysis. *Natl. Renew. Energy Lab. Doc.* **2012**, *1*, 1–40.
48. OSIT. *Kurdistan Province Statistical Yearbook*; Kordistan Management and Planning Institute: Sananadaj, Iran, 2016; p. 168.
49. Marion, W.; Wilcox, S. *Solar Radiation Data Manual for Buildings*; National Renewable Energy Lab.(NREL): Golden, Co, USA, 1995; p. 255.
50. Rahpeyma, S. Steps for natural gas demand saving. Available online: http://www.mashal.ir/ (accessed on 9 September 2019).

© 2020 by the authors. Licensee MDPI, Basel, Switzerland. This article is an open access article distributed under the terms and conditions of the Creative Commons Attribution (CC BY) license (http://creativecommons.org/licenses/by/4.0/).

 *sustainability*

Article

Enhancing Integrated Power and Water Distribution Networks Seismic Resilience Leveraging Microgrids

Javad Najafi [1], Ali Peiravi [1] and Amjad Anvari-Moghaddam [2,3,*]

[1] Department of Electrical Engineering, Ferdowsi University of Mashhad, Mashhad, Iran; javad.najafi@mail.um.ac.ir (J.N.); peiravi@um.ac.ir (A.P.)
[2] Department of Energy Technology, Aalborg University, 9220 Aalborg, Denmark
[3] Faculty of Electrical and Computer Engineering, University of Tabriz, 5166616471 Tabriz, Iran
* Correspondence: aam@et.aau.dk

Received: 30 January 2020; Accepted: 9 March 2020; Published: 11 March 2020

Abstract: An earthquake, as one of the natural disasters, can damage vital infrastructures including the power distribution network (PDN) and water distribution network (WDN). The dependency of WDN on PDN is the other challenge that can be highlighted after the earthquake. In this paper, the resilience improvement planning of integrated PDN and WDN against earthquakes is solved through stochastic programming. Power lines and substation hardening in PDN and water pipes rehabilitation with better material are the candidate strategies to minimize the expected inaccessibility value of loads to power and water as the resilience index and to minimize the cost of strategies. The proposed model is tested on the modified IEEE 33-bus PDN with a designed WDN and its performance is evaluated under different cases where the impacts of using distributed generations (DG) in PDN, equipping the water pumps to back-up generators, and the value of loads accessibility to water on the system resilience are investigated.

Keywords: earthquake; power distribution network; resilience improvement planning; water distribution network; microgrid

---

1. Introduction

Long inaccessibility of cities and societies to critical infrastructures including power and water after earthquakes has highlighted the importance of resilience improvement planning of the power distribution network (PDN) and water distribution network (WDN). Traditionally, *N-1* or *N-k* security indices guaranteed the reliability of energy networks. However, low-probability but high-impact events that can trigger the interdependencies of infrastructures may surpass the traditional reliability methods.

Simultaneous resilience improvement planning of PDN and WDN has different advantages over the individual system planning. The dependency of WDN on PDN due to water pumps operation as one important reason for the inaccessibility of loads to water is ignored in the resilience study of an individual WDN. Although, a number of water pumps might be equipped to emergency generators. But, these units are not efficient in long emergency conditions resulting from natural disasters due to fuel limitation [1]. Furthermore, with simultaneous analysis and improvement of the resilience of PDN and WDN, the allocated budget is spent in such a way that the accessibility of all loads to power and water will be improved. In other words, the problem detects the priority of PDN and WDN to enhance resilience considering the budget limits [2].

1.1. Background and Previous Works

The previous works can be reviewed in three categories: 1—Resilience study of PDN, 2—Resilience study of WDN, and 3—Resilience study of joint PDN and WDN.

1.1.1. Resilience Study of PDN

Three types of strategies can be adapted for a PDN to enhance resilience including pre-natural disasters actions, operational, and planning types [3]. The first type can be implemented only for predictable natural disasters such as hurricanes. Crews or mobile generators pre-positioning is among the first type of resilience improvement strategies, which aims to minimize the expected outage duration of disconnected loads [4,5]. An earthquake is an unpredictable natural disaster and this type of strategy is not efficient to enhance the resilience of PDN against earthquakes.

Operational strategies as the second type of resilience improvement strategies that can be efficient against any natural disasters refer to actions after natural disasters occurrence to help the PDN to deal better with emergency conditions. Microgrids formation by distributed generation (DG) is one of the well-known operational resilience improvement strategies. This strategy partitions the PDN into some smaller self-sustained networks [6]. Another operational strategy for enhancing the resilience of PDN is modeling the microgrids as emergency sources for PDN [7]. Microgrids are independent AC or DC energy networks that can be operated in an isolated or connected mode [8,9].

Planning strategies are the last type of aim to improve the resilience of PDN for a long time. Some kinds of these strategies try to physically enhance the network such as pole hardening [10,11] and strengthening the substations with anchored components [12] to decrease the damaged probability of components against natural disasters. Another strategy toward this type is to place distributed means of generation or energy storage devices in the PDN [13–15].

According to the above categories, this paper belongs to the third type of resilience improvement of PDN which can be named resilience improvement planning of PDN. In this kind of problem, it is important to determine which kind of resilience improvement strategies should be implemented in the PDN as candidate strategies, how they will be modeled in the problem, and which kind of optimization tool will be implemented to solve the problem. It is worth mentioning that the best candidate strategies for improving the resilience of PDN should be chosen based on the conditions and the type of natural disasters threatening the PDN. For example, undergrounding the cables can increase the resilience of PDN against hurricanes, however, it might not be an efficient solution against floods or earthquakes. The optimization tool can impact the modeling of strategies in the problem. For example, when robust optimization is implemented in [10] to solve the problem, if a power distribution line is chosen to be hardened, that line will not be vulnerable against natural disasters anymore. Another limitation of robust optimization is that when different candidate strategies (line hardening and DG placement) are implemented such as in [10], reconfiguration cannot be implemented to restore the network. Due to these limitations of robust optimization, stochastic optimization has been used to solve the problem in recent works such as [1,16]. In stochastic programming, which is implemented in [1,16], the final decision will be obtained by studying a set of scenarios that are produced by uncertain parameters of the problem.

1.1.2. Resilience Study of WDN

Earthquake and flood are two high potential natural disasters that can severely damage the WDN. The 2011 Japan Earthquake disconnected the accessibility of 2,300,000 households to water [17]. The most vulnerable components which are forming most of the WDN are the pipes [18]. Numerous works with different models such as stochastic framework or fuzzy TOPSIS technique have tried to identify the critical pipes [19–21]. Other components of WDN such as water tanks and water resources are less vulnerable and they can be hardened well as only a small number of them exist in WDN. As mentioned in [22], four factors including pipe diameter, pipe material, topography, and soil liquefaction determine the vulnerability of a water pipe against an earthquake. Therefore, to enhance the resilience of WDN, some of the mentioned factors should be improved. In other words, the WDN should be rehabilitated based on the conditions of the WDN and the region where it is located. In [23], two resilience improvement strategies, including replacement of water pipes with better materials and different diameters, are implemented to enhance the resilience of WDN against an earthquake. Due to the limitation of the budget, only a few numbers of critical pipes in a WDN can be rehabilitated.

1.1.3. Simultaneous Resilience Study of PDN and WDN

Each network is composed of nodes and links. Topological and flow-based are two available analysis methods to study the vulnerability of a network. In the topological method, graph theory and structural indices are used to investigate the vulnerability of a network. In contrast, the flow-based method utilizes physics-based equations to assess the vulnerability of a network. The serviceability of the network can be predicted with high accuracy using a flow-based analysis compared to the topological method [19,24]. Most of the works, such as [25–27], that studied the simultaneous resilience of power and water distribution networks have implemented the topological analysis method. However, in this works, the technical equations related to each network such as power flow equations in PDN and water flow equations in WDN are not considered. This issue has been resolved in [1,16] by proposing a joint model of PDN and WDN where power/water flows can be analyzed throughout the network, but the vulnerability of WDN to natural disasters has been neglected in that study. In other words, the mentioned works solve the resilience improvement planning problem of PDN and WDN against hurricanes and the only reason for loads inaccessibility to water is the dependency of WDN on PDN.

1.2. Contribution of The Paper

In light of the reviewed literature, this paper enhances the resilience of PDN and WDN by using technical equations related to each network against earthquakes which can damage both networks. Substation and power lines hardening in PDN and water pipes rehabilitation with better material in WDN as three candidate strategies are implemented through a stochastic framework to maximize the resilience of integrated PDN and WDN and to minimize the planning cost.

The proposed method can identify vulnerable parts in both PDN and WDN and do the needed contingency analysis following an event to determine the most efficient resilience improvement strategy. By using the proposed method, the impact of DGs in PDN, emergency generators of water pumps in WDN and the value of loads accessibility to water on the resilience level of the networks can also be investigated.

The rest of the paper is organized as follows. The general framework of the proposed stochastic programming including uncertain parameters, scenarios generation, and scenario reduction method is explained in Section 2. Section 3 formulates the resilience improvement planning of integrated PDN/WDN and presents the solution methodology. Numerical case studies are presented in Section 4. Finally, Section 5 concludes the paper.

2. General Framework of the Proposed Stochastic Programming

In the resilience improvement planning of integrated power/water distribution networks, a number of uncertainties might exist. These main factors include: (1) the severity of earthquakes, (2) time of event occurrence in a day, (3) operational states of substation and power lines in PDN and water pipes in WDN, (4) repair time of damaged component, (5) the accessibility of a load to water, and (6) value of loads inaccessibility to power and water. It is worth mentioning that only some of these factors with more importance are considered as uncertain parameters in the proposed model. If the uncertain parameters increase in a stochastic problem, it is vital to generate more scenarios for studying and it means the computational burden of the problem will be increased significantly. To well address such uncertainties in an operational planning problem, stochastic programming techniques for analysis of potential future events can be effectively used. To generate such events (also called scenarios), the uncertainty of each parameter is essential to be modeled.

In this paper, the historical seismic data of an area is used to generate stochastic earthquake events. According to [21], the uncertainty of the earthquake magnitude in each scenario is formulated as (1) by using a cumulative density function and inverse transform sampling:

$$F_M = 1 - \frac{e^{(-\beta M)} - e^{(-\beta M_{\max})}}{e^{(-\beta M_{\min})} - e^{(-\beta M_{\max})}},$$

(1)

where $\beta = d \ln 10$ and $d = 0.8$ are obtained by analyzing the historical data of earthquakes [17]. To determine the magnitude of the earthquake in a scenario, a random number between 0 and 1 is generated by uniform distribution and then it is projected into the cumulative distribution function. In order to assess the intensity of an earthquake, in addition to the magnitude, another factor as the location of the earthquake is also important. Therefore, by using the location and magnitude, PGA (Peak Ground Acceleration) of the earthquake as (2) can be calculated, which is needed to evaluate the vulnerability of components in the networks. There are different equations for calculating the PGA. In this paper, Equation (2) is obtained from [28]. It should be noted that the PGA equation can be changed based on the region where the PDN and WDN are located. The used PGA in this paper belongs to rock and soil grounds such as Iran, Tehran:

$$\ln \mathrm{PGA} = 4.15 + 0.623\left(\frac{M + 0.38}{1.06}\right) - 0.96 \ln(\Delta).$$

(2)

The occurrence time of the earthquake in a day is modeled by a random number between 1 and 24. This parameter shows the time that the restoration of the networks will be triggered.

The failure probability of a structure against the severity of an earthquake is called fragility function [29]. By using different fragility functions, the operational states of substation which supply the PDN, power lines in PDN, and water pipes in WDN should be determined. According to (3), cumulative normal distribution function with various parameters is used to show the fragility function of a power pole or substation. In this paper, a power distribution line is out of service if one of its' power pole is damaged against an earthquake:

$$p_f^{\text{pole or sub}}(\mathrm{PGA}) = \Phi[\ln(\mathrm{PGA}/m^{\text{pole or sub}})/\xi^{\text{pole or sub}}].$$

(3)

The fragility function of a water pipe is considered as below [17]. According to (4), in addition to the length of a pipe and PGA of an earthquake, four other factors affect the vulnerability of a pipe against earthquakes. These factors depend on the water pipe characteristics (pipe diameter (C_1) and pipe material (C_2)), surrounding soil characteristics and topology (topography (C_3) and liquefaction (C_4)):

$$p_f^{\text{pipe}} = 1 - \exp(-0.00187 C_1 C_2 C_3 C_4 \mathrm{PGA} \times L).$$

(4)

Now, in order to determine the operational states of the substation, power lines, and water pipes in each scenario, a random number between 0 and 1 is generated for each component and compared with the failure probability of that component which is obtained from the fragility function. If the random number is less than the failure probability, then the component is considered to be damaged, otherwise it will be operated.

A set of scenarios will be generated and due to computational burden, only a few of them will be analyzed. Backward scenario reduction as a well-known method will be applied to obtain the final scenarios set [30].

3. Problem Formulation and Solution Methodology

3.1. Problem Formulation

The problem is formulated in the form of multi-objective stochastic programming. The first objective function takes the resilience index into consideration and aims at minimizing the expected inaccessibility value of loads to power and water due to the earthquakes:

$$OF_1 = \min \sum_{s=1}^{N_s} \rho_s \sum_{l=1}^{N_l} \sum_{t=t_s^0}^{t_s^0+T_s} \left(IVP_{l,t}\alpha_{s,l,t} + IVW_{l,t}\beta_{s,l,t} \right), \tag{5}$$

where α is a binary variable that is indicated in (6).

$$\alpha = \begin{cases} 1 & \text{if load is connected} \\ 0 & \text{if load is disconected} \end{cases}, \tag{6}$$

where unlike the binary accessibility of a load to power, the accessibility of a load to water is proportional to water pressure head in that node:

$$\beta = \begin{cases} 1 & ph \geq ph^{\text{required}} \\ \left(\dfrac{ph}{ph^{\text{required}}} \right)^2 & ph < ph^{\text{required}} \end{cases}. \tag{7}$$

The second objective function of the problem considers the minimization of the total cost of planning including power lines and substation hardening and water pipes rehabilitation. In this paper, it is assumed that the life span of all candidate strategies is equal. Ignoring this assumption, the model also should consider the life cycle cost of candidate strategies:

$$OF_2 = \min \sum_{pl=1}^{N_{pl}^{pw}} \Omega_{pl} C_{pl}^H + \sum_{wp=1}^{N_{wp}^{wt}} \Psi_{wp} C_{wp}^R + \sum_{sb=1}^{N_{sb}} \lambda_{sb} C_{sb}^H. \tag{8}$$

For each scenario, the restoration and recovery of the networks will be solved in such a way that the accessibility of loads to power and water be maximized. A disconnected load in PDN can be restored by reconfiguration or by microgrid formation using DGs. If the disconnected loads include water pumps, by analyzing the WDN considering the water pipes damages, the level accessibility of loads to water will be assessed and then it will be decided that the disconnected water pumps be restored or not.

Throughout the entire PDN, constraints (9)–(13) must be satisfied. Equations (9)–(10) denote the power balance constraints:

$$P_i^{u,s,t} = \left| V_i^{u,s,t} \right| \sum \left| V_j^{u,s,t} \right| \left(G_{ij} \cos \theta_{ij}^{u,s,t} + B_{ij} \sin \theta_{ij}^{u,s,t} \right) \quad \begin{array}{l} s \in \{1,2,\ldots,N_s\}, u \in \{1,2,\ldots,U_{s,t}+1\}, \\ i \in \{1,2,\ldots,N_b^{u,s,t}\}, t \in [t_s^0, t_s^0 + T_s] \end{array}, \tag{9}$$

$$Q_i^{u,s,t} = \left| V_i^{u,s,t} \right| \sum \left| V_j^{u,s,t} \right| \left(G_{ij} \sin \theta_{ij}^{u,s,t} - B_{ij} \cos \theta_{ij}^{u,s,t} \right) \quad \begin{array}{l} s \in \{1,2,\ldots,N_s\}, u \in \{1,2,\ldots,U_{s,t}+1\}, \\ i \in \{1,2,\ldots,N_b^{u,s,t}\}, t \in [t_s^0, t_s^0 + T_s] \end{array}. \tag{10}$$

In any operating mode, bus voltages and line flows must be kept within a safe range indicated by (11)–(12), respectively:

$$|V_{\min}| \leq \left| V_i^{u,s,t} \right| \leq |V_{\max}| \quad s \in \{1,2,\ldots,N_s\}, u \in \{1,2,\ldots,U_{s,t}+1\}, i \in \{1,2,\ldots,N_b^{u,s,t}\}, t \in [t_s^0, t_s^0 + T_s], \tag{11}$$

$$\left| I_{ij}^{u,s,t} \right| \leq \left| I_{ij}^{\max} \right| \quad s \in \{1,2,\ldots,N_s\}, u \in \{1,2,\ldots,U_{s,t}+1\}, i \in \{1,2,\ldots,N_b^{u,s,t}\}, t \in [t_s^0, t_s^0 + T_s]. \tag{12}$$

Equation (13) guarantees the radiality of each network:

$$N_b^{u,s,t} = N_{line}^{u,s,t} + 1 \quad s \in \{1, 2, \ldots, N_s\}, u \in \{1, 2, \ldots, U_{s,t} + 1\}, t \in [t_s^0, t_s^0 + T_s]. \tag{13}$$

Furthermore, in each intentional islanded network (microgrid) which includes one or more DGs, Equations (14)–(15) as constraints must be satisfied. Equations (14) and (15) show that the active/reactive of each DG should not exceed the maximum active/reactive capacity of that DG:

$$P_{DG_g}^{u,s,t} \leq P_{DG_g}^{\max} \quad s \in \{1, 2, \ldots, S\}, u \in \{1, 2, \ldots, U_{s,t}\}, g \in \{1, 2, \ldots, N_{DG}^{u,s,t}\}, t \in [t_s^0, t_s^0 + T_s], \tag{14}$$

$$Q_{DG_g}^{u,s,t} \leq Q_{DG_g}^{\max} \quad s \in \{1, 2, \ldots, S\}, u \in \{1, 2, \ldots, U_{s,t}\}, g \in \{1, 2, \ldots, N_{DG}^{u,s,t}\}, t \in [t_s^0, t_s^0 + T_s]. \tag{15}$$

To calculate the accessibility of loads to water, it is necessary to formulate the hydraulic model of the WDN. According to [31], there are three fundamental equations. The first equation explained in (16) is the mass conservation that must be satisfied at the nodes except fixed-head nodes such as reservoirs of WDN:

$$\sum_{wp \in LK_n} f_{wp,t,s} + F_{n,t} = 0, s \in \{1, 2, \ldots, N_s\} \quad n \in \{1, 2, \ldots, N_n^{water} - NR^{water}\}, t \in [t_s^0, t_s^0 + T_s]. \tag{16}$$

Moreover, according to (17), energy conservation must be satisfied in each simple loop of the water network:

$$\sum_{wp=1}^{wp_{ls}} h_{wp,ls,t,s} = 0 \quad ls \in \{1, 2, \ldots, LS\}, t \in [t_s^0, t_s^0 + T_s], s \in \{1, 2, \ldots, N_s\}. \tag{17}$$

The last equation represents the hydraulic head loss. This equation shown in (18) indicates the head loss of a pipe as a function of the flow through the pipe:

$$h_{wp} = x f_{wp}{}^y, \tag{18}$$

where x and y are coefficients which are determined based on the Hazen–Williams model.

3.2. Solution Methodology

Greedy search as an iterative algorithm is utilized in this paper to solve the problem. In order to implement the greedy search algorithm, the aforementioned objective functions are mapped into the following mixed-objective function:

$$OF = \max \frac{OF_1^{itr-1} - OF_1^{itr}}{\cos t\_st_{st}} \quad st \in \{1, 2, \ldots, N_{st}\}. \tag{19}$$

In each iteration of the greedy search algorithm, the problem is solved considering the objective function in (19) which is indicating the difference of resilience improvement (expected inaccessibility of loads to power and water) compared to the previous iteration (itr-1) per cost of each chosen strategy. This iterative procedure will be continued until the maximum budget (determined by the planner) is exhausted. Another advantage of this approach is that the priority of each strategy can be determined in enhancing the resilience of PWN and WDN. In other words, the problem can be solved to choose a specific number of strategies to yield the maximum resilience improvement per cost of strategies.

In order to assess the accessibility of loads to power and water of each scenario, both PDN and WDN should be analyzed based on the problem formulation. The PDN and WDN analyses are done in MATLAB and EPANET, respectively. EPANET is a WDN modeling software package that can facilitate the extended-period simulation of hydraulic behavior within pressurized pipe networks with high precision. To enable the joint operation of PDN and WDN while capturing systems interactions, an interface for data exchange between simulation platforms (i.e., MATLAB and EPANET) is established. To solve the restoration problem of PDN reconfiguration option to reroute the power from the substation and microgrids formation by DG(s) are available. To this end, the number of autonomous islanded networks (clusters) that could be formed for serving local loads should be defined. In this paper, graph theory and the modified Viterbi algorithm, which is proposed in [32] and improved in [1], is utilized to solve the restoration problem. Veterbi is like dynamic algorithms and in each stage, some candidate switching states are checked until the maximum disconnected loads be restored. In the improved version of the Veterbi algorithm, the number of candidate switching states is decreased. The effectiveness of this method in solving the problem is shown in [1]. It should be noted that if the substation in the main network is also damaged, the disconnected loads can be restored only by microgrid formation. By using the method in [33], the restoration trees will be constructed for each DG and the best path will be chosen based on the objective function of the problem. If there exist power distribution lines in a region to feed the load through one or more DGs locally, then those DGs and corresponding local loads can be merged into a single microgrid. The accessibility of loads to water also will be obtained by EPANET considering the water pipes damages. If disconnected loads include water pumps, by using EPANET, the impact of restoration of each disconnected water pump on the accessibility of loads to water will be calculated.

4. Simulations and Results

To illustrate the effectiveness of the proposed method, the modified IEEE 33 bus PDN with connected DGs and its related designed WDN as shown in Figure 1 is studied.

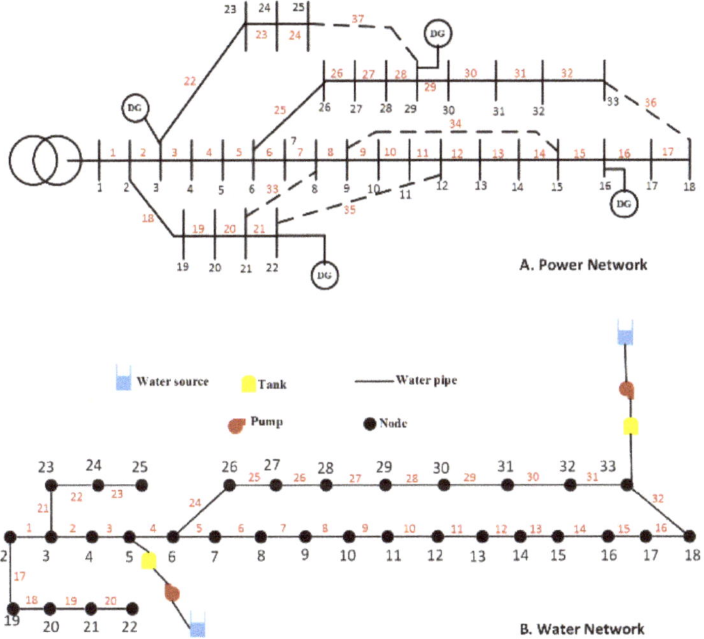

Figure 1. PDN and related designed WDN.

The capacity of each DG is assumed to equal to 150 kW. The active/reactive power peak demands of PDN and water peak demand of WDN are depicted in Figure 2.

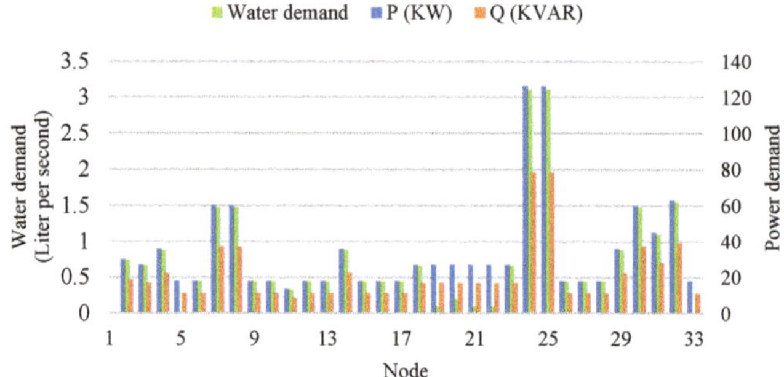

Figure 2. Water, active, and reactive power peak demands of networks.

To obtain these values at different hours of a day, they should be multiplied in the related hourly multipliers which are illustrated in Figure 3.

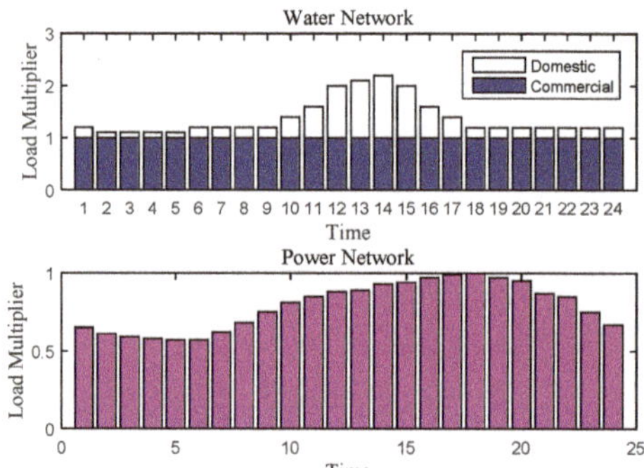

Figure 3. Load multipliers of PDN and WDN.

The multipliers of active/reactive power demands show the normalized loads values based on the maximum load, while the hourly multipliers in WDN are denoting the hourly normalized water needs based on the minimum demand in a day. The water demand of commercial loads is considered constant during a day.

The value of load inaccessibility to power and water is depicted in Figure 4. It is assumed that the loads in nodes (19–22) are commercial, while the rests are residential. Power accessibility is more important than water accessibility for a commercial load. Therefore, according to Figure 4, the inaccessibility value of commercial loads to power is much more than the inaccessibility value to water. Loads in nodes 5 and 33 of the distribution network are water pumps. Thus, the value of load inaccessibility to water of these nodes is zero. It should be noted that the dynamic value of each water pump for restoration will be determined in the restoration problem. In order to determine the importance of one pump, the accessibility function of loads will be obtained with EPANET and will be compared with the state in which the water pump is restored.

The minimum and maximum magnitudes of the earthquake are considered 3 and 7 Richter, respectively. The distance of networks from focus of an earthquake is assumed to be 10 km. The fragility function of each power pole and substation before and after hardening are depicted in Figure 5.

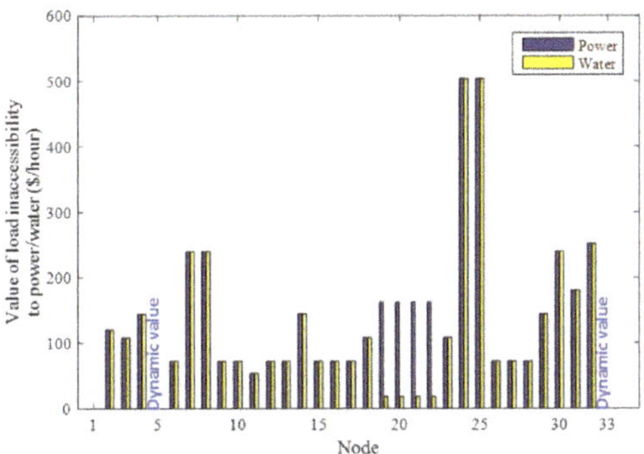

Figure 4. Value of loads inaccessibility to power and water.

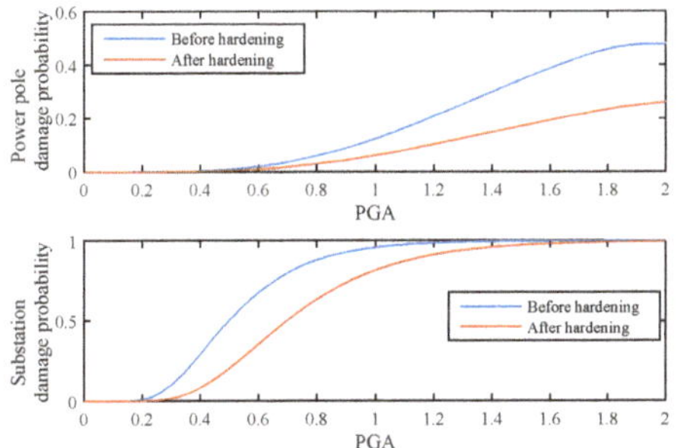

Figure 5. Fragility function of a power line and substation before and after hardening.

Cost of each power pole and substation hardening are assumed equal to 5000$ and 56,000$, respectively. As mentioned before, in order to rehabilitate the water pipes as a resilience improvement strategy, the material of a water pipe will be improved. Therefore, by implementation of this strategy, parameter C2 (pipe material correction factor) will be changed from 1 to 0.3. The cost of this strategy for water pipes with diameters of 80, 100, 150, 200, and 250 mm are 28.6$, 28.6$, 35.24$, 52.8$, and 95.26$ per meter, respectively [34]. C1 as water pipe diameter is obtained from [20] and other factors (C3 and C4) are assumed to be 1.

The repair time of the damaged power line, substation, and water pipe considering the PGA of an earthquake is tabulated in Table 1.

Table 1. Discrete probability distribution function of damaged components.

		Component Repair Time (Hours)		
		Power Line	**Substation**	**Water Pipe**
	[0,0.5)	4	6	6
PGA	[0.5,1)	6	8	8
	[1,1.5)	8	10	10
	[1.5,2]	10	12	12

Fifty scenarios are produced to solve the scenarios. The PGA of each scenario is shown in Figure 6.

Furthermore, Figure 7 depicts that each component including the power line, water pipe, and substation experience how many failures in the scenarios set.

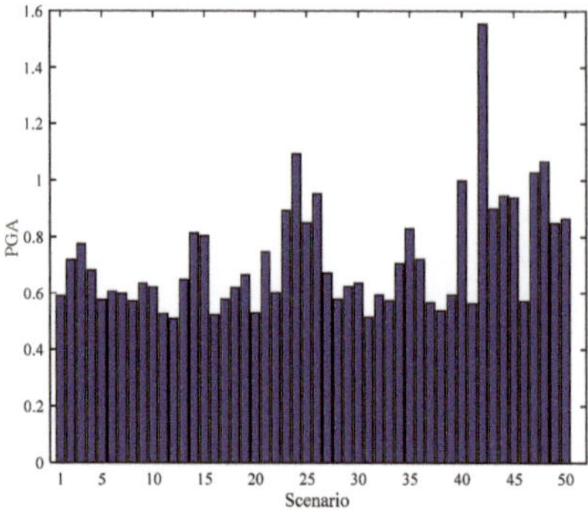

Figure 6. The PGA of each scenario.

Case 1: In this case, the resilience improvement planning of PDN/WDN with DGs in PDN is solved for budget constraint equal to 100,000$. The results are shown in Table 2.

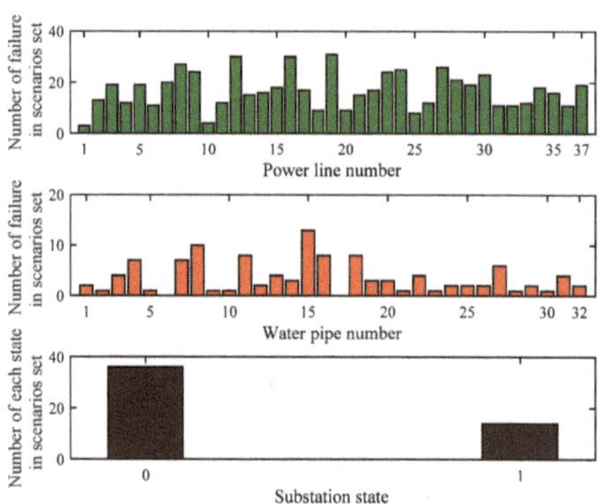

Figure 7. The failure number of each power line, water pipe, and substation in the scenarios set.

Table 2. The results of resilience improvement planning in Case 1.

Step	Strategy	EP[1] ($)	EW[2] ($)	EPW[3] ($)	Cost($)
0	No strategy	19,872.4	15,091.4	34,963.8	0
1	Power line 3–4	19,387.2	11,430.7	30,817.9	30,000
2	Water pipe 6–7	19,387.2	10,922.5	30,309.7	4370
3	Water pipe 4–5	19,387.2	10,416.5	29,803.7	8951
4	Substation	15,631.2	10,035.8	25,667	56,000
Sum					99,321

(1) EP: Expected loads inaccessibility value to power; (2) EW: Expected loads inaccessibility value to wat; (3) EPW: Expected loads inaccessibility value to power and water.

According to Table 2, step 0 evaluates the resilience level of PDN and WDN before solving the problem. In other words, the expected inaccessibility value of loads to power and water against an earthquake is 36,963.8$. In step 1 and with objective function in (20), the best and most efficient strategy for improving the resilience of PDN and WDN is to harden line 3–4 in PDN. This strategy can decrease the expected inaccessibility value of loads to power only by 2.44%. However, this strategy decreases the expected inaccessibility value of loads to water by 24.3%. In other words, this strategy improves the resilience of WDN by decreasing the dependency of WDN on PDN. By hardening power line 3–4, the path between DG in node 3 and water pump in node 5 is more robust against an earthquake and backup power could be provided once an emergency case is triggered. In step 2, water pipe 6–7 is chosen to be rehabilitated with better material. Maybe, based on the topological method, it seems that water pipe 5–6 has more priority than water pipe 6–7 to be rehabilitated. But, based on WDN physics equations, the pressure head of nodes (6)–(33) has better conditions with the water pump in node 33 as long as the exiting loop in WDN can be maintained. In step 3, the accessibility of these loads to water can be improved with the rehabilitation of the water pipe 4–5 which serves as the main link between nodes (2)–(5) and (19)–(25) and the water pipe in node 5. In the last step, the accessibility of loads to power is improved significantly by hardening the substation which is the main source of PDN. Finally, by spending 99,321\$, the expected inaccessibility value of loads to power and water against an earthquake is decreased from 34,963.8\$ to 25,667\$ which denotes a great resilience improvement (nearly 27%). In order to understand the impact of DGs in PDN on the resilience level of the networks, in Case 2, the resilience improvement planning is solved again, but without DG.

Case 2: The goal of this case is to investigate the resilience of PDN and WDN without DGs. The results are expressed in Table 3. In this case, to compare the priority of chosen strategies with the previous case, the budget constraint is ignored and the problem is solved for obtaining four strategies.

Table 3. The results of resilience improvement planning in Case 2 (without DGs in PDN).

Step	Strategy	EP(\$)	EW(\$)	EPW(\$)	Cost(\$)
0	No strategy	30,853.7	24,155.8	55,009.5	0
1	Substation	25,167.1	20,956.4	46,123.5	56,000
2	Power line 1–2	24,820.6	20,232.4	44,753	10,000
3	Power line 2–3	23,420.4	19,651.8	43,072.2	40,000
4	Power line 3–4	22,971.1	17,707.4	40,678.5	30,000
Sum					136,000

According to Table 3, the expected inaccessibility value of loads to power and water is 55,009.5\$. In other words, resilience is decreased by 57.3% compared to the previous case when DGs exist and form the microgrid to restore the disconnected loads. Since the only power source of PDN in this case is the main substation, substation hardening is chosen as the first action plan in step 1. Unlike the previous case, this strategy is chosen with high priority. By doing so, not only the accessibility of loads to power improve, but also, the dependency of WDN on PDN is significantly decreased. In other steps including 2, 3, and 4, hardening of power lines 1–2, 2–3, and 3–4 are chosen to enhance the resilience of PDN and WDN through making a reliable path between the only power source (main substation) and the loads in PDN. It should be noted that these strategies can also decrease the dependency of WDN on PDN. According to Figure 7, the number of failures of line 1–2 is less than the other power lines in the scenarios set. But, due to the importance of this line that plays an important role in delivering the energy to the loads, the hardening of this line can be efficient in improving the resilience of PDN and WDN. As the results show, the DGs in PDN can significantly enhance the accessibility of loads to power and also decrease the dependency of WDN on PDN.

Case 3: In this case, it is assumed that all the water pumps are equipped to backup DGs during the whole emergency period. So, the only reason for loads inaccessibility to water is the water pipes damages. The results are tabulated in Table 4.

Table 4. The results of resilience improvement planning in Case 3 (water pumps equipped with back-up generators).

Step	Strategy	EP($)	EW($)	EPW($)	Cost($)
0	No strategy	19,647.4	4876.5	24,523.9	0
1	Substation	16,618.8	4876.5	21,495.3	56,000
2	Water pipe 6–7	16,618.8	4693.5	21,312.3	4370
3	Water pipe 4–5	16,618.8	3777.7	20,386.5	8951
4	Power line 1–2	16,304.2	3777.7	20,081.9	10,000
Sum					**79,321**

According to Tables 2 and 4, the expected loads inaccessibility to water is decreased from 15,091.4$ to 4876.5$. This means the dependency of WDN on PDN is an important reason in the inaccessibility of loads to water. With the DGs in the PDN and equipping the water pumps to the emergency generators, the substation is chosen again in step 1 to be hardened. This strategy improves the expected accessibility of loads to power by 15.4%. In steps 2 and 3, two water pipes, 6–7 and 4–5, are chosen to be rehabilitated with better material to enhance the resilience of WDN. With these two strategies, the expected accessibility of loads to water is improved by 22.5%. Finally, in step 4, power line 1–2 hardening is chosen to improve the resilience of PDN. Considering the results shown in this case and the previous cases, the dependency of WDN on PDN can be the main reason for the inaccessibility of loads to water.

Case 4: This case is designed to investigate the impact of the value of loads inaccessibility to power and water on the planning results. The only difference of this case and Case 1 is that the loads inaccessibility values for residential loads are considered fifty times bigger than Case 1 (Figure 4). In other words, the residential loads accessibility to water is very important than to the power in this system. According to Table 5, which shows the results of this case, all the chosen strategies are in such a way to improve the accessibility of loads to water. Therefore, chosen strategies in steps 1 and 4 make the path between the substation and the water pump in node 5 more robust and decrease the dependency of WDN on PDN. Other chosen strategies in steps 2 and 4 with the rehabilitation of the water pipes, decrease the vulnerability of WDN against earthquakes.

Table 5. The results of resilience improvement planning in Case 4 (accessibility to water is more important than to power for residential loads).

Step	Strategy	EP($)	EW($)	EPW($)	Cost($)
0	No strategy	19,872.5	739,190.4	759,062.9	0
1	Power line 3–4	19,387.2	573,021.1	592,408.3	30,000
2	Water pipe 6–7	19,387.2	535,115.2	554,502.4	4370
3	Power line 4–5	19,064.4	450,448.5	469,512.9	30,000
4	Water pipe 4–5	19,064.4	427,540.3	446,604.7	8951
Sum					**73,321**

5. Conclusions

In this paper, the resilience improvement planning of integrated PDN and WDN with candidate strategies including substation and power line hardening in PDN and water pipes rehabilitation with better material in WDN was solved. The results showed that the proposed method could determine the priority and kind of strategies based on the conditions of both PDN and WDN. The summary of the highlighted results is below:

- Substation hardening as the most expensive but efficient strategy can enhance the resilience of PDN and WDN, significantly. When there is no DG in PDN and substation is the only power source, this strategy is chosen with high priority to improve resilience. This strategy can also decrease the dependency of WDN on PDN.

- Rehabilitation of some water pipes can significantly improve the accessibility of loads to water. It should be noted that the identification of such water pipes for rehabilitation cannot be done based on a simple approach such as the topological method. Therefore, it is vital to assess the WDN based on a physics-based equation in WDN.

- DGs are vital energy sources in PDN that can decrease the expected loads inaccessibility value to power and water by forming local microgrids. When there is no DG in PDN, strategies which enhance the path between substation and PDN will be chosen with high priority. Furthermore, in order to decrease the dependency of WDN to PDN with DGs, some strategies can be chosen to make a more robust path between DGs and important water pumps.

- The main reason for loads inaccessibility to water is the dependency of water pumps on power outages. This challenge should be considered in the resilience improvement problem.

- The importance of loads accessibility to power or water in a system affects the chosen strategies of the problem.

Author Contributions: J.N. designed the model, did the simulation studies and accomplished writing of the paper. A.P. and A.A.-M. supervised the whole work, revised the manuscript and checked the results. All authors have read and agreed to the published version of the manuscript.

Funding: This research received no external funding.

Acknowledgments: A.A.-M. acknowledges the support by "HeatReFlex-Green and Flexible Heating/Cooling" project (www.heatreflex.et.aau.dk) funded by Danida Fellowship Centre and the Ministry of Foreign Affairs of Denmark under the grant no. 18-M06-AAU.

Conflicts of Interest: The authors declare no conflict of interest.

Nomenclature

Indices

$g, N_{DG}^{u,s,t}$	Index and the number of DGs in microgrid u in scenario s at hour t
i, j	Bus indices
l, N_l	Index and number of loads
ls, LS	Index and number of loops in WDN
n, N_n^{water}	Index and number of nodes in WDN
pl, N_{pl}^{pw}	Index and number of power lines
s, N_s	Index and number of scenarios
t	Time index
sb, N_{sb}	Index and number of substations
u	Index for power network including the main network and intentionally islanded networks
wp, N_{wp}^{wt}	Index and number of water pipes

Parameters and variables

$C_{pl}^H, C_{wp}^R, C_{sb}^H$	Cost of power line pl hardening, cost of water pipe wp rehabilitation and cost of substation sb hardening, respectively.
C_1, C_2, C_3, C_4	Correction factors for the pipe diameter, pipe material, topography and liquefaction, respectively.
F_m	Poisson cumulative distribution of earthquake magnitude
$F_{n,t}$	Node n demand at time t
$f_{wp,t,s}$	Pipe wp flow rate at hour t in scenario s
G_{ij}, B_{ij}	Conductance and susceptance of the line which connects bus i and j,
$h_{wp,ls,t,s}$	Hydraulic head loss of pipe wp in loop ls at hour t in scenario s
$IVP_{l,t}, IVW_{l,t}$	Inaccessibility value of load l to power and water at hour t, respectively.

L	length of the pipe
LK_n	Set of all links (pipes) connected to node n in WDN
M	Specific magnitude magnitudes of the earthquake.
m, ξ	Mean and standard deviation of cumulative normal distribution function
$N_{line}^{u,s,t}$	Number of power lines in network u in scenario s at hour
$N_b^{u,s,t}$	Number of buses in network u in scenario s at hour t
NR^{water}	Number of fixed-head nodes in WDN
OF_1, OF_2	First and the second objective function, respectively.
$p_f^{pipe}, p_f^{pole}, p_f^{sub}$	Fragility function of a water pipe, power pole and substation, respectively.
$P_{DG_g}^{max}, Q_{DG_g}^{max}$	Maximum active/reactive capacity of DG g
$P_{DG_g}^{u,s,t}, Q_{DG_g}^{u,s,t}$	Active/reactive power of DG g in network u in scenario s at hour
$P_{loss,l}^{u,s,t}, Q_{loss,l}^{u,s,t}$	Active/reactive loss of distribution line l in microgrid u in scenario s at hour t
$P_i^{u,s,t}, Q_i^{u,s,t}$	Active and reactive power injected into bus i in network u in scenario s at hour t
$ph, ph^{required}$	Water pressure head and required water pressure head for full accessibility of a load to water, respectively.
t_s^0, T	The initial hour in each scenario (when the recovery of the networks starts right after the earthquake), the final hour in each scenario (when all the loads will access to power and water)
$U_{s,t}$	Number of microgrids at hour t in scenario s
Δ	Distance from the earthquake focus
ρ_s	Probability of scenario s
$\alpha_{s,l,t}, \beta_{s,l,t}$	Accessibility state of load l to power and water at hour t in scenario s, respectively.
$\Omega_{pl}, \Psi_{wp}, \lambda_{sb}$	Binary variables that determine the power line pl is hardened or not, the water pipe wp is rehabilitated or not and the substation sb is hardened or not, respectively.
$\theta_{ij}^{u,s,t}$	Difference phase voltage angle between bus i and j
$\left\|V_i^{u,s,t}\right\|$	Voltage magnitude at bus i in network u in scenario s at hour t
$\|V_{min}\|, \|V_{max}\|$	Minimum and maximum allowable voltage magnitude in the PDN
$\left\|I_{ij}^{u,s,t}\right\|$	The line flow between bus i and j in network u in scenario s at hour t
$\left\|I_{ij}^{max}\right\|$	Maximum allowable line current capacity between bus i and j

References

1. Najafi, J.; Peiravi, A.; Guerrero, J.M. Power distribution system improvement planning under hurricanes based on a new resilience index. *Sustain. Cities Soc.* **2018**, *39*, 592–604. [CrossRef]
2. Anvari-Moghaddam, A.; Mohammadi-Ivatloo, B.; Asadi, S.; Guldstrand Larsen, K.; Shahidehpour, M. Sustainable Energy Systems Planning, Integration, and Management. *Appl. Sci.* **2019**, *9*, 4451. [CrossRef]
3. Wang, Y.; Chen, C.; Wang, J.; Baldick, R. Research on resilience of power systems under natural disasters—A review. *IEEE Trans. Power Syst* **2016**, *31*, 1604–1613. [CrossRef]
4. Kavousi-Fard, A.; Wang, M.; Su, W. Stochastic Resilient Post-Hurricane Power System Recovery Based on Mobile Emergency Resources and Reconfigurable Networked Microgrids. *IEEE Access* **2018**, *6*, 72311–72326. [CrossRef]
5. Lei, S.; Wang, J.; Chen, C.; Hou, Y. Mobile emergency generator pre-positioning and real-time allocation for resilient response to natural disasters. *IEEE Trans. Smart Grid* **2018**, *9*, 2030–2041. [CrossRef]
6. Zhu, J.; Yuan, Y.; Wang, W. An exact microgrid formation model for load restoration in resilient distribution system. *Int. J. Elec. Power Energy Syst.* **2020**, *116*, 105568. [CrossRef]
7. Najafi, J.; Peiravi, A.; Anvari-Moghaddam, A.; Guerrero, J.M. An efficient interactive framework for improving resilience of power-water distribution systems with multiple privately-owned microgrids. *Int. J. Elec. Power Energy Syst.* **2020**, *116*, 105550. [CrossRef]
8. Chowdhury, D.; Hasan, A.S.M.K.; Khan, M.Z.R. Scalable DC microgrid architecture with phase shifted full bridge converter based power management unit. In Proceedings of the 2018 10th International Conference on Electrical and Computer Engineering (ICECE), Dhaka, Bangladesh, 20–22 December 2018; pp. 22–25.
9. Hasan, A.S.M.K.; Chowdhury, D.; Khan, M.Z.R. Scalable DC microgrid architecture with a one-way communication based control Interface. In Proceedings of the 2018 10th International Conference on Electrical and Computer Engineering (ICECE), Dhaka, Bangladesh, 20–22 December 2018; pp. 265–268.

10. Lin, Y.; Bie, Z. Tri-level optimal hardening plan for a resilient distribution system considering reconfiguration and DG islanding. *Appl. Energy* **2018**, *210*, 1266–1279. [CrossRef]

11. Ma, S.; Li, S.; Wang, Z.; Qiu, F. Resilience-oriented design of distribution systems. *IEEE Trans. Power Syst.* **2019**, *34*, 2880–2891. [CrossRef]

12. Salman, A.M.; Li, Y. A probabilistic framework for multi-hazard risk mitigation for electric power transmission systems subjected to seismic and hurricane hazards. *Struct. Infrastruct. Eng.* **2018**, *14*, 1499–1519. [CrossRef]

13. Dong, J.; Zhu, L.; Su, Y.; Ma, Y.; Liu, Y.; Wang, F.; Tolbert, L.M.; Glass, J.; Bruce, L. Battery and backup generator sizing for a resilient microgrid under stochastic extreme events. *IET Gener. Transm. Distrib.* **2018**, *12*, 4443–4450. [CrossRef]

14. Najafi, J.; Peiravi, A.; Anvari-Moghaddam, A.; Guerrero, J.M. Power-Heat Generation Sources Planning in Microgrids to Enhance Resilience against Islanding due to Natural Disasters. In Proceedings of the 2019 IEEE 28th International Symposium on Industrial Electronics (ISIE), Vancouver, BC, Canada, 12–14 June 2019; pp. 2446–2451.

15. Xie, H.; Teng, X.; Xu, Y.; Wang, Y. Optimal energy storage sizing for networked microgrids considering reliability and resilience. *IEEE Access* **2019**, *7*, 86336–86348. [CrossRef]

16. Najafi, J.; Peiravi, A.; Anvari-Moghaddam, A.; Guerrero, J.M. Resilience improvement planning of power-water distribution systems with multiple microgrids against hurricanes using clean strategies. *J. Clean. Prod.* **2019**, *223*, 109–126. [CrossRef]

17. Cimellaro, G.P.; Solari, D.; Bruneau, M. Physical infrastructure interdependency and regional resilience index after the 2011 Tohoku Earthquake in Japan. *Earthq. Eng. Struct. D.* **2014**, *43*, 1763–1784. [CrossRef]

18. Kongar, I.; Esposito, S.; Giovinazzi, S. Post-earthquake assessment and management for infrastructure systems: learning from the Canterbury (New Zealand) and L'Aquila (Italy) earthquakes. *B. Earthq. Eng.* **2017**, *15*, 589–620. [CrossRef]

19. Pudasaini, B.; Shahandashti, S.M. Identification of Critical Pipes for Proactive Resource-Constrained Seismic Rehabilitation of Water Pipe Networks. *J. Infrastruct. Syst.* **2018**, *24*, 04018024. [CrossRef]

20. Salehi, S.; Jalili Ghazizadeh, M.; Tabesh, M. A comprehensive criteria-based multi-attribute decision-making model for rehabilitation of water distribution systems. *Struct. Infrastruct. Eng.* **2018**, *14*, 743–765. [CrossRef]

21. Yoo, D.G.; Jung, D.; Kang, D.; Kim, J.H.; Lansey, K. Seismic hazard assessment model for urban water supply networks. *J. Water Resour. Plan. Manag.* **2016**, *142*, 04015055. [CrossRef]

22. Yoo, D.G.; Jung, D.; Kang, D.; Kim, J.H. Seismic-reliability-based optimal layout of a water distribution network. *Water* **2016**, *8*, 50. [CrossRef]

23. Farahmandfar, Z.; Piratla, K.; Andrus, R. Flow-based modeling for enhancing seismic resilience of water supply networks. Proceedings of Pipelines 2015, Baltimore, ML, USA, 23–26 August 2015; pp. 756–765.

24. Farahmandfar, Z.; Piratla, K.R. Comparative Evaluation of Topological and Flow-Based Seismic Resilience Metrics for Rehabilitation of Water Pipeline Systems. *J. Pipeline Syst. Eng. Pract.* **2018**, *9*, 04017027. [CrossRef]

25. Almoghathawi, Y.; Barker, K.; Albert, L.A. Resilience-driven restoration model for interdependent infrastructure networks. *Reliab. Eng. Syst. Safe.* **2019**, *185*, 12–23. [CrossRef]

26. González, A.D.; Dueñas-Osorio, L.; Sánchez-Silva, M.; Medaglia, A.L. The interdependent network design problem for optimal infrastructure system restoration. *Comput.-Aided Civ. Inf. Eng.* **2016**, *31*, 334–350. [CrossRef]

27. Zhang, Y.; Yang, N.; Lall, U. Modeling and simulation of the vulnerability of interdependent power-water infrastructure networks to cascading failures. *J. Syst. Sci. Syst. Eng.* **2016**, *25*, 102–118. [CrossRef]

28. Nazemi, M.; Moeini-Aghtaie, M.; Fotuhi-Firuzabad, M.; Dehghanian, P. Energy storage planning for enhanced resilience of power distribution networks against earthquakes. *IEEE Trans. Sustain. Energy* **2019**. [CrossRef]

29. Panteli, M.; Pickering, C.; Wilkinson, S.; Dawson, R.; Mancarella, P. Power system resilience to extreme weather: Fragility modelling, probabilistic impact assessment, and adaptation measures. *IEEE Trans. Power Syst.* **2017**, *32*, 3747–3757. [CrossRef]

30. Growe-Kuska, N.; Heitsch, H.; Romisch, W. Scenario reduction and scenario tree construction for power management problems. In Proceedings of the 2003 IEEE Bologna Power Tech Conference Proceedings, Bologna, Italy, 23–26 June 2003. [CrossRef]

31. Zhang, H.; Cheng, X.; Huang, T.; Cong, H.; Xu, J. Hydraulic analysis of water distribution systems based on fixed point iteration method. *Water Resour. Manag.* **2017**, *31*, 1605–1618. [CrossRef]

32. Yuan, C.; Illindala, M.S.; Khalsa, A.S. Modified Viterbi algorithm based distribution system restoration strategy for grid resiliency. *IEEE Trans. Power Deliv.* **2017**, *32*, 310–319. [CrossRef]

33. Xu, Y.; Liu, C.-C.; Schneider, K.P.; Tuffner, F.K.; Ton, D.T. Microgrids for service restoration to critical load in a resilient distribution system. *IEEE Trans. Smart Grid* **2018**, *9*, 426–437. [CrossRef]
34. Soto, R.; Crawford, B.; Misra, S.; Monfroy, E.; Palma, W.; Castro, C.; Paredes, F. Constraint programming for optimal design of architectures for water distribution tanks and reservoirs: a case study. *Tehnički Vjesnik* **2014**, *21*, 99–105.

 © 2020 by the authors. Licensee MDPI, Basel, Switzerland. This article is an open access article distributed under the terms and conditions of the Creative Commons Attribution (CC BY) license (http://creativecommons.org/licenses/by/4.0/).

 *sustainability*

Article

Stochastic Operation of a Solar-Powered Smart Home: Capturing Thermal Load Uncertainties

Esmaeil Ahmadi [1,2]**, Younes Noorollahi** [2]**, Behnam Mohammadi-Ivatloo** [3,4,]*** and Amjad Anvari-Moghaddam** [3,5,]*

[1] Energy Economics Laboratory, Department of Socio-Environmental Energy Science,
    Graduate School of Energy Science, Kyoto University, Kyoto 606-8501, Japan;
    esmaeil.ahmadi.53s@st.kyoto-u.ac.jp
[2] Renewable Energy and Environmental Department, Faculty of New Sciences and Technologies,
    University of Tehran, Tehran 15119-43943, Iran; Noorollahi@ut.ac.ir
[3] Faculty of Electrical and Computer Engineering, University of Tabriz, Tabriz 5157944533, Iran
[4] Institute of Research and Development, Duy Tan University, Da Nang, 550000, Vietnam
[5] Department of Energy Technology, Aalborg University, 9220 Aalborg, Denmark
*   Correspondence: bmohammadi@tabrizu.ac.ir (B.M.-I.); aam@et.aau.dk (A.A.-M.)

Received: 18 May 2020; Accepted: 16 June 2020; Published: 22 June 2020

Abstract: This study develops a mixed-integer linear programming (MILP) model for the optimal and stochastic operation scheduling of smart buildings. The aim of this study is to match the electricity demand with the intermittent solar-based renewable resources profile and to minimize the energy cost. The main contribution of the proposed model addresses uncertainties of the thermal load in smart buildings by considering detailed types of loads such as hot water, heating, and ventilation loads. In smart grids, buildings are no longer passive consumers. They are controllable loads, which can be used for demand-side energy management. Smart homes, as a domain of Internet of Things (IoT), enable energy systems of the buildings to operate as an active load in smart grids. The proposed formulation is cast as a stochastic MILP model for a 24-h horizon in order to minimize the total energy cost. In this study, Monte Carlo simulation technique is used to generate 1000 random scenarios for two environmental factors: the outdoor temperature, and solar radiation. Therefore in the proposed model, the thermal load, the output power of the photovoltaic panel, solar collector power generation, and electricity load become stochastic parameters. The proposed model results in an energy cost-saving of 20%, and a decrease of the peak electricity demand from 7.6 KWh to 4.2 KWh.

Keywords: smart home; solar renewable; thermal load; stochastic operation; energy storage

---

1. Introduction

Concerns such as global warming, unfair distribution of fossil sources, and fluctuating fossil fuel prices have caused governments to consider alternative strategies like applying renewable energy supplies [1]. In the future, renewables will have to compete with other energy resources without demand-side policy support [2]. Under a regime with no support storage, system profitability will widely hinge on suitable energy management and operation. The optimal energy management of the residential sector has also attracted more attention because 30 to 40 percent of the world's primary energy consumption is spent on buildings [3,4].

Smart homes, as a chief component of the Internet of Things (IoT) and smart grids, is the network of physical devices that provide electronic, sensor, software, and network connectivity inside a home [5]. A smart home is a domain of IoT, which are automated buildings with installed detection and control devices, such as air conditioning, heating, and ventilation. Smart homes enable the energy systems of the building to operate as an active load in smart grids. How to control these loads, forecast the energy consumption and production of the system, and manage the integrated system, are vital factors for smart buildings to serve effectively in energy systems aimed at energy-saving, load peak shaving, load shifting, and compensating for the fluctuations due to variable renewable power production.

Different methods such as load disaggregation algorithms [6] or forecasting methods [7] can be used to determine consumption of different loads in residential buildings. In the residential energy management, considering an integrated energy system including heating and electrical demands is a reasonable strategy. Previous studies [8,9] indicate that this integration improves the efficiency of the energy systems. Authors of [10] developed an MILP model to minimize the energy costs and greenhouse gas emissions of 30 smart homes under two different tariff schemes including real-time price and peak-demand charge. The results showed that the proposed model in this study tunes a trade-off between the cost of greenhouse gas emissions and the electricity bill. Authors of [11] used MILP model to investigate the impacts of various demand response programs on the residential customers' profit. Another study [12] introduced an MILP model to minimize the energy operating costs of a smart house. Lighting, heating, ventilation, and air conditioning loads were included as controllable loads to participate in its demand response plans. The developed model achieved up to 7 percent reduction in energy costs. MILP model is used in [13] for smart home energy management using hybrid stochastic optimization and robust optimization.

Variable renewable resources (wind and solar supplies) impose fluctuations on energy systems. Separating renewable energy supplies from the grid by applying energy storage systems could solve the issues such as fluctuating power generation to avoid instability because of an imbalance between the supply-side and demand-side. Furthermore, these energy storages could reduce the cost of clean energies through load shifting, peak shaving, etc. Battery units and thermal storages (water tank) are the most common energy storage devices used in the residential sector [14,15]. A study [16] considered heat pumps and thermal storage as flexible loads to integrate fluctuating renewable power resources into the electricity grid. Instead of modeling heat pump energy conversion as linear for simplification, this study proposed a heuristic method for scheduling a smart home with a photovoltaic system aiming to minimize the costs of electricity exchange with the grid.

Controllable loads in the smart buildings mainly include electric water heater, heating, ventilation and air-conditioning, washing machines and dryers, electric vehicles, dishwashers, thermal storage devices, and batteries. Among all these devices, thermostatically controlled loads [17,18] such as electric water heaters and heating, and ventilation and air-conditioning, have received substantial attention in recent years. However, the majority of models proposed for optimal scheduling of smart buildings in previous studies were either too complicated or had limited applications. For instance, the electric water heater models introduced in studies [19,20] require too many measurements for scheduling. Another study [21] developed a mixed-integer nonlinear model to find the optimal operation of conventional cooling chillers and decrease the energy costs. Thermal loads such as heating, ventilation and air conditioning, and the water heating system make a significant contribution to the power consumption of a typical residential load, which is over 50% of the total residential energy consumption [22,23].

The optimal energy management in a building is affected by various factors, such as technical, economic, social, and environmental parameters, which are inherently ambiguous and uncertain. Studies [24,25] that dealt with stochastic optimization problems in smart grids, including renewables, clearly showed that taking uncertainties into account in the process of optimizing the energy building operation leads to more reliable and accurate results. Uncertainty can be captured using sensitivity analysis methods or optimization under stochastic methods in this field [26,27]. The sensitivity analysis methods are able to identify the most influencing uncertain parameters while they cannot determine the optimal operation under conditional uncertainties [28].

It is common among studies [29–33] to only consider one uncertainty aspect (demand-side or supply-side) in the optimization problems of scheduling smart buildings. To the best of the authors' knowledge, none of the previous studies in this field consider at the same time both the uncertainties of thermal demand for heating, and also the energy demand for the hot water required by the washing machine and dishwasher in the optimal scheduling of smart buildings. In the proposed model in this study, the uncertainties of the supply-side (the intermittent energy generation of photovoltaics and solar collectors) and the demand-side (heat demand and electricity load) are represented by the Monte Carlo (MC) [34] method as different scenarios. In the current study, the proposed method makes the uncertain parameters used for scenario generation effective.

This study proposes a stochastic model in order to find the optimal scheduling of a smart home using the flexibility offered by smart controllable devices, and match the load with intermittent variable renewable resources profile, thereby mitigating the uncertainties in both demand and supply sides. This study includes three steps: (1) system modeling and gathering historical data; (2) generating scenarios and choosing best scenarios, and; (3) optimal stochastic scheduling for one day.

2. Model Description

This study develops a mixed-integer linear programming (MILP) model for the optimal and stochastic operation scheduling of smart homes. The rest of this section is organized to describe the stochastic scheduling optimization model in detail.

2.1. Uncertainty Factors

Monte Carlo simulation technique is used to generate scenarios for one day in January, the coldest month of the year (as the worst scenario for the amount of thermal demand). Environmental conditions, namely the outdoor temperature and solar radiation, have been considered as uncertainty factors, as a result of which, thermal load, output power of photovoltaic panel, solar collector power generation and electricity load become stochastic parameters. Finally 100 scenarios combining 10 solar irradiance and 10 outdoor temperature have been used to model the scheduling problem.

2.1.1. Solar Radiation

The solar irradiation is an uncertainty factor depending on the weather conditions. Its uncertainty is assumed to follow the beta distribution as (1).

$$f_{I_t}(I_t) = \frac{\Gamma(a_{I,t} + b_{I,t})}{\Gamma(a_{I,t})\Gamma(b_{I,t})} (I_t)^{a_{I,t}-1} (1 - I_{t,t})^{b_I-1} \tag{1}$$

where a_I and b_I are beta distribution parameters, Γ is the gamma function and I_t refers to solar irradiance in each hour.

2.1.2. Outdoor Temperature

The outdoor temperature is assumed to follow the Gaussian distribution as (2).

$$f_{T_o,t}(T_o, t) = \frac{1}{\sqrt{2\sigma_t^2 \pi}} e^{\frac{(T_{o,t} - \mu_t)^2}{2\sigma_t^2}} \tag{2}$$

where σ_t and μ_t are the parameters of the Gaussian distribution and T_o, t is the outside temperature in each hour.

2.2. Photovoltaic Array

Photovoltaic electricity generation varies according to the solar radiation depending mainly on the site location and weather conditions. Output power of the photovoltaic (PV) system has been obtained as Equation (3) from studies [35,36].

$$P_{pv}(s,t) = \eta_{pv}(s,t) A_{pv} I(s,t) \tag{3}$$

The efficiency of the PV panel is expressed as Equation (4) from study [35].

$$\eta_{pv}(s,t) = \eta_{pv,ref}\left[1 - \alpha(T_o(s,t) + I(s,t)\frac{NOCT - 20}{0.8} - T_{ref})\right] \tag{4}$$

2.3. Solar Collector

Similar to photovoltaic resources, the power generation of solar collector hinges on solar irradiance. The thermal energy produced by solar collector, has been obtained by applying the Equation (5) from study [35].

$$Q_{sc}(s,t) = \eta_{sc}(s,t) A_{sc} I(s,t) \tag{5}$$

The efficiency of solar collector is generally not a fixed value and depends on the outdoor temperature and radiation parameter. The efficiency $\eta_{sc}(s,t)$ is calculated as Equation (6) from study [35].

$$\eta_{sc}(s,t) = \eta_{sc0} - \frac{a_1}{G}(T_{sc}(s,t) - T_o(s,t)) - \frac{a_2}{G}(T_{sc}(s,t) - T_o(s,t))^2 \tag{6}$$

2.4. Optimal Scheduling

The amount of stored energy in the thermal storage is obtained as the summation of the remaining thermal energy from the previous hour (by considering the thermal loss), thermal power generation of solar collector and the electric heater in that hour, minus the hourly thermal demand. To obtain power consumption of electrical boiler, Equation (7) and constraints (8) and (9) are considered. The thermal energy waste or loss of the thermal storage (water tank) is considered as 0.03% per hour [37].

$$Q_{tank}(s,t+1) = \eta_{tank} \times Q_{tank}(s,t) + Q_{eb}(t) + Q_{sc}(s,t) - Q_l(s,t) \tag{7}$$

The electric boiler has a limitation for generating thermal power (Q_{eb}) which is its designed capacity (Cap_{eb}) and considered in the model as:

$$Q_{eb}(t) \leq Cap_{eb} \tag{8}$$

The stored energy in the thermal storage (Q_{tank}) cannot exceed its designed capacity (Cap_{tank}), which is described as:

$$Q_{tank}(s,t) \leq Cap_{tank} \tag{9}$$

To reach a balance between the supply-side and the demand-side and avoid electricity shortage, the electricity demand (p_l) must be met by photovoltaic electricity generation (p_{pv}), the battery discharge (dch_{bat}) and the purchasing power from the grid (p_{grid}) minus the battery charge (ch_{bat}) and the power sold to the grid (p_{sold}). The electrical power's exchange with the grid has been calculated as in Equation (10). The amount of PV power production in excess is sold to the grid as in Equation (11).

$$p_{grid}(s,t) = p_l(t) + p_{sold}(s,t) + ch_{bat}(t) - p_{pv}(s,t) - dch_{bat}(t) \tag{10}$$

$$p_{sold}(s,t) = \begin{cases} p_{pv}(s,t) - p_l(st) & \text{if} \quad p_{pv}(s,t) + dch_{bat}(t) \geq p_l(t) + ch_{bat}(t) \\ 0 & \text{if} \quad p_{pv}(s,t) + dch_{bat}(t) \leq p_l(t) + ch_{bat}(t) \end{cases} \tag{11}$$

where the electricity demand ($p_l(t)$) has been obtained from Equation (12). The electricity demand refers to uncontrollable loads (p_{l0}), such as electricity consumption of the electric boiler and other controllable loads such as the dishwasher.

$$p_l(t) = \frac{Q_{eb}(t)}{\eta_{eb}} + p_{l0}(t) + \sum_i Sl(i,t) \times En(i) \quad \forall i \in \{wsh, dry, dsh, pmp\} \tag{12}$$

Allowed operation time of each component has been controlled by constraint (13).

$$Sl(i,t) = \begin{cases} 0 \quad \text{or} \quad 1 & \text{if} \quad t \in T_i \\ 0 & \text{if} \quad t \notin T \end{cases} \quad \forall i \in \{ws, dry, dsh, pmp\}, EOT_i \leq T_i \leq LOT \tag{13}$$

Constraint (13) represents the particular preferences of households, namely their preferred time intervals for operating the smart appliances. These tasks must be completed within and cannot be operated outside of their specific time intervals. The required length of operation for each component is reached by applying constraint (14). This second constraint is applied to satisfy the required operation time of each smart appliance.

$$\sum_{t=1}^{24} Sl(i,t) = \text{rot}(i) \quad \forall i \in \{ws, dry, dsh, pmp\} \tag{14}$$

The dryer machine needs to operate after completing the washing procedure, which is considered in the model as (15).

$$Sl(dry,t) = 0 \quad \text{if} \quad \sum_{n=1}^{t-1} Sl(ws,n) < \text{rot}(ws) \tag{15}$$

The amount of stored energy in the battery is calculated as Equation (16).

$$E_{bat}(t) = \eta_{bat} \times E_{bat}(t-1) + ch\_bat(t) - dch\_bat(t) \tag{16}$$

where E_{bat} is the hourly stored energy and η_{bat} represents the efficiency of the battery. The hourly stored energy cannot exceed its designed capacity which is described as:

$$E_{bat}(t) \leq Cap_{bat} \tag{17}$$

The hourly battery charge and discharge are limited, which are considered as constraints (18) and (19). It is assumed that the battery has no stored energy at the beginning of the day. The battery must discharge the stored electricity to its initial state by the end of each day.

$$ch\_bat(t) \leq r_{bat} \times Cap_{bat} \tag{18}$$

$$dch\_bat(t) \leq r_{bat} \times Cap_{bat} \tag{19}$$

where r_{bat} is the maximum charge/discharge rate of the battery storage which is defined as a share of the battery capacity [10,38].

Finally, minimizing the expected electricity payment for the entire day as objective function has been formulated as (20).

$$\min J = \sum_{s=1}^{100} \sum_{t=1}^{24} \alpha(s) \times [p_{sold}(s,t) \times c_{sold} - p_{grid}(s,t) \times c_{elec}(t)] \tag{20}$$

where $\alpha(s)$ is the probability of each scenario with summation of 1.

Thermal Load

The thermal energy from the solar collector and the boiler is stored into the water tank for providing hot water and heating and ventilation demands. Hourly thermal load for each scenario has been calculated as Equation (21).

$$Q_l(s,t) = Q_h(s,t) + Q_{wt}(s,t) \tag{21}$$

where $Q_h(s,t)$ refers to the required thermal energy for heating system and $Q_{wt}(s,t)$ shows the hot water energy demand.

It is assumed that the indoor temperature should be maintained in a predetermined range $22°C \leq T_{r,t}^s \leq 24°C$ when the household is occupied based on ASHRAE Standard 55-2013 for 30% humidity. It is assumed indoor temperature follows Equation (22) from study [35].

$$T_r(s,t+1) = (1 - \frac{\tau}{R_a C_r})T_r(s,t) + \frac{\tau}{R_a C_r}T_o(s,t) + \frac{\eta_w A_w \tau}{C_r}I(s,t) + \frac{Q_h(s,t)}{C_r} \tag{22}$$

where $Q_h(s,t)$ depicts the required thermal demand for the heating and ventilation load for each scenario.

In this model, it is assumed that the daily average hot water demand for the benchmark is a linear function of the bedrooms number [39], which is summarized in Table 1. Hot water energy demand is classified to controllable loads (thermal energy consumption of washing machine and dishwasher) and uncontrollable loads (thermal energy needed for shower, sink, and bath) as Equation (23).

$$Q_{wt}(s,t) = Q_{wt,0}(s,t) + Q_{wt,ws}(s,t) + Q_{wt,dsh}(s,t) \tag{23}$$

Thermal energy consumption of washing machine and dishwasher are assumed as variable factors in the optimal scheduling problem as Equations (24) and (25).

$$Q_{wt,ws}(s,t) = 0.0439 \times (49 - T_0(s,t)) \times Sl(ws,t) \tag{24}$$

$$Q_{wt,dsh}(s,t) = 0.0146 \times (49 - T_0(s,t)) \times Sl(dsh,t) \tag{25}$$

Table 1. Benchmark water temperature and volume for each hot water end-use [39].

End-Use	End-Use Temperature (°F)	Water Demand (gal/day)
Washing machine	120	$7.5 + 2.5 \times N_{br}$
Dishwasher	120	$2.5 + 0.833 \times N_{br}$
Shower	105	$14 + 4.67 \times N_{br}$
Bath	105	$3.5 + 1.17 \times N_{br}$
Sink	105	$12.5 + 4.16 \times N_{br}$

3. Simulation Results and Discussion

A two-bedroom smart home in Shiraz city, Iran was considered as the case study. The household was equipped with solar collector (SC), photovoltaic array (PV), thermal storage tank, electrical boiler, battery storage (see Figure 1) and other loads of different types such as electric washing machine, washing dryer, dishwasher, water supply pump, and uncontrollable loads. The boiler and solar collector provide thermal energy for heating and ventilation system, and also deliver hot water for washing machine, dishwasher, sink, bath, shower, and other hot water outlets with a hot water storage tank.

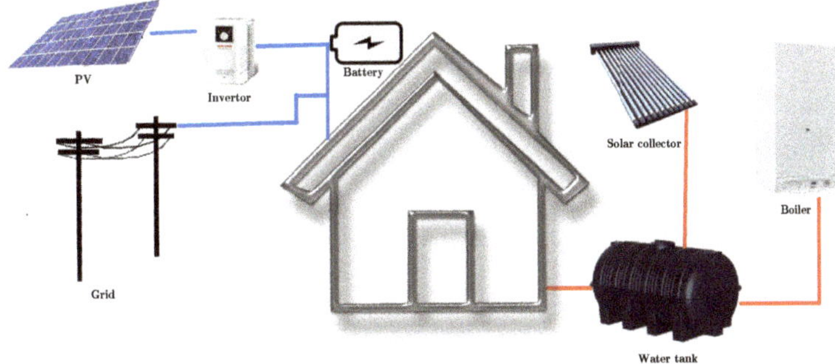

Figure 1. The proposed energy supply for the case study.

3.1. Input Data

Shiraz temperature data for every 10 minutes at the height of 10 m in 2008 are available on [40]. Temperature of each hour for Shiraz during one year can be seen in Figure 2. It is assumed that the smart home was equipped with 14.6 m^2 PV panels, 3.12 m^2 solar collector and 1 KWh lithium-ion battery. The photovoltaic panel considered in this study was SPR-E20-327 made by SunPower and the modeled solar collector was assumed to be APSE-10 made by APRICUS company.

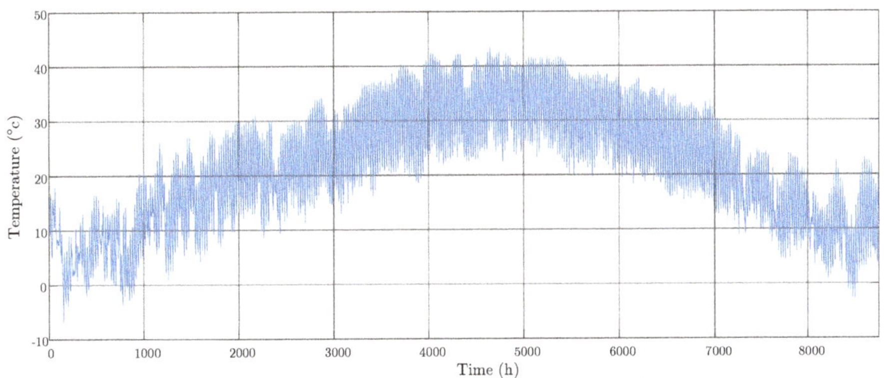

Figure 2. Temperature data for Shiraz city, Iran.

Solar radiation data for this case study were taken from the Iranian Renewable Energy Organization (SUNA) database [40] shown in Figure 3 and average daily radiation was equal to 5.48 KWh/m^2/d. Electricity price was considered as $0.044 /KWh for (1–19) time interval and $0.053 /KWh for (20–24) time interval [41]. It was assumed that the feed-in tariff for solar electricity was equal to $0.23/KWh [42].

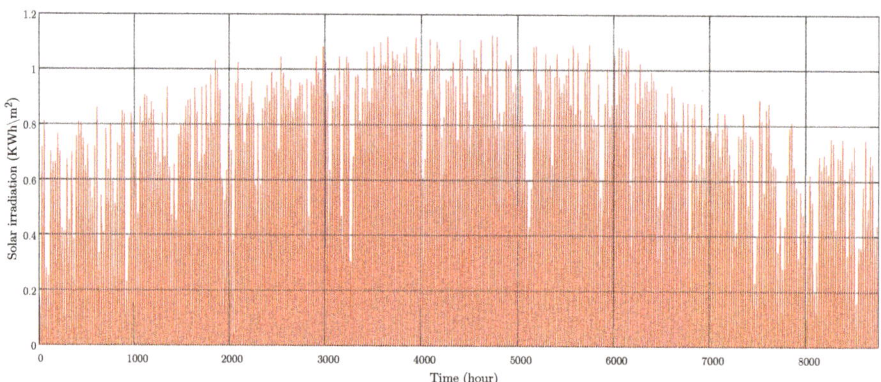

Figure 3. Solar radiation in Shiraz city, Iran.

Electric loads were classified into two categories: controllable loads and uncontrollable loads. A constant load (Figure 4), was considered to represent uncontrollable loads in the proposed model. Controllable loads include washing machine, cloth dryer, dishwasher, and water pump. The rated power consumption of controllable smart appliances, the allowed interval time for operation, and the required operation time during one day in winter are summarized in Table 2. The required thermal parameters of the smart home are given by detail in Table 3.

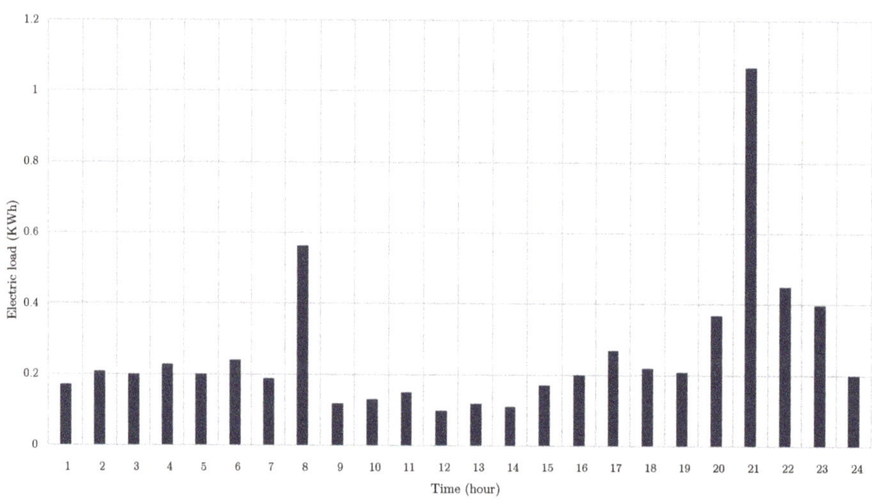

Figure 4. Uncontrollable electricity demand.

Table 2. Controllable electric devices.

Component	Rated Power (KW)	Required Operation (h)	Allowed Operation Time Interval
Washing machine	1	2	$[9-15]$
Dryer	1.3	1	$[16-23]$
Dish washer	0.5	2	$[9-23]$
Water pump	0.7	3	$[1-24]$

Table 3. The smart home thermal parameters.

C_r (KWh/$^\circ$C)	A_w (m^2)	R_a ($^\circ$C/KW)
8.188	20.203	37.984

3.2. Uncertainty

Historical data of solar radiation for 30 days of January (Figure 3) were used to obtain the beta PDF parameters by curve fitting coded into MATLAB software. In the next step, 1000 scenarios were generated for each hour based on the obtained beta PDF. The possibility of each hourly datum was calculated based on that hour's specific beta distribution. As shown in Figure 5, 10 final solar irradiance scenarios were chosen using SCENRED2 coded into GAMS software.

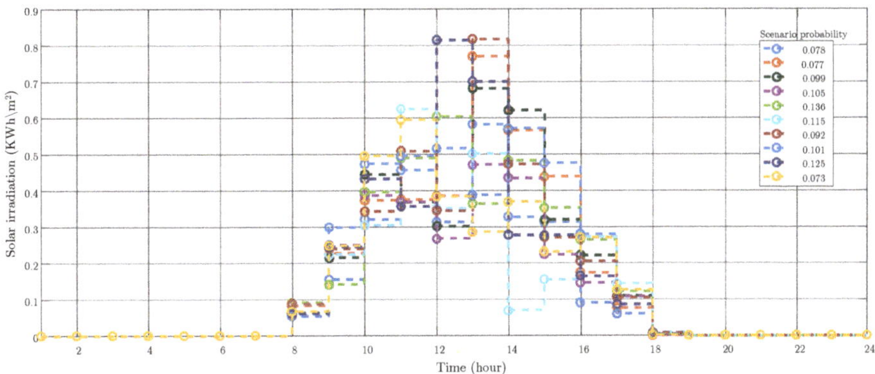

Figure 5. The final scenarios for solar irradiance.

Similar to solar irradiance, the parameters of the Gaussian distribution were estimated using 30 historical data in January 2008 for Shiraz (see Figure 2). Then, 1000 scenarios were generated for each hour based on the obtained PDF. The probability of each hourly datum was calculated based on that hour's specific Gaussian distribution. Finally, 10 outdoor temperature scenarios were chosen using SCENRED2 in GAMS software as illustrated in Figure 6.

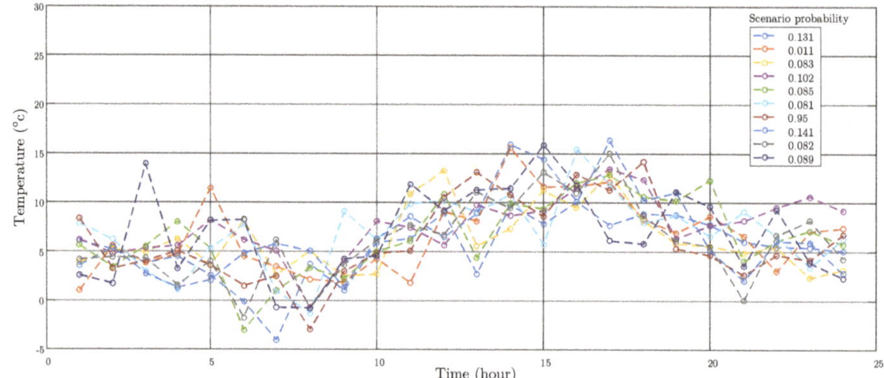

Figure 6. The final scenarios for outdoor temperature.

3.3. Optimal Scheduling

The described stochastic scheduling problem was coded into GAMS software and was solved as a mixed-integer linear programming (MILP) model. For the proposed scheduling, the weighted average cost of optimal scheduling of 100 scenarios for one day in January was equal to $0.05 , which was 20% less than the operation without scheduling ($0.06). This cost reduction mostly came from shifting controllable loads from higher-cost on-peak periods to lower-cost off-peak time intervals.

As mentioned in previous section, the indoor temperature should be maintained in a predetermined range 22 °C $\leq T_{r,t}^s \leq$ 24 °C when the household is occupied. The weighted average indoor temperature of 100 scenarios are depicted in Figure 7, which shows the indoor temperature was set about 22 °C for reducing energy consumption.

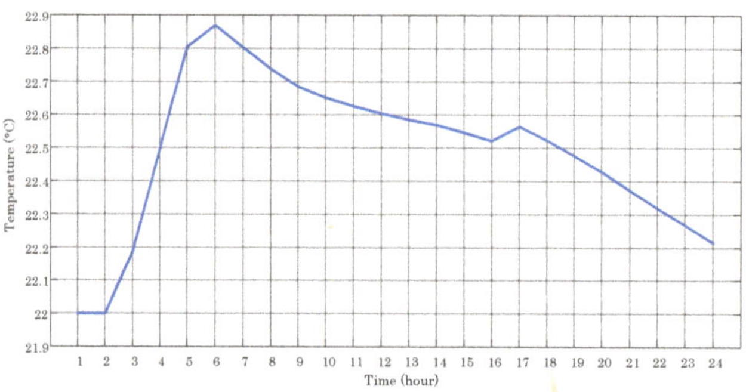

Figure 7. Average indoor temperature.

Figure 8a depicts the electricity consumption and the optimal schedule for operating the dishwasher set to operate between 9:00 am to 11:00 pm. The required operating period for the dishwasher was 2 h. To decrease the energy costs, the dishwasher operated during off-load periods. The operation of the dryer depended on the operation of the washing machine. It was assumed that the dishwasher had the capacity to wash dishes of three meals, which took 2 h. It meant the dishwasher washed the dishes of the dinner the previous evening, breakfast, and lunch. The allowed operation interval for the washing machine, which was between 9:00 am to 3:00 pm, forced the model to run the washing machine from 1:00 pm to 2:00 pm. As shown in Figure 8, the operation time of the dishwasher, dryer, and the electric heater (partially) was shifted to early morning and late afternoon to increase the sold electricity to the grid and profit from the feed-in tariff. The electric pump benefited from freedom in operation time throughout the studied 24-h period. This device needed to operate for 3 h to meet the demand, which is illustrated in Figure 8c. Figure 8b shows the optimal timing for the operation of these devices. Figure 9 illustrates the optimal operation of the battery storage. The battery was charged early in the morning to cover a part of electricity demand after sunrise and increase the amount of electricity that was fed into the grid. In the afternoon, the battery was recharged to provide electricity during the on-peak period and thereby decrease the overall costs.

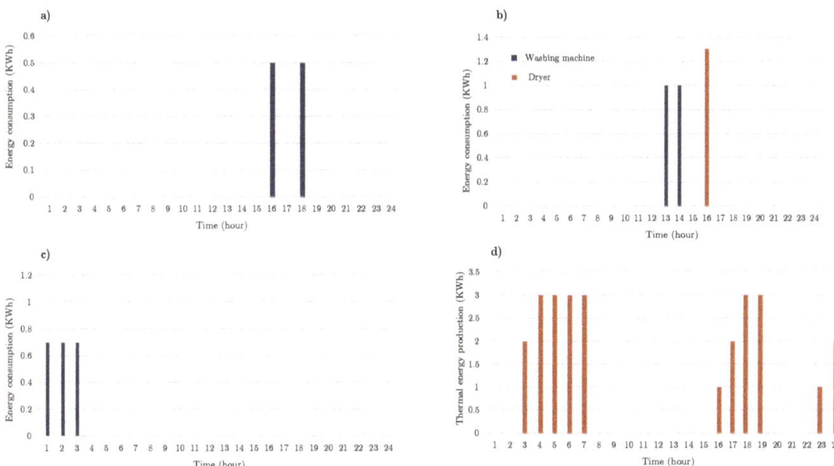

Figure 8. The optimal schedule for operation of devices: (**a**) Dishwasher. (**b**) Washing and drying machine. (**c**) Water pump. (**d**) Electric heater.

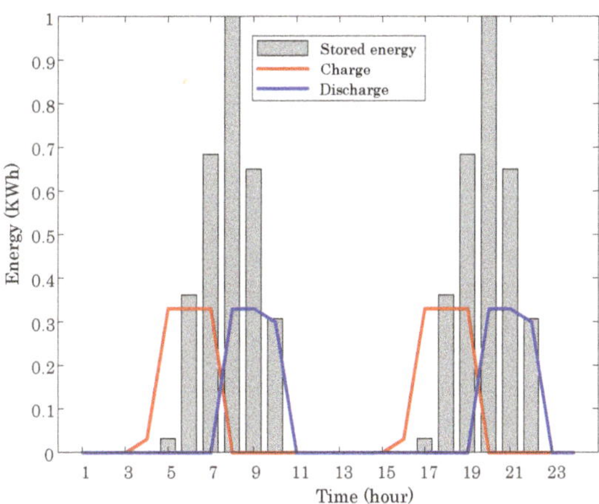

Figure 9. The optimal schedule for the battery operation.

In the proposed model, the thermal load was categorized into two types: (1) the thermal demand for heating and ventilation, and (2) the thermal demand for the hot water. These thermal loads are illustrated in Figure 10a,c as the average of the 100 proposed scenarios (weighted average based on the probability of each scenario). The peak hot water demand occurred at 8 am with thermal demand of 3.2

KWh, while the demand for heating and ventilation reached 3 KWh maximum at 4 am. The weighted average amount of stored thermal energy in the thermal storage for the 100 scenarios is described in Figure 10b. Solar collector with 3.2 KWh thermal power generation (see Figure 11b)provided around 10% of the total thermal energy demand.

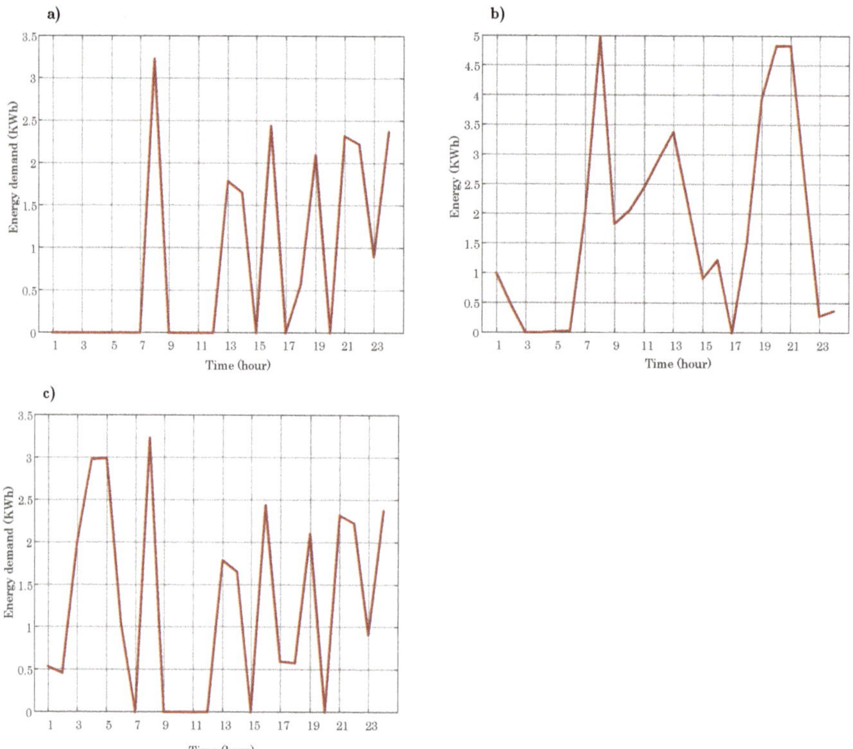

Figure 10. The average thermal energy: (**a**) demand for the required hot water. (**b**) stored in the thermal storage. (**c**) total thermal energy demand.

The optimal operation schedule of the electric boiler is depicted in Figure 8d. The electric boiler operated from 3:00 am to 7:00 am with its full capacity in order to let the user benefit from the feed-in tariff after sunrise by selling photovoltaic electricity to the grid. In the late afternoon, the electric boiler was rerun to its the full capacity at 6:00 pm and 7:00 pm to avoid the higher electricity cost after 7:00 pm (on-peak period with higher-cost). The required thermal energy for the dishwasher and the washing machine are illustrated in Figure 12a,b. The results indicate in weighted average for the 100 proposed scenarios the thermal demand for the dishwasher (1.1 KWh) and washing machine (3.4 KWh) accounted for 15% of the total thermal energy demand. Running the washing machine (7:00–8:00 pm) and dishwasher (9:00–10:00 pm) in the evening, as in the operation without scheduling, increased the thermal energy demand of these devices (5% for the washing machine and 13% for the dishwasher) due to different outdoor temperature. Ignoring the thermal energy demand of these devices in optimal scheduling of smart appliances could cause inaccurate results.

The weighted average power generation for the 100 proposed scenarios of the photovoltaic panels and the solar collectors are illustrated in Figure 11a,b. The maximum power generation culminated at 1:00 pm. The exchange electricity between the grid and the proposed energy system is shown in Figure 13. In the proposed operation schedule, the peak electricity demand was 4.2 KWh at 6:00 pm, which was around 45% less compared to the peak for the operation without scheduling (7.6 KWh at 9 pm). The results indicate that the total electricity demand during on-peak hours (20–24) with the optimal scheduling was 4.7 KWh, while this value was equal to 15.6 KWh without scheduling (see Figure 13).

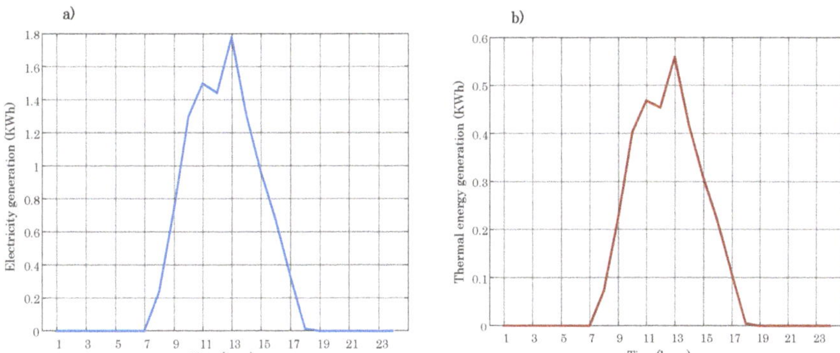

Figure 11. (**a**) The electricity generation of the photovoltaic panels. (**b**) The thermal energy generation of the solar collector.

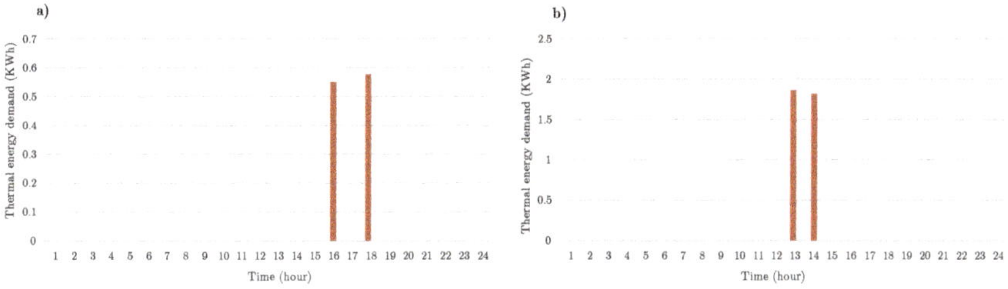

Figure 12. The thermal energy consumption of: (**a**) dishwasher. (**b**) washing machine.

As discussed in the introduction, variable renewable resources, including solar and wind power, are non-dispatchable and generate fluctuating power due to their inherent intermittency. Mismatches between the variable renewable power generation and electric demand can cause instability in electricity systems. The solutions to avoid imbalance between the supply-side and the demand-side, such as energy storage and demand response, play a chief role in the deployment of variable renewable resources [38]. The proposed model in the current study shifted a portion of the controllable loads from the on-peak time intervals to off-peak time intervals. Due to this shift and the described constraints for the allowed operation periods (see Table 2), the renewable power self-consumption increased compared to the base scenario without scheduling. The portion of photovoltaic electricity generation, which was consumed by the smart house, is defined as self-consumption. The excess photovoltaic electricity generation decreased from 78% (8.1 KWh) for the base scenario without scheduling to 67% (6.9 KWh) for the optimal scheduling, which was beneficial for the stability of the electricity system. The excess renewable generation refers to a share of the renewable power generation exceeding the load demand which was fed into the electric grid. Because of renewable feed-in tariffs, a higher share of the excess renewable generation increased the profit margin of the households with renewable generators. On the other hand, this excess renewable generation could cause instability in the electricity system and force additional costs on the system.

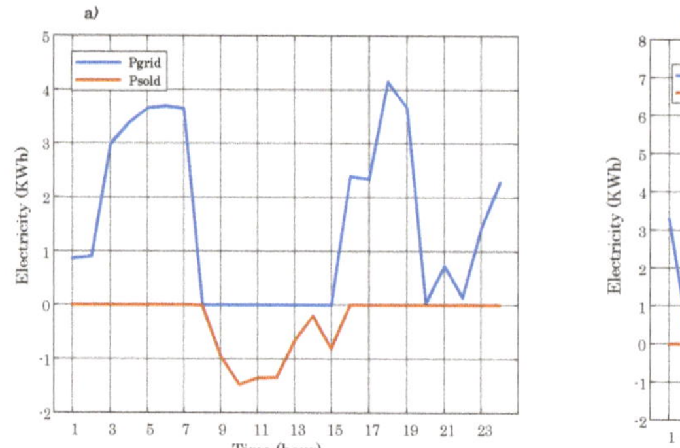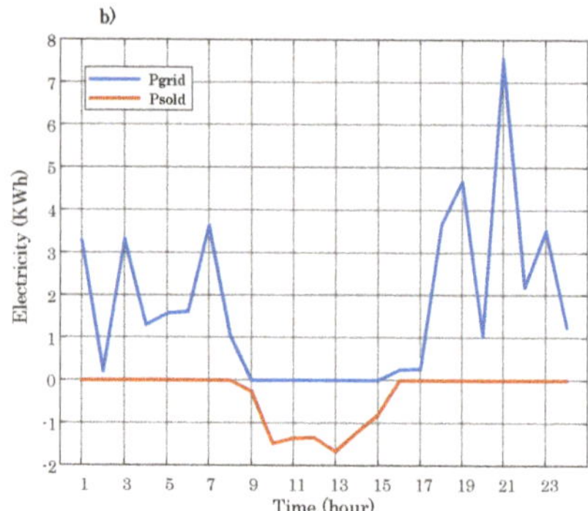

Figure 13. The exchange electricity with the electricity grid: (**a**) with the optimal scheduling; (**b**) without the optimal scheduling.

4. Conclusions

In smart grids, buildings are no longer passive consumers, but they are controllable loads, which can be used for demand-side energy management. In this study, a stochastic mixed-integer linear programming model is used to find the optimal scheduling of a smart home using the flexibility offered by smart controllable devices, in order to minimize the total energy cost and address the uncertainties in both demand and supply sides. The weighted average cost of optimal scheduling of the 100 final scenarios for one day in January was obtained as $0.05, which shows 20 percent cost-saving compared to the operation without scheduling. In the meantime, the proposed model in this study shifts the load from on-peak periods to off-peak periods to decrease the energy costs and the peak demand. The results indicate that the proposed model has the potential to decrease the peak electricity demand from 7.6 KWh to 4.2 KWh which is about 45% peak shaving.

To address the uncertainties in the scheduling of smart homes, Monte Carlo simulation technique is used to generate 1000 random scenarios for the two relevant environmental factors, namely the outdoor temperature and solar radiation. As a result, thermal load, output power of the photovoltaic panel, solar collector power generation, and electricity load become stochastic parameters. Modeling the thermal load revealed that around 15% of thermal demand goes to the washing machine and dishwasher. Ignoring the captured uncertainties of thermal demand in smart building scheduling could cause unrealistic and inaccurate results. Furthermore, the thermal energy demand of the washing machine and the dishwasher depends on their operation time and varies during the day. Considering this factor in the scheduling of smart buildings assists in energy saving and in avoiding unnecessary costs. For the current case study, the proposed operation schedule saved one percent of thermal energy by shifting these loads.

Although the model improved the share of the excess renewable generation to 67% from a 78% excess share for the operation without scheduling, a further investigation is required to consider the stability of the electricity grid. Future studies should explore the trade-offs between the share of excess renewable production (the household's profit) and the stability of the power system (the additional costs of the electricity system).

Author Contributions: E.A. designed, conceptualized, and developed the simulation model, analyzed the simulation results, visualized, and wrote the manuscript. Y.N., B.M.-I. and A.A.-M. gave guidance, provided the materials, and helped to improve the quality of the work. All authors have read and agreed to the published version of the manuscript.

Funding: This research received no external funding.

Conflicts of Interest: The authors declare no conflict of interest.

Nomenclature

t	Set of time.
i	Set for the smart appliances.
s	Set for the scenarios.
α	Scenario probability.
N_{br}	Number of bedrooms.
T_r	Indoor temperature (°C).
τ	Time interval (hour).
R_a	Room equivalent thermal resistance (°C/KW).
C_r	Room thermal capacity (KWh/°C).
T_o	Outdoor temperature (°C).
A_W	Room effective window (m²).
I	Solar irradiance (KWh/m²).
Q_h	Heating and ventilation thermal energy demand (KWh).
Q_l	Total thermal load (KWh).
Q_{wt}	Hot water thermal load (KWh).
P_{pv}	Photovoltaic electricity generation (KWh).
eta_{pv}	Efficiency of photovoltaic panel.
eta_{sc}	Efficiency of solar collector.
A_{pv}	Photovoltaic panel size (m²).
A_{sc}	Solar collector size (m²).
Q_{tank}	Stored thermal energy in the water tank (KWh).
p_{sold}	Injected electricity to the grid (KWh).
p_{grid}	Purchased electricity form the grid (KWh).
p_l	Electricity demand (KWh).
Q_{eb}	Thermal energy generation by eclectic boiler (KWh).
p_{eb}	Electricity consumption of eclectic boiler (KWh).
C_{sold}	Sold electricity price to the grid ($/KWh).
C_{elec}	Purchased electricity cost from the grid ($/KWh).
f	Probability density function (PDF).
a_l, b_l	Beta distribution parameters.
σ, μ	Normal distribution parameters.
$Q_{wt,ws}$	Thermal energy consumption of the washing machine (KWh).
$Q_{wt,dsh}$	Thermal energy consumption of the dishwasher (KWh).
$Q_{wt,0}$	Uncontrollable hot water energy consumption (KWh).
En	Rated power of smart appliances (KW).
rot	Required operation time of smart appliances (h).
EOT	Earliest operation time of smart appliances.
LOT	Latest operation time of smart appliances.
Sl	State of smart appliance (binary variable).
dsh	Dishwasher.
pmp	Water supply pump.
dry	Dryer (cloth).
ws	Washing machine.
E_{bat}	Energy stored in the battery.
η_{bat}	Efficiency of the battery storage.
ch_{bat}	Electricity charge of battery storage.
dch_{bat}	Electricity discharge of battery storage.
Cap_{bat}	Capacity of battery storage.
r_{bat}	Battery charge/discharge rate.

References

1. Tanaka, K.; Uchida, K.; Ogimi, K.; Goya, T.; Yona, A.; Senjyu, T.; Funabashi, T.; Kim, C.H. Optimal Operation by Controllable Loads Based on Smart Grid Topology Considering Insolation Forecasted Error. *IEEE Trans. Smart Grid* **2011**, *2*, 438–444. [CrossRef]
2. Noorollahi, Y.; Shabbir, M.S.; Siddiqi, A.F.; Ilyashenko, L.K.; Ahmadi, E. Review of two decade geothermal energy development in Iran, benefits, challenges, and future policy. *Geothermics* **2019**, *77*, 257–266. [CrossRef]
3. Akikur, R.; Saidur, R.; Ping, H.; Ullah, K. Comparative study of stand-alone and hybrid solar energy systems suitable for off-grid rural electrification: A review. *Renew. Sustain. Energy Rev.* **2013**, *27*, 738–752. [CrossRef]
4. Baljit, S.; Chan, H.Y.; Sopian, K. Review of building integrated applications of photovoltaic and solar thermal systems. *J. Clean. Prod.* **2016**, *137*, 677–689. [CrossRef]
5. Alaa, M.; Zaidan, A.; Zaidan, B.; Talal, M.; Kiah, M. A review of smart home applications based on Internet of Things. *J. Netw. Comput. Appl.* **2017**, *97*, 48–65. [CrossRef]
6. Moradzadeh, A.; Sadeghian, O.; Pourhossein, K.; Mohammadi-Ivatloo, B.; Anvari-Moghaddam, A. Improving Residential Load Disaggregation for Sustainable Development of Energy via Principal Component Analysis. *Sustainability* **2020**, *12*, 3158. [CrossRef]
7. Roy, S.S.; Samui, P.; Nagtode, I.; Jain, H.; Shivaramakrishnan, V.; Mohammadi-ivatloo, B. Forecasting heating and cooling loads of buildings: A comparative performance analysis. *J. Ambient Intell. Humaniz. Comput.* **2020**, *11*, 1253–1264. [CrossRef]
8. Heiskanen, E.; Nissilä, H.; Lovio, R. Demonstration buildings as protected spaces for clean energy solutions—The case of solar building integration in Finland. *J. Clean. Prod.* **2015**, *109*, 347–356. [CrossRef]
9. Yang, L.; Entchev, E.; Rosato, A.; Sibilio, S. Smart thermal grid with integration of distributed and centralized solar energy systems. *Energy* **2017**, *122*, 471–481. [CrossRef]
10. Pooranian, Z.; Abawajy, J.H.; P, V.; Conti, M. Scheduling Distributed Energy Resource Operation and Daily Power Consumption for a Smart Building to Optimize Economic and Environmental Parameters. *Energies* **2018**, *11*, 1348. [CrossRef]
11. Oskouei, M.Z.; Mohammadi-Ivatloo, B.; Abapour, M.; Anvari-Moghaddam, A.; Mehrjerdi, H. Practical implementation of residential load management system by considering vehicle-for-power transfer: Profit analysis. *Sustain. Cities Soc.* **2020**, *60*, 102144. [CrossRef]
12. Baek, K.; Ko, W.; Kim, J. Optimal Scheduling of Distributed Energy Resources in Residential Building under the Demand Response Commitment Contract. *Energies* **2019**, *12*, 2810. [CrossRef]
13. Akbari-Dibavar, A.; Nojavan, S.; Mohammadi-Ivatloo, B.; Zare, K. Smart home energy management using hybrid robust-stochastic optimization. *Comput. Ind. Eng.* **2020**, *143*, 106425. [CrossRef]
14. Mehrpooya, M.; Mohammadi, M.; Ahmadi, E. Techno-economic-environmental study of hybrid power supply system: A case study in Iran. *Sustain. Energy Technol. Assessments* **2018**, *25*, 1–10. [CrossRef]
15. Mohammadi, M.; Ghasempour, R.; Astaraei, F.R.; Ahmadi, E.; Aligholian, A.; Toopshekan, A. Optimal planning of renewable energy resource for a residential house considering economic and reliability criteria. *Int. J. Electr. Power Energy Syst.* **2018**, *96*, 261–273. [CrossRef]
16. Sánchez, C.; Bloch, L.; Holweger, J.; Ballif, C.; Wyrsch, N. Optimised Heat Pump Management for Increasing Photovoltaic Penetration into the Electricity Grid. *Energies* **2019**, *12*, 1571. [CrossRef]
17. Wu, Z.; Zhou, S.; Li, J.; Zhang, X. Real-Time Scheduling of Residential Appliances via Conditional Risk-at-Value. *IEEE Trans. Smart Grid* **2014**, *5*, 1282–1291. [CrossRef]
18. Ali, M.; Jokisalo, J.; Siren, K.; Lehtonen, M. Combining the Demand Response of direct electric space heating and partial thermal storage using LP optimization. *Electr. Power Syst. Res.* **2014**, *106*, 160–167. [CrossRef]
19. Booysen, M.J.; Cloete, A.H. Sustainability through Intelligent Scheduling of Electric Water Heaters in a Smart Grid. In Proceedings of the 2016 IEEE 14th International Conference on Dependable, Autonomic and Secure Computing, 14th International Conference on Pervasive Intelligence and Computing, 2nd International Conference on Big Data Intelligence and Computing and Cyber Science and Technology Congress(DASC/PiCom/DataCom/CyberSciTech), Auckland, New Zealand, 8–12 August 2016; pp. 848–855.
20. Nel, P.J.C.; Booysen, M.J.; van der Merwe, B. A Computationally Inexpensive Energy Model for Horizontal Electric Water Heaters With Scheduling. *IEEE Trans. Smart Grid* **2018**, *9*, 48–56. [CrossRef]

21. Shi, E.; Jabari, F.; Anvari-Moghaddam, A.; Mohammadpourfard, M.; Mohammadi-Ivatloo, B. Risk-Constrained Optimal Chiller Loading Strategy Using Information Gap Decision Theory. *Appl. Sci.* **2019**, *9*, 1925. [CrossRef]
22. Khan, J.; Arsalan, M.H. Solar power technologies for sustainable electricity generation—A review. *Renew. Sustain. Energy Rev.* **2016**, *55*, 414–425. [CrossRef]
23. Luthander, R.; Widén, J.; Nilsson, D.; Palm, J. Photovoltaic self-consumption in buildings: A review. *Appl. Energy* **2015**, *142*, 80–94. [CrossRef]
24. Pickering, B.; Choudhary, R. District energy system optimisation under uncertain demand: Handling data-driven stochastic profiles. *Appl. Energy* **2019**, *236*, 1138–1157. [CrossRef]
25. Onishi, V.C.; Antunes, C.H.; Fraga, E.S.; Cabezas, H. Stochastic optimization of trigeneration systems for decision-making under long-term uncertainty in energy demands and prices. *Energy* **2019**, *175*, 781–797. [CrossRef]
26. Afzali, S.F.; Cotton, J.S.; Mahalec, V. Urban community energy systems design under uncertainty for specified levels of carbon dioxide emissions. *Appl. Energy* **2020**, *259*, 114084. [CrossRef]
27. Zhang, B.; Hu, W.; Cao, D.; Huang, Q.; Chen, Z.; Blaabjerg, F. Deep reinforcement learning–based approach for optimizing energy conversion in integrated electrical and heating system with renewable energy. *Energy Convers. Manag.* **2019**, *202*, 112199. [CrossRef]
28. Gabrielli, P.; Fürer, F.; Mavromatidis, G.; Mazzotti, M. Robust and optimal design of multi-energy systems with seasonal storage through uncertainty analysis. *Appl. Energy* **2019**, *238*, 1192–1210. [CrossRef]
29. Testi, D.; Urbanucci, L.; Giola, C.; Schito, E.; Conti, P. Stochastic optimal integration of decentralized heat pumps in a smart thermal and electric micro-grid. *Energy Convers. Manag.* **2020**, *210*, 112734. [CrossRef]
30. Urbanucci, L.; Testi, D. Optimal integrated sizing and operation of a CHP system with Monte Carlo risk analysis for long-term uncertainty in energy demands. *Energy Convers. Manag.* **2018**, *157*, 307–316. [CrossRef]
31. Yin, Y.; Liu, T.; He, C. Day-ahead stochastic coordinated scheduling for thermal-hydro-wind-photovoltaic systems. *Energy* **2019**, *187*, 115944. [CrossRef]
32. Bashir, A.A.; Pourakbari-Kasmaei, M.; Contreras, J.; Lehtonen, M. A novel energy scheduling framework for reliable and economic operation of islanded and grid-connected microgrids. *Electr. Power Syst. Res.* **2019**, *171*, 85–96. [CrossRef]
33. Gomes, I.; Pousinho, H.; Melício, R.; Mendes, V. Stochastic coordination of joint wind and photovoltaic systems with energy storage in day-ahead market. *Energy* **2017**, *124*, 310–320. [CrossRef]
34. Wu, L.; Shahidehpour, M.; Li, T. Stochastic Security-Constrained Unit Commitment. *IEEE Trans. Power Syst.* **2007**, *22*, 800–811. [CrossRef]
35. Nguyen, H.T.; Nguyen, D.T.; Le, L.B. Energy Management for Households With Solar Assisted Thermal Load Considering Renewable Energy and Price Uncertainty. *IEEE Trans. Smart Grid* **2015**, *6*, 301–314. [CrossRef]
36. Ahmadi, E.; McLellan, B.; Ogata, S.; Tezuka, T. Modelling the Water-Energy-Nexus to Assist the Design of Economic and Regulatory Support Instruments towards Sustainability. In Proceedings of the Chemeca 2019: Chemical Engineering Megatrends and Elements, Sydney, Australia, 29 September–2 October 2019; p. 550.
37. Sioshansi, R.; Denholm, P. The Value of Concentrating Solar Power and Thermal Energy Storage. *IEEE Trans. Sustain. Energy* **2010**, *1*, 173–183. [CrossRef]
38. Ahmadi, E.; McLellan, B.; Ogata, S.; Mohammadi-Ivatloo, B.; Tezuka, T. An Integrated Planning Framework for Sustainable Water and Energy Supply. *Sustainability* **2020**, *12*, 4395. [CrossRef]
39. Hendron, R.; Burch, J. Development of standardized domestic hot water event schedules for residential buildings. In Proceedings of the ASME 2007 Energy Sustainability Conference. American Society of Mechanical Engineers Digital Collection, Long Beach, CA, USA, 27–30 June 2016; pp. 531–539.
40. Iranian Renewable Energy and Energy Efficiency Organization, Shiraz Temperature and Solar Radiation Data. 2008. Available online: http://www.satba.gov.ir/ (accessed on 31 October 2019).

41. Tehran Regional Electric Energy Distribution Portal, Electricity Price. 2014. Available online: https://www.tvedc.ir/en/ (accessed on 22 September 2019).
42. Iranian Renewable Energy and Energy Efficiency Organization, Guaranteed Electricity Purchase Tariff for Renewables in Iran. 2015. Available online: http://www.satba.gov.ir/en/guaranteed-Guaranteed-Renewable-Electricity-Purchase-Tariffs (accessed on 22 September 2019).

 © 2020 by the authors. Licensee MDPI, Basel, Switzerland. This article is an open access article distributed under the terms and conditions of the Creative Commons Attribution (CC BY) license (http://creativecommons.org/licenses/by/4.0/).

Article

A Deep Neural Network-Assisted Approach to Enhance Short-Term Optimal Operational Scheduling of a Microgrid

Fatma Yaprakdal [1,*], M. Berkay Yılmaz [2], Mustafa Baysal [1] and Amjad Anvari-Moghaddam [3,4]

[1] Faculty of Electrical and Electronics Engineering, Yildiz Technical University, Davutpasa Campus, 34220 Esenler, Istanbul, Turkey; baysal@yildiz.edu.tr
[2] Computer Engineering Department, Akdeniz University, Antalya Campus, Dumlupinar Boulevard, 07058 Antalya, Turkey; berkayyilmaz@akdeniz.edu.tr
[3] Department of Energy Technology, Aalborg University, 9220 Aalborg East, Denmark; aam@et.aau.dk
[4] Faculty of Electrical and Computer Engineering, University of Tabriz, Tabriz 5166616471, Iran
* Correspondence: f4913026@std.yildiz.edu.tr

Received: 6 February 2020; Accepted: 14 February 2020; Published: 22 February 2020

Abstract: The inherent variability of large-scale renewable energy generation leads to significant difficulties in microgrid energy management. Likewise, the effects of human behaviors in response to the changes in electricity tariffs as well as seasons result in changes in electricity consumption. Thus, proper scheduling and planning of power system operations require accurate load demand and renewable energy generation estimation studies, especially for short-term periods (hour-ahead, day-ahead). The time-sequence variation in aggregated electrical load and bulk photovoltaic power output are considered in this study to promote the supply-demand balance in the short-term optimal operational scheduling framework of a reconfigurable microgrid by integrating the forecasting results. A bi-directional long short-term memory units based deep recurrent neural network model, DRNN Bi-LSTM, is designed to provide accurate aggregated electrical load demand and the bulk photovoltaic power generation forecasting results. The real-world data set is utilized to test the proposed forecasting model, and based on the results, the DRNN Bi-LSTM model performs better in comparison with other methods in the surveyed literature. Meanwhile, the optimal operational scheduling framework is studied by simultaneously making a day-ahead optimal reconfiguration plan and optimal dispatching of controllable distributed generation units which are considered as optimal operation solutions. A combined approach of basic and selective particle swarm optimization methods, PSO&SPSO, is utilized for that combinatorial, non-linear, non-deterministic polynomial-time-hard (NP-hard), complex optimization study by aiming minimization of the aggregated real power losses of the microgrid subject to diverse equality and inequality constraints. A reconfigurable microgrid test system that includes photovoltaic power and diesel distributed generators is used for the optimal operational scheduling framework. As a whole, this study contributes to the optimal operational scheduling of reconfigurable microgrid with electrical energy demand and renewable energy forecasting by way of the developed DRNN Bi-LSTM model. The results indicate that optimal operational scheduling of reconfigurable microgrid with deep learning assisted approach could not only reduce real power losses but also improve system in an economic way.

Keywords: day-ahead operational scheduling; reconfigurable microgrid; DRNN Bi-LSTM; aggregated load forecasting; bulk photovoltaic power generation forecasting

---

1. Introduction

Traditional power systems require innovation to bridge the gap between demand and supply while also overcome essential challenges such as grid reliability, grid robustness, customer electricity cost minimization, etc. Accordingly, the so-called smart grids have been developed based on the recent integration of modern communication technologies and infrastructures into conventional grids [1]. A smart grid can be defined in short as the computerization of the electrical networks with the primary objective of decreasing costs to the consumer while improving the reliability and quality of the power supply. Even though the use of computers and digital technology as part of the electrical grid has existed for at least a few decades, this technology has primarily been used for Supervisory Control and Data Acquisition (SCADA) rather than the autonomous intelligent control which is what smart grid paradigms aim for [2]. The smart grid concept is mainly comprised of microgrids (MGs) as key components [3]. These are parts of a central grid that can operate independently from the central utility grid [4,5]. A point of common coupling (PCC) is used for ensuring interactions with a central grid where the microgrid is connected to a central grid. A PCC located on the primary side of the transformer defines the separation between the central grid and the microgrid. In addition to providing local customers with their thermal and electricity needs, microgrids also improve local reliability while reducing emissions thus resulting in lower energy supply costs. Hence, microgrids are frequently utilized and accepted in the utility power industry, due mostly to their environmental and economic benefits. When the external grid suffers from disturbances or grid-connected mode, an MG system can be used in either islanded mode with the external grid supporting part of the power consumption and consisting of distributed energy resources (DERs), power conversion circuits, storage units and adjustable loads thereby providing sustainable energy solutions [6]. An MG system can be operated in either islanded mode in case the external grid suffers from disturbances or grid-connected mode, where the external grid supports part of the power consumption, and consists of distributed energy resources (DERs), power conversion circuits, storage units and adjustable loads thereby providing sustainable energy solutions [7–9]. There are a wide range of distributed generation (DG) units such as wind turbines (WTs), photovoltaics (PVs), and distributed storage (DS) units such as batteries as part of DERs [7]. Power generation in the MG system is generally utilized through the use of DERs or/and conventional power generators, such as diesel generators [10]. Power from in-plant generators has to be utilized for critical loads either to complement the grid or as an emergency source which can tolerate very little or no interruptions. The simplicity and ease of maintenance of diesel generators make them a perfect match for use under these circumstances. External supply assistance is not required to start them and they come in a wide range of ratings [11].

Additional interest has sparked recently for the use and development of renewable energy resources due to factors related to global warming and the energy crisis over the past few decades [12,13]. Minimum fuel cost has been the dominant strategy for electric power dispatching until now; however, environmental concerns have to be taken into consideration. Hereof, future demands of power grids can be supplied via microgrids that can meet these requirements [14]. A significant number of microgrid demonstration projects have been put forth in various countries including the U.S., E.U., Japan and Canada where microgrids have been integrated with the development agenda of future electric grids [15]. The penetration of microgrids has thereby increased rapidly, gradually reaching significant levels. Microgrids make up low voltage (LV) networks with distributed generation units complete with energy storage units and controllable loads (e.g. water heaters, air condition) offering considerable control capabilities over the power system operation. Extensive complications arise in the operation of an LV grid due to the operation of micro-sources in the power system, but in the meantime, it can also provide noticeable benefits to the overall network performance when managed and coordinated in an efficient manner [16]. Wind turbines and photovoltaic panels which are currently the important appliances for extracting solar energy are typical non-dispatchable DERs used in MGs for overcoming issues related to alternative, sustainable, and clean energy [10,17]. Solar energy is considered among the most promising renewable resources for bulk power generation in addition to being an infinite, eco-friendly form of Energy. Nonetheless, it is highly dependent on temperature and solar radiation, which have direct effects on the solar power output thus resulting in the intermittent and variable characteristic of solar power [18]. Another dimension of uncertainty is present in the load forecast with the adoption of renewable distributed generation technologies [19]. Forecast needs are underlined by the presence of a de-regulated environment, especially for distribution networks. Load forecasting is essential for the convenient operation of the electrical industry [20]. Reliable short-term (hour-ahead, day-ahead) load forecasts under service constraints are required for actions such as network management, load dispatch, and network reconfiguration. Short-term load forecasting algorithms are among methodologies that aim to increase the effectiveness of planning, operation, and conduction in electric energy systems [21–25].

Network reconfiguration that is considered as a significant solution for microgrids has been used for optimal operation management. It can be defined as operational schemes that alter the network topology by modifying the on/off status of remotely controlled sectionalizing switches (normally closed switches) and tie switches (normally open switches) of active distribution networks thereby enabling the controlling of power flow from substation to power consumers with additional benefits such as load balancing, real power loss reducing, optimizing the load sharing between parallel circuits by directing power flow along contractual paths [26]. However, the majority of the network reconfiguration (NR) studies aim to decrease power losses on the grids [27]. Real power losses in active distribution networks can be decreased by way of two essential methods: the NR and the optimal dispatch (OD) of diesel DG units. The NR technique can only partially mitigate the losses due to the distribution systems. The OD of DGs is a significant contributor to obtain greater power loss reduction [28]. The sizing of DGs and NR has been implemented in [29–31] either sequentially or simultaneously to attain further reductions in power loss. Optimal NR and distributed generation allocation have been carried out simultaneously in the distribution network. Nonetheless, in [28] it has been observed that significant improvements such as voltage profile improvement and reduced energy production cost in the entire system can be attained via the simultaneous application of NR and OD of the DGs techniques during the analysis. Moreover, the proposed joint approach of the PSO and SPSO methods displayed a higher performance in power loss reduction in comparison with the other methods in the literature [29–32].

The optimal operational scheduling problem should be resolved for attaining the minimum loss under the new profile if the net load profile keeps changing. The net load profile is calculated as a sum of electrical power consumption and renewable power generation for each period. Therefore, the dynamic NR and dispatch of diesel DGs can be pre-performed in response to load and renewable power output changes with the forecasted electrical power consumption and renewable power generation profiles [33]. Various articles have been published in the literature on the scope of general power system operational scheduling and planning. However, load and renewable power generation output estimates have been used in very few. In [33], the data-driven NR based on the 1-h-ahead load forecasting is solved in a dynamic and pre-event manner with the utilization of support vector regression (SVR) and parallel parameters optimization based on short-term load forecasting results as an input to the reconfiguration algorithm. An SVR-based short-term load forecasting approach is used in [34] to cooperate with the NR for minimizing the system loss. These papers only consider NR within the operational planning studies and machine learning-based 1-h-ahead load forecasting results are utilized as input to the optimization frameworks. Since the next days' power generation must be scheduled every day in power systems, a day-ahead short-term load forecasting is an obligatory daily task for power dispatch. The economic operation and reliability of the system are significantly affected by its accuracy. An artificial neural network-based method for forecasting the next day's load as well as the economic scheduling for that particular load is put forth in [35]. The economic dispatch problem integrated with artificial neural network-based short-term load forecasting is taken into consideration in [36] regarding studies on operational scheduling with power dispatch. Reference [37] contributes to the optimal load dispatch of a community MG with deep learning-based solar power and load forecasting. A two-stage dispatch model based on the day-ahead scheduling and real-time scheduling to optimize the dispatch of MGs is presented in [38]. A two-phase approach for short-term optimal scheduling operation integrating intermittent renewable energy sources for sustainable energy consumption is proposed in [39]. A day-ahead energy acquisition model is developed in the first phase while the second phase presents real-time scheduling coordination with hourly NR. Reference [40] puts forth a method for optimal scheduling and operation of load aggregators with electric energy storage in power markets to schedule the imported power in each period of the next day with day-ahead forecasted price and load. However, it has been observed upon examining the power system operational scheduling and planning studies that the simultaneous approach of NR and OD of DG units has not been performed as in [28]. Moreover, one of the studies in the literature has made use of DL-based load and solar power forecasting results and it is concluded that the proposed LSTM-based deep RNN model has a great potential for more accurate short-term forecasting results to promote economic power dispatch on a microgrid as an optimal operational work [37]. The results of this study have encouraged us in terms of microgrid optimal operational studies. Therefore, the present study is supported by DL-based prediction studies and is put forth as a continuation of the study in [28]. Furthermore, DRNN-LSTM is mostly used among the deep learning approaches for short-term and aggregated level forecasting studies as can be seen in [37,41–44]. Here, A DRNN model based on Bi-LSTM units has been developed as a significant contribution of this study to forecast the aggregated electrical power load and bulk PV power output of reconfigurable MG during a short-term period and the net load profile is considered as a sum of demand consumption and PV power generation for each hour in a day.

The rest of the study is organized as follows: Section 2 provides a detailed description of the DRNN Bi-LSTM model for forecasting aggregated power load and PV power output and the optimization model for determining the optimal day-ahead operational scheduling of the reconfigurable MG. The simulation results and discussions are given in Section 3 while the conclusions are presented in Section 3.

2. The Architecture of the Proposed Approach

It is of significant importance with regard to developing economies to put into practice power system operational planning practices for employing the already existing capacity in the best possible manner [45]. Optimal generation scheduling is especially important for microgrid operation [46]. As such, putting forth the least-cost dispatch among the DGs that minimize the total operating cost, while meeting the electrical load and satisfying various technical, environmental, and operating constraints is part of the daily operation of a microgrid [47]. Contrarily, real load data measured at that instant in a conventional dynamic power system reconfiguration work is used for putting forth the optimal topology at each scheduled time point. An accurate prediction of the load power is possible by way of the development of load forecasting technique which takes place at a future time and provides more information on load changes. Optimal topology with the incorporation of load forecasting can be resolved subject to forecasted load conditions during a longer time period rather than the use of a snapshot of the energy consumption at the time when the reconfiguration actualizes; thereby, this information can be used by the distribution network operator for an improved system reconfiguration operation as well as for bringing about the desired optimal solutions [33]. Moreover, it is possible to operate the smart grids in an economical manner only via accurate forecasting of solar power/irradiance [12].

In this context, this study develops a DRNN Bi-LSTM model to forecast hourly power load and the hourly PV power output over a short-term horizon (24-h) respectively. Afterwards, the optimal operational scheduling model of the reconfigurable MG which contains aggregated power load, bulk PV arrays, and diesel DGS is established under different scenarios. In the optimal operational scheduling model, the aggregated load and the PV power output are obtained from the estimation results of the DRNN Bi-LSTM model [37].

2.1. Day-Ahead Load and Solar Power Output Forecasting

These forecasting studies are generally carried out at aggregated and individual levels and are classified based on the forecasting horizons as follows:

- Very Short-Term Load Forecasting (VSTLF): ranging from seconds or minutes to several hours [48],

- Short-Term Load Forecasting (STLF): ranging from hours to weeks [49],
- Medium and Long-Term Load Forecasting (MTLF/ LTLF): ranging from months to years [50].

It is critical for power companies to put forth an accurate forecast of load profile for the next 24 h as it can have a direct impact on the optimal hourly scheduling of the generation units in addition to their participation in different energy markets. Interestingly, the number of values to predict can also be used for classifying the forecasting models. There are two main groups: the first group consists of those that forecast only one value (next hour's value, next day's peak value, next day's total value, etc.); the second group is comprised of forecasts with multiple values, such as next hours or even next day's hourly forecast. Single-value forecasts are used for on-line operation and optimization of load flows, whereas multiple-value forecasts are utilized for generator scheduling and economic dispatching. Energy demand forecasting has been an important field in order to allow generation planning and adaptation. In addition to demand forecasting, electrical generation forecasting models have also attracted increased attention recently, especially with regard to renewable generation sources that depend on the forecasting of a particular energy resource (solar radiation, wind, etc.) [50].

Forecasts, usually 24-h ahead, should be used for anticipating the electricity demand to be met with sufficient energy and thus it will be apparent whether it is necessary to buy energy in the market (energy defect) or sell it (excess energy). This is known as STLF which helps in planning the operation of generators and energy-related systems owned by the utility [51]. In the meantime, there are many factors such as calendar type, weather, climate, and special activity that have an impact on load consumption. Similarly, the majority of the forecasting approaches applied for power forecasting are available for load forecasting solutions [10]. Load forecasting is essentially a time series forecasting problem. Autoregressive moving averages models [52], auto-regressive integrated moving average [53], linear regression (LR) [54], iteratively re-weighted least-squares (IRWLS) [55]; nonlinear methods such as artificial neural network [56], multi-layer perceptron (MLP) [57], regression decision tree machine learning algorithm [58], general regression neural network and support vector machine (SVM) [59] have recently been utilized for this kind of forecasting study. However, deep neural networks (DNNs), a type of artificial neural networks (ANN) with multiple hidden layers of neurons between the input and output [8], have recently been increasing in popularity as the latest developed subset of machine learning techniques for time series electrical load forecasting problems [60]. The short-term and aggregated level has the highest number of studies on electrical load estimation with LSTM-RNN as the most commonly used deep learning method in these studies.

PV panels are used on-grid or off-grid to provide electricity to individual buildings, aggregated settlements, and commercial and industrial areas. The intermittent nature of solar energy makes it very difficult to establish a balance between electricity generation and consumption. The successful integration of solar energy into the power grid requires an accurate PV power prediction thereby reducing the impact of the uncertainty of output power of PV panels leading to a more stable system [61]. Power forecasting for PV power generation has especially become one of the fundamental technologies for improving the quality of operational scheduling and reducing spare capacity reserves [62]. The methods used to estimate the PV generation which is influenced by atmospheric conditions such as temperature, cloud amount, dust and relative humidity are generally divided into three main categories; time-series statistical methods, physical methods, and combined methods. Nonetheless, solar forecasting studies frequently utilize artificial intelligence (AI) techniques due to their capacity to solve complex and nonlinear data structures. Deep learning algorithms have especially been used in solar power prediction studies that outperform traditional methods as a sub-branch of AI methods. The most widely used deep learning method is deep LSTM-RNN [44] and the combination of LSTM with other DL algorithms [63,64].

2.1.1. Forecasting Model

DNN is broadly accepted to model complex non-linear systems in engineering. Besides, the computation of DNN only includes basic algebraic equations, thus providing a fast computation speed [8]. Recent forecasting studies have put forth that improved accuracies are attained by DL systems in comparison with conventional methods. CNN is a type of feed-forward artificial neural network in the field of machine learning research where a structure is formed among artificial neurons inspired by the organization of human neurons [65]. CNN is most utilized in cases related to tasks in which data have high local correlation such as visual imagery, video prediction, and text categorization. It can capture when the same pattern appears in different regions [66]. A CNN architecture consists of a stack of distinct layers that transform the input data into an output volume. The network structure of the CNN model is comprised of distinct types of layers such as convolution layers and pooling layers [65]. In addition, CNNs require multi-dimensional inputs to attain a high prediction accuracy. Time series data, e.g., the energy consumption data, forecasting poses a significant challenge even for deep learning technologies when the desired output is another time series, namely the 24 h of the next day. Contrary to such traditional feed-forward networks (FFNs) where all inputs and outputs are assumed to be independent of each other, RNN maintains a memory about the history of all past inputs using the internal state. RNN contains feedback connections to ensure the flow of activations in a loop. RNN can be considered more like a human brain because of the recurrent connectivity found in the visual cortex of the brain. Therefore, an RNN architecture is more appropriate for time series, as the case of the present work. It proposes an RNN consisting of a Bi-LSTM unit to learn the sequential flow of various measurements through consecutive days and hours, predicting electrical energy consumption and bulk PV power generation values through the next 24 h at an aggregated level for day-ahead operational scheduling of reconfigurable MG. Information flow is multidirectional with the Bi-LSTM unit.

All hyper-parameters (number of layers, number of hidden units, length of input feature sequence, etc ...) are chosen according to the tests applied during the dates of 08.01.2016–31.12.2016, with models trained on a small subset of the previous samples. A few experiments have been conducted for that purpose, without fine-tuning of the hyper-parameters.

The proposed RNN consists of a sequence input layer of size 11 (one sequence input for each feature), followed by a Bi-LSTM layer of 150 hidden units, a fully connected layer that outputs a single sequence of either PV power generation or electrical demand, and finalized by a regression layer. Adam optimizer was used and the training was stopped earlier to prevent overfitting to training data. The initial learning rate is 0.005.

The test data used represents electrical energy consumption and PV power generation measurements, along with a number of other observations in the Czech Republic from 01.01.2012 to 31.12.2016. There are 24 measurements per day with 1-h resolution. 5-fold cross-validation is applied over the entire dataset to train and test (training on the first 4 years and testing on the 5th year, trainining on first 3 plus the last year and testing on 4th year, etc ...).

Two different models are trained to forecast daily PV power generation and electrical demand values separately. The input features for daily power load forecasting model are previous 6 days' (hourly) month, day, hour, PV power generation, electrical energy consumption with pumping load, electrical energy consumption, wind speed, temperature, direct horizontal radiation, diffusion horizontal radiation; while these features are previous 6 days' (hourly) month, day, hour, PV power generation, wind speed, temperature, direct horizontal radiation, diffusion horizontal radiation for daily PV power generation forecasting model. Each feature input is a sequence (past 6 days) of length $24 \times 6 = 144$. RNN outputs either a sequence of forecasted PVPP or forecasted load values. The proposed RNN is shown in Figure 1. All available input features are used without any handcrafted modification or selection, to allow the deep model to capture all necessary information from raw data.

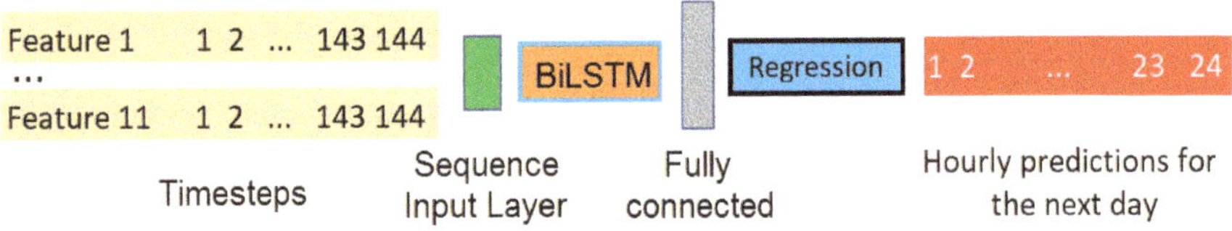

Figure 1. The proposed RNN model.

During training and testing, only the feature sequences starting just at the start of a new day at time 00:00 are considered. During training, it is also possible to use intermediate sequences (such as previous 144 hours' at time 14:00 as input, with a corresponding ground truth output sequence for the next 24 h until next day's 14:00) however, such intermediate sequences are skipped for simplicity.

As an example; when the first-ever year in the dataset (2012) is used in training, input time-steps for the first-ever training sample will thus be the first 6 days' (from 01.01.2012 to 06.01.2012) hourly feature measurements (of length 144 for each feature), with the target as either the hourly electrical demand or PVPP values for 07.01.2012.

To normalize the data; month, day and hour are divided by their maximum values plus one, i.e., 13 for the month, 32 for day and 25 for the hour. With the help of this normalization; maximum values become just less than one, easier to converge to with most activation functions. Other features are normalized by subtracting their mean and dividing to their standard deviation so that these features become zero mean and unit variance. Let x_i be some observation of feature type x (Can be wind power plant (WPP) generation, PVPP, electrical load consumption with pumping, electrical load consumption, wind speed, temperature, direct horizontal radiation or diffusion horizontal radiation). x_i is normalized as:

$$x_i^n = (x_i - \mu(x))/\sigma(x) \tag{1}$$

$\mu(x)$ and $\sigma(x)$ are calculated from the complete dataset. Performances (root mean square error (RMSE)) of the proposed RNNs during the learning phase for power demand and PV power generation with the number of training iterations are shown in Figures 2 and 3 accordingly, trained with the first 80% of the days (first 4 years).

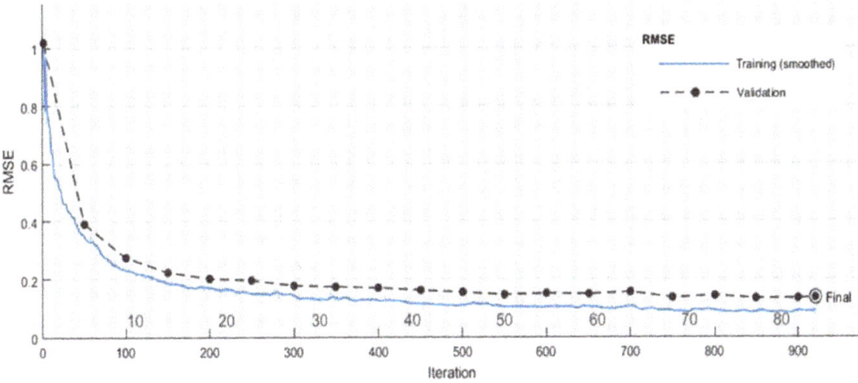

Figure 2. Performance (RMSE) of the proposed RNN during the learning phase for power demand, trained with the first 80% of the days (first 4 years).

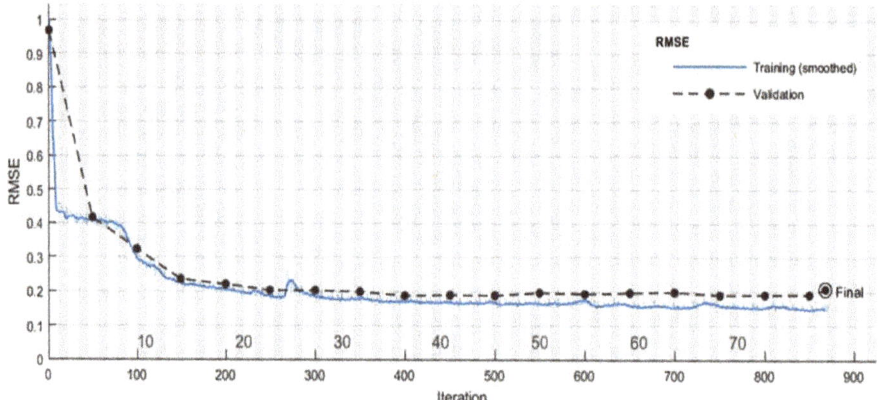

Figure 3. Performance (RMSE) of the proposed RNN during the learning phase for PV power generation, trained with the first 80% of the days (first 4 years).

2.1.2. Baseline Feed-Forward Neural Network

RNNs have significantly stronger abilities in modelling complex processes and learning temporal behaviors rather than a normal feedforward network [67]. However, a feed-forward neural network (FFNN) which accepts inputs of feature hour sequences all stacked as a single 1D feature vector (length $11 \times 144 = 1584$) and outputs 24 hourly forecasts for the following day is utilized for comparison here. There are 12 hidden units in the hidden layer, chosen on a small validation set similar to that of the RNN. The baseline FFNN model is shown in Figure 4.

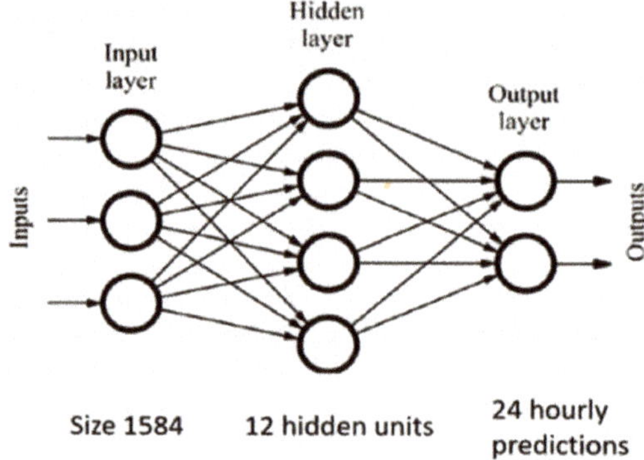

Figure 4. The baseline FFNN model.

2.1.3. Error Metrics

In the following equation:

$$e_j = a_j - p_j \tag{2}$$

is the error. Several error metrics are reported in this study for comparison.

- Scale-dependent measures:

Mean absolute error (MAE) is formulated as:

$$MAE = \frac{1}{n} \sum_{j=1}^{n} |e_j|, \ 0 \leq MAE < \infty \tag{3}$$

and MAE is easy to interpret and heavily used in regression and time-series problems.

Root mean square error (RMSE) is defined as:

$$RMSE = \sqrt{\frac{1}{n} \sum_{j=1}^{n} e_j^2}, \ 0 \leq RMSE < \infty \tag{4}$$

and RMSE is a quadratic error metric, representing the standard deviation of errors. RMSE exaggerates bigger errors, being more sensitive to outliers than MAE.

- Measures based on percentage errors:

Percentage errors are mostly used to compare forecast performance across several data sets as they are scale-independent. Mean absolute percentage error (MAPE) is defined as:

$$MAPE = \frac{100}{n} \sum_{j=1}^{n} \frac{|e_j|}{|a_j|} \tag{5}$$

and MAPE is frequently used in regression and time-series problems to measure the accuracy of predictions [68]. If there are zero values in the data, MAPE cannot be calculated. Similarly, MAPE values become large if there are small values in the data. There is no upper limit on the MAPE value. MAPE values are biased in the sense that they systematically reward the method that is predicting smaller values.

Symmetric mean absolute percentage error (sMAPE) is defined as:

$$sMAPE = \frac{100}{n} \sum_{j=1}^{n} \frac{2|e_j|}{|a_j| + |p_j|} \tag{6}$$

and sMAPE is an alternative to MAPE if there are data points at zero or close to zero. Such problems can be less severe for sMAPE. sMAPE may still involve division by a number close to zero.

Large (or infinite) errors can be avoided by excluding the non-positive data or that of less than one [65]. However, this solution is ad hoc and is impossible to apply in practical applications. Moreover, it leads to the problem of how outliers can be removed. The exclusion of outliers may result in the loss of information when the data involve numerous small a_j's.

Because the underlying error distributions of percentage errors have only positive values and no upper bound, percentage errors are highly prone to right-skewed asymmetry in practice [68]. Percentage errors are neither resistant nor robust measures because a few outliers can dominate and they will not be close in value for many distributions. This work proposes to get rid of the undefined percentage errors by adding a constant positive number α to each data point a_j and prediction p_j so that all data points are positive and much greater than zero. MAPE and sMAPE thus use $a_j + \alpha$. and $p_j + \alpha$ values in the corresponding formulas. It is further proposed that scale-dependent measures better represent the magnitude of the error than percentage errors when the data is normalized to be unit variance and zero mean.

Both the errors on normalized and unnormalized actual values $a_j^n \sigma(x) + \mu(x)$ and predictions $p_j^n \sigma(x) + \mu(x)$ are reported. In some sense, the proposed process of adding a fixed constant is inevitable as for the unnormalized data the following MAPE value is calculated:

$$MAPE = \frac{100}{n} \sum_{j=1}^{n} \frac{\left| a_j^n \sigma(x) + \mu(x) - (p_j^n \sigma(x) + \mu(x)) \right|}{\left| a_j^n \sigma(x) + \mu(x) \right|} \tag{7}$$

Simplifying to:

$$MAPE = \frac{100}{n} \sum_{j=1}^{n} \frac{|e_j|}{\left| a_j^n + \mu(x)/\sigma(x) \right|} \tag{8}$$

leads to smaller MAPE values as shown in the results. sMAPE is similarly affected when unnormalized data is used.

Mean absolute scaled error (MASE) is the mean absolute error of the forecast values, divided by the mean absolute error of the one-step naive forecast of training data. MASE is scale-invariant and it has predictable behaviour even when the data values are close to 0. In this work we calculate the naive forecast as the value at the same hour of the previous day, utilizing the seasonal time series formula:

$$MASE = \frac{MAE}{\frac{1}{T-24} \sum_{t=25}^{T} |Y_t - Y_{t-24}|} \tag{9}$$

where MAE is calculated according to Equation (3).

2.2. Optimal Operational Scheduling Problem

This section of the study focuses on the optimal operational scheduling problem of reconfigurable MGs in which the optimal radial topology of the balanced medium-voltage reconfigurable MG system as well as the optimum power generation level of diesel DGs has to be determined by the system operator to minimize real power losses. The aforementioned non-linear combinatorial problem which can be considered as a single-objective optimization problem is represented by the mathematical formulation given below [69]:

$$x = [x_1, x_2, \ldots, x_{dv}] \tag{10}$$

$$\min(f_1(x), f_2(x), \ldots f_N(x)) \tag{11}$$

$$s.t. h_i(x) = 0; i = 1, \ldots, p \tag{12}$$

$$g_i(x) \leq 0; i = 1, \ldots, q \tag{13}$$

In this paper, the optimization problem is a minimization problem with its equality and inequality constraints given in the following sections.

• Objective Function

Power loss reduction and voltage profile improvement for the whole system are significantly influenced by NR techniques as well as OD of DG units which eventually determine the direction of power flow in an MG. Therefore, the main objective of our specific problem is the minimization of the sum of active power losses in all branches as given in the following equation [28]:

$$\min \left\{ P_L = \sum_{i=1}^{b} |I_i|^2 R_i \right\} \tag{14}$$

• Equality and Inequality Constraints

Various equality and inequality constraints of the reconfigurable MG have to be taken into consideration during the simultaneous application of NR and OD of DGs in reconfigurable MG. The following constraints have been considered for this study:

(1) Equality Constraint

Power balance constraint has to be met based on the following equation:

$$P_{EP} + \sum_{j=1}^{N_{DG}} P_{DGj} - \sum_{i=1}^{N_{MGL}} P_{MGLi} - P_L = 0 \tag{15}$$

(2) Inequality Constraints

The constraint for the maximum and minimum active power generation of dispatchable units can be represented as given below:

$$P_{DGi}^{min}(t) \leq P_{DGi}(t) \leq P_{DGi}^{max}(t) \tag{16}$$

The amount of the flowing current I_i at the ith branch should not exceed it's maximum thermal value I_i^{max} [28]:

$$|I_i| \leq I_i^{max} \tag{17}$$

The bus voltage values, V_i, should vary between the minimum and maximum values after reconfiguration and slack bus voltage is taken as following [70]. The limits in the present study have been set to $V_{min} = 0.90$ p.u. and $V_{max} = 1.10$ p.u., respectively:

$$V_{min} \leq V_i \leq V_{max}; V_{slack} = 1 \tag{18}$$

• Radiality Constraint

All possible MG configurations have to be in radial condition throughout the NR process. Moreover, there must not be any loops and all loads must be connected to the main power supply in the topological structure of MG which can be expressed using the following formula [28]:

$$\sum_{b}^{N_b} \beta_b = m - N_{sub} \tag{19}$$

• Operational Cost Calculation

Total of purchasing power cost from the main grid and the production cost of dispatchable diesel DGs is considered as the total operational cost here in this study:

$$TotalCost = Cost_{MG} + Cost_{DG} \tag{20}$$

In case it is not possible to meet the total energy demand via distributed DGs, reconfigurable MG has to purchase power from the upstream grid. The following formula is used total active power purchase for this case is calculated [71]:

$$Cost_{MG} = \sum_{t=1}^{24} v^b(t) P^b(t) \tag{21}$$

The mathematical relation given below is used for calculating the generation cost of diesel DGs on a daily basis which is comprised of the fuel cost [72].

$$Cost_{DG} = \sum_{t=1}^{24} a + b \times P_{DG}(t) + c \times (P_{DG}(t))^2 \tag{22}$$

2.2.1. Overview of PSO and SPSO

The majority of the researchers have used Particle Swarm Optimization (PSO) for solving optimization related problems in power systems. The behavior of clustered social animals such as fish and birds are used for creating the PSO method. Birds or fish move towards food at certain speeds or positions. Each particle part of the population of n particles in D-dimensional space represents a possible solution for PSO defined by two parameters as position (p) and velocity (v) that are initially chosen randomly. The movement of population members will depend on their own experience and experience from other 'friends' in the group Pbest and Gbest. The parameters are updated based on the following model:

$$v_{iD}^{k+1} = w \times v_{iD}^k + c_1 \times rand \times \left(p_{best-i} - x_i^k\right) + c_2 \times rand \times \left(g_{best-i} - x_i^k\right) \tag{23}$$

$$x_i^{k+1} = x_i^k \times v_i^{k+1} \tag{24}$$

and here, w is calculated by the following formula:

$$w = w_{max} - (w_{max} - w_{min}) \times \left(\frac{k}{k_{max}}\right) \tag{25}$$

The velocities are confined in the range of [0,1] via sigmoid transformation on the velocity parameters in binary PSO, thus ensuring that the particle position values are either 0 or 1:

$$sig\left(v_{iD}^{k+1}\right) = \frac{1}{1 + \exp\left(-v_{iD}^{k+1}\right)} \tag{26}$$

$$x_{iD}^{k+1} = \begin{cases} 1, & if \ \sigma < sig\left(v_{iD}^{k+1}\right) \\ 0, & if \ \sigma \geq sig\left(v_{iD}^{k+1}\right) \end{cases} \tag{27}$$

Khalil and Gorpinich suggested a minor change to binary PSO, SPSO by keeping the search in the selected search space. The search space in SPSO at each D dimension SD = [SD1, SD2, ... , SDN] is comprised of a set of DN positions where DN represents the number of selected positions in dimension D. A fitness function is described in SPSO as is the case for the basic PSO; which maps at each D dimension from DN positions of the selective space SD leading to alter the position of each particle from being in real-valued space to be a point in the selective space, thereby changing the sigmoid transformation as per (20):

$$v_{iD}^{k+1} = \begin{cases} rand \times v_{iD}^{k+1}, & if \ \left|v_{iD}^{k+1}\right| < \left|v_{iD}^k\right| \\ v_{iD}^{k+1}, & otherwise \end{cases} \tag{28}$$

The number of tie switches in the MG indicates the dimension of the reconfiguration problem. When all tie switches are closed some loops are present in the network with the number of loops equal to the number of switches. All branches in the loop that define the dimension make up the search space in a certain dimension. The optimization algorithm does not take into account the branches out of any loop. The common branch that belongs to more than one loop should be placed in just one loop in the dimension. The optimum configuration can be determined via SPSO following the determination of the dimensions and search space for each dimension [28].

2.2.2. PSO&SPSO Method

The framework discussed in this study is NR in parallel with OD of DGs with an objective of minimizing real power losses with some constraints on the MG. The PSO algorithm has been selected for solving the problem due to its improved potential in the solution of discrete, nonlinear and complex optimization problems. The fact that PSO and SPSO algorithms combine the advantages of both PSO approaches can be considered as the motivation for their selection.

There are many equality and inequality constraints for the non-linear optimization problem of OD which puts forth the optimal power output of DGs to meet the estimated electrical consumptions from an economic perspective. Traditional optimization algorithms may not be sufficient for solving such problems due to local optimum solution convergence, while metaheuristic optimization algorithms, and specially PSO, have succeeded unusually in solving such types of OD problems during the last decades.

MG reconfiguration comprises the combinational part of the whole optimization problem. To attain the suitable arrangement of power and radial configuration for every load, distribution system planners operate with numerous switches. The sectionalizing switches (which are normally closed) along with the tie switches (which are normally open) to maintain radiality. An accurate switching operation plan can be attained through the various switch. The combinatorial nature of the constrained optimization problem can be easily overcome by embedding selective operators into the standard PSO.

The optimization problem turns out to be even more complex when time-sequence variation in load, power market price and output power of DGs are considered. High computation time for larger systems hindering real-time operation can be indicated as a problem for the majority of meta-heuristic methods. Hence, PSO is preferred to overcome the complexity of the optimization problem due to its faster convergence rate, accuracy, parallel calculation, and ease of application.

It is of significant importance to associate the MG parameters with optimization parameters for the simultaneous MG reconfiguration and OD problem. Two decision variable sets make up the particles in the proposed PSO&SPSO. The dimension of search space equals the number of diesel DGs regarding the OD part while it is equal to the number of tie switches in the MG about the NR part throughout the combined algorithm. This method is used for carrying out the OD for the DG units via the basic PSO, while switch positions are determined by applying the SPSO method simultaneously at every iteration [28].

2.3. The Test System Features

The standard 33-bus test system is taken into consideration as a reconfigurable MG in the present study by integrating three diesel DGs and a bulk PV generation unit as can be seen in Figure 5. The detailed characteristic information about the test system can be found in [24].

It is considered that a bulk PV energy generation system is integrated into bus number 6 on the reconfigurable MG for this short-term operational scheduling study. As put forth in Table 1, diesel DGs with 4 MW total maximum real power capacity operated at a unity power factor has been installed on different buses.

Table 1. The features of dispatchable DGs.

Dispatchable Units	Bus Number	Cost Function Coefficients of Dispatchable Units		
		a (£)	b (£/MW)	c (£/MW^2)
Diesel DG-1	14	25	87	0.0045
Diesel DG-2	18	28	92	0.0045
Diesel DG-3	32	26	81	0.0035

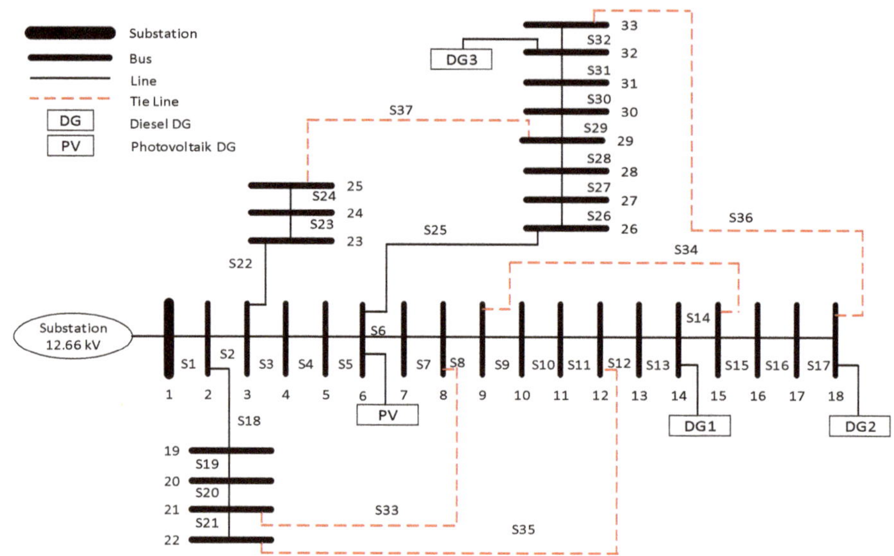

Figure 5. The standard 33-bus test system considered as a reconfigurable MG.

2.4. Forecasting Results for the Optimal Operational Scheduling Problem

The results in normalized and unnormalized data are reported in this section. The results measured on peak hours of each day instead of all 24 h are also reported. "N" and "U" denote the normalized and unnormalized data respectively. "P" denotes only peak-hour unnormalized data. Error calculating results for Load and PV power forecasting are presented in Tables 2 and 3 for the proposed RNN, Table 4 for the baseline FFNN, Table 5 for random guess results. Note that, MG optimal operational scheduling results are provided only when the forecast is performed on the last 20% of the days (from 08.01.2016 to 31.12.2016) with the proposed RNN, explicitly shown in Table 2.

Table 2. Power load and generation forecast errors with the proposed RNN tested only on the last 20% of the days (from 08.01.2016 to 31.12.2016).

Forecast Errors	Test Data	MAE	RMSE	MAPE	sMAPE	MASE
	N	0.1361	0.2029	4.68%	4.68%	0.3778
Power Demand	U	173.71	258.93	1.18%	1.19%	0.3778
	P	259.55	470.83	1.47%	1.49%	0.5361
	N	0.2055	0.3861	14.96%	13.49%	0.9316
PV Power Generation	U	77.74	146.06	14.00%	12.71%	0.9316
	P	216.59	269.83	25.53%	24.02%	0.8090

Table 3. Power load and generation forecast errors with the proposed RNN, 5-fold cross-validation results.

Forecast Errors	Test Data	MAE	RMSE	MAPE	sMAPE	MASE
	N	0.1287 ± 0.0064	0.1850 ± 0.0115	$5.59 \pm 0.62\%$	$5.45 \pm 0.49\%$	0.3543 ± 0.0191
Power Demand	U	164.20 ± 8.16	236.14 ± 14.72	$1.21 \pm 0.07\%$	$1.21 \pm 0.07\%$	0.3543 ± 0.0191
	P	195.36 ± 40.49	288.60 ± 103.75	$1.19 \pm 0.21\%$	$1.20 \pm 0.21\%$	0.3958 ± 0.0868
	N	0.1984 ± 0.0076	0.3821 ± 0.0123	$15.95 \pm 1.34\%$	$14.01 \pm 0.86\%$	0.8990 ± 0.0408
PV Power Generation	U	75.03 ± 2.88	144.54 ± 4.66	$14.78 \pm 1.15\%$	$13.10 \pm 0.75\%$	0.8990 ± 0.0408
	P	217.37 ± 9.85	273.29 ± 14.24	$27.72 \pm 2.65\%$	$25.07 \pm 1.16\%$	0.8165 ± 0.0483

Table 4. Power load and generation forecast errors with the baseline FFNN, 5-fold cross-validation results.

Forecast Errors	Test Data	MAE	RMSE	MAPE	sMAPE	MASE
Power Demand	N	0.1948 ± 0.0176	0.2788 ± 0.0672	$8.61 \pm 1.15\%$	$8.50 \pm 1.06\%$	0.5361 ± 0.0475
	U	248.57 ± 22.51	355.74 ± 85.76	$1.84 \pm 0.17\%$	$1.84 \pm 0.17\%$	0.5361 ± 0.0475
	P	334.22 ± 196.39	554.18 ± 557.28	$1.62 \pm 0.78\%$	$1.67 \pm 0.87\%$	0.6795 ± 0.4102
PV Power Generation	N	0.2549 ± 0.0166	0.4424 ± 0.0236	$25.78 \pm 4.11\%$	$23.45 \pm 3.79\%$	1.1549 ± 0.0764
	U	96.40 ± 6.28	167.35 ± 8.91	$23.37 \pm 3.51\%$	$21.40 \pm 3.29\%$	1.1549 ± 0.0764
	P	197.15 ± 26.09	276.35 ± 31.02	$22.06 \pm 5.47\%$	$25.83 \pm 6.83\%$	0.7401 ± 0.0977

Table 5. Power load and generation forecast errors with random guesses drawn from normal distribution $N(\mu(x),\ \sigma(x))$ where x is either power demand or PV generation, 5-fold cross-validation results (5 random repetitions).

Forecast Errors	Test Data	MAE	RMSE	MAPE	sMAPE	MASE
Power Demand	N	1.1429 ± 0.0400	1.4202 ± 0.0481	$54.94 \pm 6.80\%$	$45.96 \pm 2.79\%$	3.1462 ± 0.1179
	U	1458.44 ± 51.10	1812.26 ± 61.39	$10.85 \pm 0.59\%$	$10.77 \pm 0.42\%$	3.1462 ± 0.1179
	P	1512.44 ± 157.48	1881.04 ± 195.12	$9.14 \pm 0.46\%$	$9.60 \pm 0.61\%$	3.0625 ± 0.3595
PV Power Generation	N	1.1110 ± 0.0162	1.4167 ± 0.0210	$177.92 \pm 5.06\%$	$105.62 \pm 1.33\%$	5.0349 ± 0.1279
	U	420.22 ± 6.12	535.87 ± 7.96	$155.78 \pm 3.73\%$	$100.44 \pm 1.22\%$	5.0349 ± 0.1279
	P	673.60 ± 44.06	812.88 ± 44.89	$68.83 \pm 1.44\%$	$94.98 \pm 2.86\%$	2.5306 ± 0.1971

MAE and RMSE results of normalized data are similar in demand and PV power generation forecasting although MAPE and sMAPE errors are much higher in PV power generation. The fact that PV power generation values are much lower than demand values generating higher MAPE and sMAPE results can be indicated as the primary reason. Regarding overall forecasting results, the proposed RNN method shows much better performance comparing with the FFNN method in all cases. However, especially with the tested data which are only on the last 20% of the days (from 08.01.2016 to 31.12.2016), it shows best ever performance when unnormalized MAPE results are benchmarked. Also, random results are the worst of all these results although they are drawn from a normal distribution where the statistics of the data are encapsulated. Only for PV power generation peak hours, FFNN provides better MAE and MAPE metrics which is at the end a one-shot prediction instead of 24-h predictions. On the other hand; peak hour demand or generation values are more variable between different days, compared to 24-h predictions.

Furthermore, it can be seen when Table 6 is observed that more accurate forecasting results can be attained with the proposed DRNN Bi-LSTM model in comparison with other methods in the literature that are based on deep architecture, in both demand and PV generation forecasting frameworks over the short-term horizon. It provides the possibility for integrating renewable energy efficiently, reducing pollutant emissions, as well as keeping the stability of power system operation.

The forecasting results of aggregated electrical energy demand and the bulk PV generation on January 8, 2016 will be used as the given experimental setup in the optimal operational scheduling of the MG, thereby contributing to promoting interaction and supply-demand balance in the grid-connected reconfigurable MG [37]. The comparative chart of the demand forecasts and actual demand values are presented in Figure 6. Solar power output estimation results are given with actual solar power generation data in Figure 7, comparatively.

Table 6. Benchmarking the error rates with other deep structured methods in the literature.

Methods	Forecasting Interval	Forecasting Level	Benchmarking Methods	Data	Load Forecasting Test MAPE	PV Power Forecasting Test MAPE
Proposed approach (DRNN-BiLSTM)	24-h (1-h resolution)	Aggregated grid power load	-	01.01.2012–31.12.2016	1.18%	14.00%
D-CNN [72]	24-h (30 min resolution)	Aggregated grid power load	Extreme Learning Machine (ELM), RNN, CNN, ARIMA	From the last week of April 2018 till the second week of July 2018	2.15%	-
Copula-DBN [73]	24-h (1-h resolution)	Aggregated grid power load	Classical NNs, SVR, ELM, and DBN	During the year of 2016	5.25%	-
DRNN-LSTM [37]	24-h (1-h resolution)	Aggregated residential power load	MLP network and SVM	01.01.2018–01.02.2018	7.43%	15.87%
Parallel CNN-RNN [41]	24-h (1-h resolution)	Aggregated grid power load	LR, SVR, DNN, CNN-RNN	10.02.2000–31.12.2012	1.405%	-

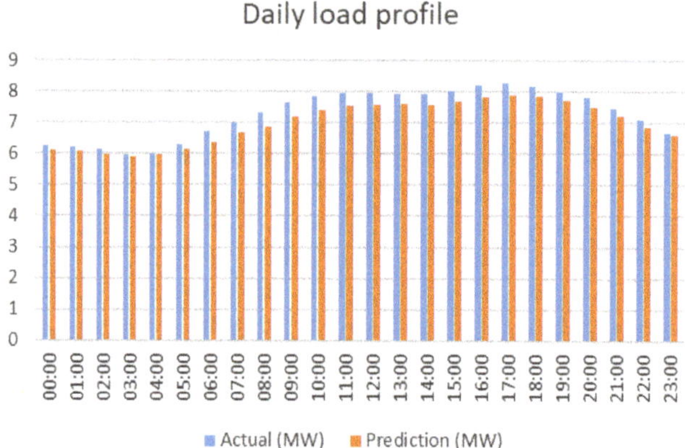

Figure 6. Day-ahead actual and predicted demand.

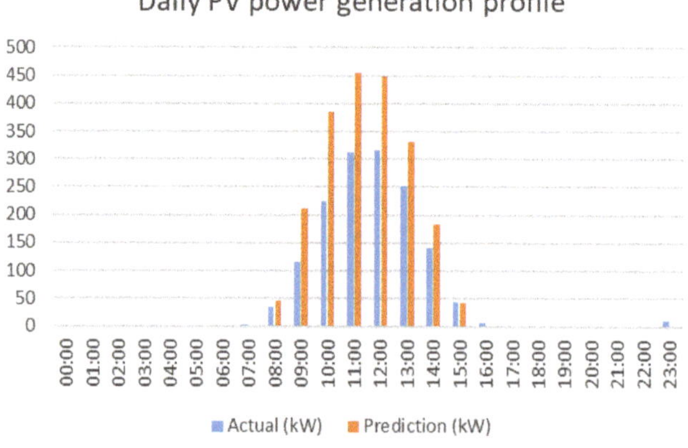

Figure 7. Day-ahead (24-h) actual and predicted PV power output.

2.5. PSO&SPSO Procedure for the Optimal Operational Scheduling Problem

The SPSO algorithm is used in this approach for determining the switch positions, whereas the OD of the DG units is performed with the basic PSO algorithm at each iteration. Both algorithms have a common objective function which is minimizing the real power loss of the whole system. In the whole PSO&SPSO algorithm, swarm population (n) is 50 and the maximum iteration number is 200. The convergence curve of the combined PSO&SPSO algorithm is presented in Figure 8 and here fitness denotes the best solution. The rest of the parameters' set values are the same as in [28].

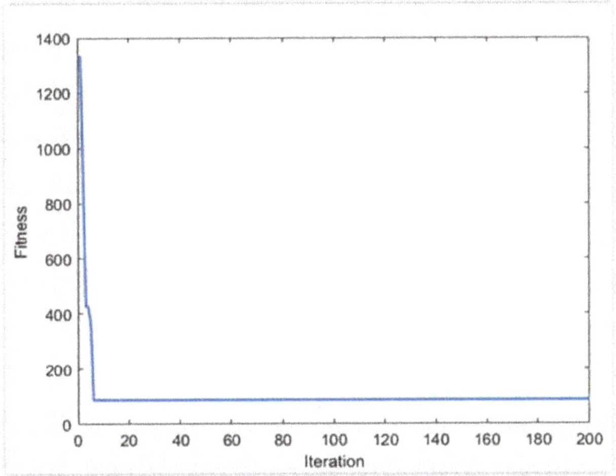

Figure 8. The convergence curve of the proposed algorithm.

The test reconfigurable MG system with all the specified sectionalizing switches and tie switches is presented in Figure 5. The dimension of the SPSO algorithm is equal to the number of loops formed when all tie switches in the reconfigurable MG are closed. Each dimension corresponds to a search space consisting of all the branches of the loop indicated with that dimension. There are five loops concerning the optimization problem in this reconfigurable MG test system once the tie switches (S33, S34, S35, S36, S37) are closed. Thus, the dimension is equal to five, and the search space in the SPSO algorithm is also represented by this dimension as five. The loops comprised of the respective branches (switches) on the reconfigurable MG test system are the same as in [24]. The connection in this case to the feeder must be maintained continuously, and the switches that are common in the loops should appear only in one loop at a time. The switches of the test system that are not in any loop do not belong to any of the search spaces and hence the optimization algorithm does not consider them It should be examined whether the test system is radial or not once the switches are selected and the connection conditions are met [74]. The optimal solution can be assigned when the radiality condition is obtained.

2.6. The Optimal Operational Scheduling Problem Test Results

Five different case studies have been performed in this section with the results presented in the following tables. The proposed single-objective problem is optimized at every time sequence in all optimization case studies presented in this study by considering the forecasted and actual hourly load demand and non-dispatchable DG unit (PV) output power profiles of the test reconfigurable MG system which has all integrated DG units (Three diesel DGs & a solar generation unit). The power market price schedule is used as presented in [24] for the economic evaluation of the operational scheduling framework. The cases that are performed for the short-term (24-h) period are presented in short as follows:

- *Case I:* Basic AC load flow analysis is performed in this case to see the test total daily real power losses of the system without performing any operational study on the reconfigurable MG system.
- *Case II:* OD of dispatchable diesel DGs is realized here by using the conventional PSO algorithm to monitor the effects of these DGs on the test system within the day.
- *Case III:* NR algorithm is applied to the test system via the SPSO method for this case.
- *Case IV:* This is a sequential study of case II and case III. Namely, NR is performed following OD of diesel DGs.
- *Case V:* Finally, in this case, NR and OD of diesel DGs are performed simultaneously by using the combined approach of conventional and selective PSO algorithms to observe the effects of both optimal operation studies at the same time.

The estimated daily total load for the MG test system used is 168,407.5 kW and the estimated daily total solar power generation is 1467.74 kW. When the basic power flow (PF) analysis is performed in line with the first case, the daily total real power loss in the system is 17681 kW with the estimated data and it corresponds to 10.49% of the estimated total energy consumption. Furthermore, the daily average minimum voltage profile on the MG test system is about 0.96 p.u. Hourly estimated energy demand values and corresponding market pricing can be found in Figure 9. Accordingly, total daily energy cost is calculated as 14,283.16 Euro with estimated data by performing basic load flow analysis on the test reconfigurable MG system.

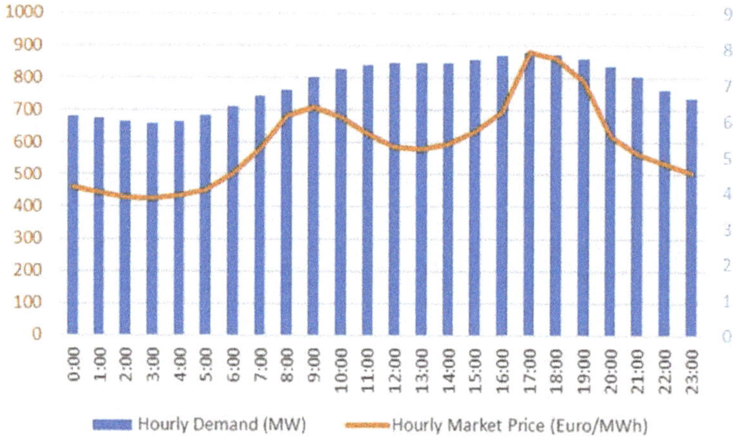

Figure 9. Hourly estimated demand and the hourly market price of the test reconfigurable MG system.

In case II, only OD of diesel DGs has performed on the test reconfigurable MG system with the estimation data and the daily total real power loss amount of 10,164 kW as seen in Table 7 which corresponds to 6.04% of the total estimated energy consumption. The average voltage level obtained by the PF study is 0.96 p.u. which has reached the level of 0.98 p.u. with the OD study as expected. With these results, total daily energy cost is calculated as 13,680.61 Euro. In the scope of the third case study, only NR framework is realized on the test system with the estimated data and it has been determined that the daily total real power loss amount is 9359 kW, which is equal to 5.56% of the total energy consumption. The contribution of the NR study to the voltage profile is observed when this study is performed with real data as seen in Table 7. When the energy cost amount is calculated with these results it corresponds to 14,288.19 Euro. It can be observed when these studies are compared through Table 7 that daily real power loss value with NR study is less than that of the OD study while the voltage profile of the OD study is better than the voltage profile of NR study.

Table 7. OD of diesel DG units and NR results.

Hour	Case II					Case III		
	Dispatch of DGs (MW)			P_L (kW)	V_{min} (p.u.)	Open Switches (Number)	P_L (kW)	V_{min} (p.u.)
	DG1	DG2	DG3					
00:00	0.580	1.209	1.318	162	0.98	21 6 14 30 26	233	0.96
01:00	0.371	0.921	1.041	252	0.98	35 6 14 30 26	226	0.95
02:00	1.147	0.423	0.263	353	0.98	11 6 14 29 26	264	0.96
03:00	0.412	0.498	1.180	269	0.98	11 7 34 30 26	209	0.95
04:00	0.226	0.222	0.725	482	0.98	21 7 14 31 26	246	0.98
05:00	0.768	0.395	1.364	225	0.98	21 6 14 30 26	235	0.96
06:00	0.631	0.936	0.121	432	0.98	11 6 14 31 26	309	0.96
07:00	0.728	0.333	0.805	433	0.98	11 6 14 30 26	353	0.96
08:00	1.240	0.600	0.529	364	0.98	11 6 14 31 26	370	0.97
09:00	0.880	0.381	0.857	461	0.98	11 6 14 30 26	435	0.97
10:00	0.077	0.971	1.259	450	0.97	35 6 34 30 26	370	0.96
11:00	0.972	0.658	1.249	343	0.97	11 6 14 31 26	473	0.96
12:00	0.379	1.024	1.017	449	0.97	21 6 14 31 26	436	0.96
13:00	0.609	0.264	1.230	532	0.97	11 6 14 30 26	511	0.97
14:00	1.257	0.068	0.911	517	0.97	11 6 14 30 26	507	0.96
15:00	1.300	0.242	1.370	366	0.97	11 7 14 30 26	473	0.96
16:00	0.716	0.642	0.517	653	0.97	21 6 14 31 26	474	0.96
17:00	1.260	0.121	0.470	712	0.97	11 6 34 31 26	466	0.96
18:00	0.784	1.008	0.465	547	0.97	11 6 14 17 26	685	0.96
19:00	1.126	1.017	0.694	388	0.97	11 6 14 30 26	532	0.96
20:00	1.095	0.808	0.985	341	0.98	21 6 14 31 26	422	0.96
21:00	0.290	0.409	0.382	781	0.98	11 6 14 30 26	442	0.95
22:00	1.254	0.121	0.9505	370	0.98	11 6 14 31 26	370	0.96
23:00	0.244	1.000	1.332	282	0.98	11 7 14 30 26	318	0.95
				10,164 (Daily total)	0.98 (Daily avr.)		9359 (Daily total)	0.96 (Daily avr.)

NR operation is actualized right after OD operation in case IV with estimated load data in this test system with the daily total real power loss amount determined as 8670 kW which corresponds to 5.15% of the total energy consumption. Furthermore, the average voltage profile has been improved to 0.97 p.u. and the total daily energy cost is calculated as 13,591.09 Euro. Thus, the lowest daily total real power loss and energy cost values have been obtained with this framework in comparison with the results of previous case studies.

It is important to notice that the results are very close to those of the NR study in terms of real power loss, while the energy cost result is quite parallel with the result OD of diesel DGs framework result as seen in Figures 10 and 11, respectively.

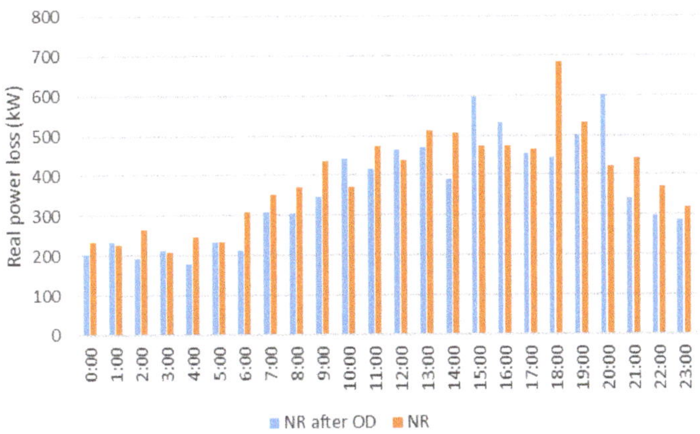

Figure 10. Daily real power loss comparative chart of case III and IV.

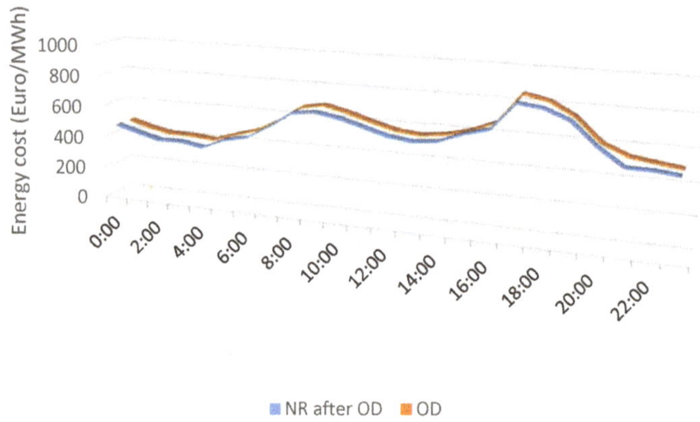

Figure 11. Daily energy cost comparative chart of case II and IV.

Consequently, NR and OD operations are performed simultaneously in case IV with the estimated load data where the daily real power loss amount is 7590 kW which corresponds to 4.51% of the total energy consumption in case 5. No improvement in the voltage profile is observed as can be seen in Table 8. However, it is still quite a good profile since it is close to the unit voltage level (1 p.u.). As a result of this case study, the lowest daily total real power loss and total energy cost amount of 13,526.1 Euro have been attained in comparison with the previous cases. The fourth case (NR after OD) has the closest results to this final case study in terms of real power loss and energy cost among all previous cases. It is worth observing the stacked chart of the daily real power loss curve for each case study since the curves of the last three studies that include NR operation are substantially parallel to each other as seen in Figure 12.

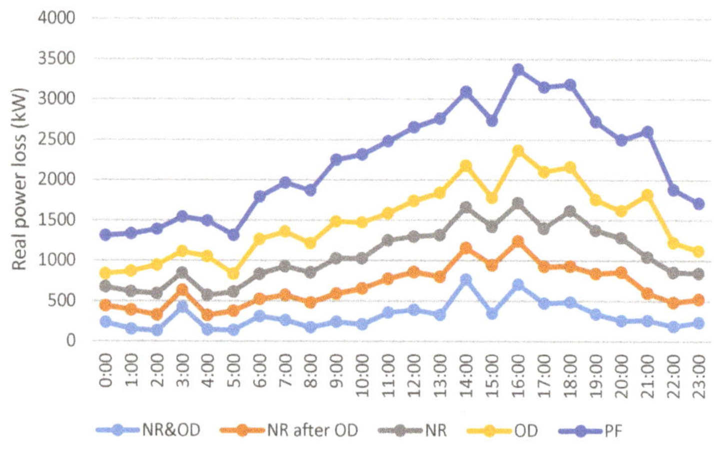

Figure 12. Daily real power loss chart of all cases.

Furthermore, the NR&OD results are calculated for every 20 days within the scope of case 5 to observe the status of each month and the differences between the actual and estimated values during the one-year test period are presented to put forth the overall impact of the forecasting method on short-term optimal operational scheduling framework in Figure 13. As can be seen from the graph, the two curves run parallel to each other or in other words, there is a very small difference between the actual and the predicted values. Total daily active power loss is 13.31 MW in the NR&OD study obtained with the test values and 12.88 MW for the same study with the estimated values as can be seen in the following figure when the mean values for this test year are calculated for these 19 measurements. Moreover, MAPE is calculated as 10.11 with these measurements.

Table 8. Simultaneous application of NR and OD of diesel DGs' results.

Hour	Open Switches (Number)	Dispatch of DGs (MW)			P_L (kW)	V_{min} (p.u.)
		DG1	DG2	DG3		
00:00	11 6 13 29 25	0.405	0.097	1.241	235	0.95
01:00	11 6 13 29 25	0.432	0.701	1.327	153	0.96
02:00	11 6 13 29 26	0.634	0.838	0.954	133	0.95
03:00	11 6 13 30 25	0.143	0.264	0.521	424	0.96
04:00	11 6 13 29 25	0.855	0.841	1.125	143	0.95
05:00	11 6 13 29 25	0.908	0.798	0.809	140	0.96
06:00	11 6 13 29 25	0.899	0.404	0.365	311	0.96
07:00	11 6 13 29 25	0.766	0.362	0.835	264	0.94
08:00	11 6 13 29 25	1.199	0.691	1.191	173	0.96
09:00	11 7 13 29 26	0.505	0.851	0.777	243	0.96
10:00	11 6 13 29 25	1.299	1.292	0.362	212	0.94
11:00	11 6 13 30 25	1.024	1.182	0.718	358	0.97
12:00	11 6 13 30 25	0.431	0.648	1.393	393	0.98
13:00	11 6 13 29 25	0.898	0.725	0.659	331	0.95
14:00	11 6 13 30 26	0.144	0.678	0.224	768	0.95
15:00	11 6 13 29 25	1.210	0.802	0.324	348	0.94
16:00	11 6 13 29 26	0.680	0.102	0.716	708	0.95
17:00	11 6 13 30 25	1.261	0.144	0.710	475	0.95
18:00	11 6 13 30 26	0.371	0.623	0.919	487	0.96
19:00	11 6 13 29 25	0.579	0.547	1.159	340	0.96
20:00	11 6 13 29 26	1.234	1.242	0.098	259	0.96
21:00	11 6 13 29 25	1.138	0.331	0.893	261	0.96
22:00	10 6 13 29 25	1.099	1.174	0.335	188	0.95
23:00	11 6 13 29 26	1.188	0.232	0.664	237	0.95
					7590 (Daily total)	0.96 (Daily avr.)

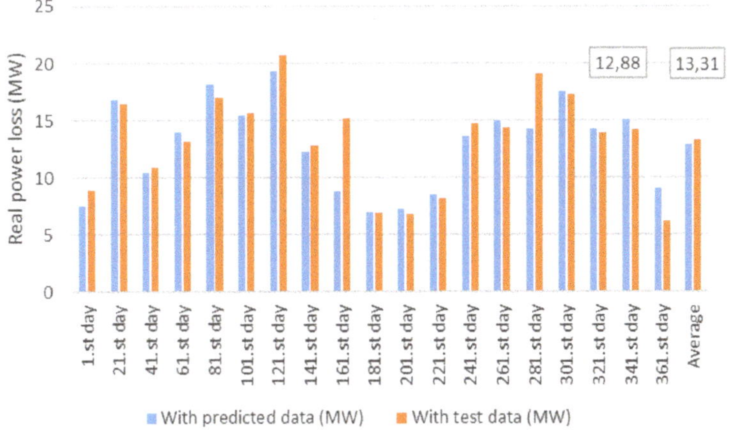

Figure 13. The daily total real power loss of every 20 days throughout the test year.

3. Conclusions

In this study, the time-varying nature of both the electrical load consumption and solar power output has been taken into account to support supply-demand balance within the scope of microgrid operational scheduling. To estimate aggregated power demand and the bulk PV output power over the short-term time period (day-ahead), a DRNN Bi-LSTM model has been proposed since conventional forecasting methods have limitations in modeling complex nonlinear problems and cannot take into consideration the time dependencies in the data set. Real-world datasets are used to test the DRNN Bi-LSTM model. They achieve excellent performance in forecasting aggregated power load and PV output power with MAPE values of 1.18% and 14%, respectively. These estimations are also performed by the FFNN method to observe the difference with traditional methods. The reason RNN mostly provides better results than the FFNN model lies behind the fact that the proposed model can learn long-term relationships between the sequential input data and the values to be predicted. Also, it has more learnable parameters compared to the FFNN model. In addition, the forecasting results put forth that the DRNN Bi-LSTM model performs better than the majority of the deep architectured methods surveyed in the literature. Therefore, the proposed DRNN Bi-LSTM model presents the potential for making more accurate short-term forecasting for day-ahead optimal operational scheduling of reconfigurable MG. The value of daily total active power loss in the reconfigurable MG test system decreases by 56% within the scope of day-ahead optimal operational scheduling with the demand and PV power data while it reduces by 57% with the estimated demand and PV power data via simultaneous NR and OD of diesel DGs' framework. The difference between the real power loss percentages obtained via the actual and forecasted data with DRNN Bi-LSTM for operational scheduling study is very small as is the difference between forecasted demand and PV power with DRNN Bi-LSTM and the actual demand and PV power data. Therefore, the proposed DRNN Bi-LSTM model has the potential for providing more accurate short-term forecasting for day-ahead optimal operational scheduling of reconfigurable MG.

Nevertheless, there are still some studies that can improve this work. The forecasting and optimization algorithms are run separately here, and the dynamic mechanism between them will be studied in future work. Afterwards, a real-time application will be added for more dynamic results in a coordinated manner with this dynamic study. To achieve more precise forecast results, shorter time periods (30 sec., 15 sec.) will be utilized instead of hourly data for real-time application studies. Besides, optimal operational scheduling of the electrical energy storage units and demand response programs will be addressed to improve the resiliency of the microgrid. Finally, development of novel algorithms that can replace the developed algorithms will be studied for improving forecasting and optimal operational scheduling studies in microgrids.

Author Contributions: F.Y. and M.B.Y. designed the model, gathered the input data, executed the simulations, and accomplished writing of the paper. A.A.-M. and M.B. supervised the entire work and edited the language. All authors have read and agreed to the published version of the manuscript.

Funding: This research was funded partially by the HeatReFlex-Green and Flexible Heating/Cooling project, (www.heatreflex.et.aau.dk) funded by Danida Fellowship Centre and the Ministry of Foreign Affairs of Denmark under the grant no. 18-M06-AAU.

Conflicts of Interest: The authors declare no conflict of interest.

Nomenclature

$\mu(x)$	the mean value of feature x
$\sigma(x)$	the standard deviation
p_j	a prediction by the proposed model
a_j	the corresponding actual value
n	is the size of the test set
x	decision vector

dv	number of decision variables
$f(x)$	optimization problem's objective function
$h_i(x)$	equality constraint that should be satisfied
$g_i(x)$	inequality constraints that should be satisfied
p	number of equality constraint
q	number of inequality constraints
P_L	total active power losses of the network
I_i	the real component of the current at branch i
R_i	branch resistance
b	sets of branches
$Cost_{MG}$	purchasing energy cost from the main grid
$Cost_{DG}$	the energy production cost of dispatchable DGs
v^b	forecasted price of the purchasing energy
p^b	value of purchased energy
P_{DG}	the output power of a distributed DG unit
a, b, c	cost function coefficients of a dispatchable DG
P_{EP}	exchanged power between the MG and the upstream grid
P_{MGL}	power consumption of each load of MG
N_{MGL}	number of MG loads
N_{DG}	number of DGs
V_i	the voltage level of each bus
V_{min}	the minimum voltage level of each bus
V_{max}	the maximum voltage level of each bus
I_i	the flowing current amount on the ith branch
I_i^{max}	thermal rating of the ith branch
β_b	a binary variable that defines a branch status (0—open, 1—closed)
N_b	set of branches (b)
m	number of network buses
N_{sub}	number of substations
S_D	selective search space at each dimension D
T	total number of samples in the training data
Y_t	training samples
k	iteration number
ω_{max}	the initial inertia weight value
ω_{min}	the final inertia weight value
DN	number of selected positions in dimension D
v_{min}	the minimum velocity of each particle
v_{max}	the maximum velocity of each particle

References

1. Ahmad, A.; Javaid, N.; Mateen, A.; Awais, M.; Khan, Z.A. Short-Term load forecasting in smart grids: An intelligent modular approach. *Energies* **2019**, *12*, 164. [CrossRef]
2. Ng E., J.; El-Shatshat, R.A. Multi-microgrid control systems (MMCS). In Proceedings of the IEEE PES General Meeting, Providence, RI, USA, 25–29 July 2010; pp. 1–6. [CrossRef]
3. Anvari-Moghaddam, A.; Seifi, A.R. A comprehensive study on future smart grids: definitions, strategies and recommendations. *J. N. C. Acad. Sci.* **2011**, *127*, 28–34.
4. Huh, J.; Seo, K. Smart Grid Framework Test Bed Using OPNET and Power Line Communication. In Proceedings of the Joint 8th International Conference on Soft Computing and Intelligent Systems (SCIS) and 17th International Symposium on Advanced Intelligent Systems (ISIS), Hokkaido, Japan, 25–28 August 2016; pp. 736–742.
5. Huh, J.; Seo, K. Hybrid Advanced Metering Infrastructure Design for Micro Grid Using the Game Theory Model. *Int. J. Softw. Eng. Appl.* **2015**, *9*, 257–268. [CrossRef]
6. Evar C., U.; Milana, T. Operational scheduling of microgrids via parametric programming. *Appl. Energy* **2016**, *180*, 672–681.

7. Shi, W.; Xie, X.; Chu, C.; Gadh, R. Distributed Optimal Energy Management in Microgrids. *IEEE Trans. Smart Grid* **2015**, *6*, 1137–1146. [CrossRef]

8. Yu J., J.Q.; Hou, Y.; Lam, A.; Li, V. Intelligent Fault Detection Scheme for Microgrids with Wavelet-based Deep Neural Networks. *IEEE Trans. Smart Grid* **2017**, *10*, 1694–1703. [CrossRef]

9. Shariatzadeh, F.; Vellaithurai, C.B.; Biswas, S.S.; Zamora, R.; Srivastava, A.K. Real-time implementation of intelligent reconfiguration algorithm for Microgrid. *IEEE Trans. Sustain. Energy* **2014**, *5*, 598–607. [CrossRef]

10. Ma, J.; Ma, X. A review of forecasting algorithms and energy management strategies for microgrids. *Syst. Sci. Control. Eng.* **2018**, *6*, 237–248. [CrossRef]

11. Kusakana, K. Optimal scheduling for distributed hybrid system with pumped hydro storage. *Energy Convers. Manag.* **2016**, *111*, 253–260. [CrossRef]

12. Wan, C.; Zhao, J.; Song, Y.; Xu, Z.; Lin, J.; Hu, Z. Photovoltaic and solar power forecasting for smart grid energy management. *CSEE J. Power Energy Syst.* **2016**, *1*, 38–46. [CrossRef]

13. Anvari-Moghaddam, A. Global Warming Mitigation Using Smart Micro-Grids. In *Global Warming—Impacts and Futur Perspective*; Bharat, R.S., Ed.; IntechOpen: London, UK, 2012; Chapter 4, pp. 119–134. [CrossRef]

14. Rabiee, A.; Sadeghi, M.; Aghaeic, J.; Heidari, A. Optimal operation of microgrids through simultaneous scheduling of electrical vehicles and responsive loads considering wind and PV units uncertainties. *Renew. Sustain. Energy Rev.* **2016**, *57*, 721–739. [CrossRef]

15. Xiao, Z.; Li, T.; Huang, M.; Shi, J.; Yang, J.; Yu, J.; Wu, W. Hierarchical MAS Based Control Strategy for Microgrid. *Energies* **2010**, *3*, 1622–1638. [CrossRef]

16. Dimeas, A.L.; Hatziargyriou, N.D. A MAS architecture for microgrids control. In Proceedings of the 13th International Conference on Intelligent Systems Application to Power Systems, Arlington, VA, USA, 6–10 November 2005; pp. 402–406.

17. Kish, G.; Lee, J.; Lehn, P. Modelling and control of photovoltaic panels utilising the incremental conductance method for maximum power point tracking. *IET Renew. Power Gener.* **2012**, *6*, 259. [CrossRef]

18. Kim, J.C.; Huh, J.H.; Ko, J.S. Improvement of MPPT Control Performance Using Fuzzy Control and VGPI in the PV System for Micro Grid. *Sustainability* **2019**, *11*, 5891. [CrossRef]

19. Xu, Y.; Xie, L.; Singh, C. Optimal scheduling and operation of load aggregators with electric energy storage facing price and demand uncertainties. In Proceedings of the North American Power Symposium, Boston, MA, USA, 4–6 August 2011.

20. Anwar, T.; Sharma, B.; Chakraborty, K.; Sirohia, H. Introduction to Load Forecasting. *Int. J. Pure Appl. Math.* **2018**, *119*, 1527–1538.

21. Pires, A.J. Short-term load forecasting based on ANN applied to electrical distribution substations. In Proceedings of the 39th International Universities Power Engineering Conference, Bristol, UK, 6–8 September 2004; Volume 1, pp. 427–432.

22. Anvari-Moghaddam, A.; Seifi, A.R. Study of forecasting renewable energies in smart grids using linear predictive filters and neural networks. *IET Renew. Power Gener.* **2011**, *5*, 470–480. [CrossRef]

23. Anvari-Moghaddam, A.; Monsef, H.; Rahimi-Kian, A.; Nance, H. Feasibility study of a novel methodology for solar radiation prediction on an hourly time scale: A case study in Plymouth, United Kingdom. *J. Renew. Sustain. Energy* **2014**, *6*, 033107. [CrossRef]

24. Lee, E.; Shi, W.; Gadh, R.; Kim, W. Design and Implementation of a Microgrid Energy Management System. *Sustainability* **2016**, *8*, 1143. [CrossRef]

25. Alani, A.Y.; Osunmakinde, I.O. Short-Term Multiple Forecasting of Electric Energy Loads for Sustainable Demand Planning in Smart Grids for Smart Homes. *Sustainability* **2017**, *9*, 1972. [CrossRef]

26. Pal, S.; Sen, S.; Sengupta, S. Power network reconfiguration for congestion management and loss minimization using Genetic Algorithm. In Proceedings of the Michael Faraday IET International Summit, Kolkata, India, 12–13 September 2015; pp. 291–296.

27. Possemato, F.; Paschero, M.; Livi, L.; Rizzi, A.; Sadeghian, A. On the impact of topological properties of smart grids in power losses optimization problems. *Int. J. Electron. Power Energy Syst.* **2016**, *78*, 755–764. [CrossRef]

28. Yaprakdal, F.; Baysal, M.; Anvari-Moghaddam, A. Optimal operational scheduling of reconfigurable microgrids in presence of renewable energy sources. *Energies* **2019**, *12*, 1858. [CrossRef]

29. Rao, R.S.; Ravindra, K.; Satish, K.; Narasimham, S.V.L. Power loss minimization in distribution system using network reconfiguration in the presence of distributed generation. *IEEE Trans. Power Syst.* **2013**, *28*, 317–325. [CrossRef]

30. Dahalan, W.M.; Mokhlis, H.; Ahmad, R.; Abu Bakar, A.H.; Musirin, I. Simultaneous Network Reconfiguration and DG Sizing Using Evolutionary Programming and Genetic Algorithm to Minimize Power Losses. *Arab. J. Sci. Eng.* **2014**, *39*, 6327–6338. [CrossRef]

31. Imran, A.M.; Kowsalya, M.; Kothari, D. A novel integration technique for optimal network reconfiguration and distributed generation placement in power distribution networks. *Int. J. Electron. Power Energy Syst.* **2014**, *63*, 461–472. [CrossRef]

32. Nguyen, T.T.; Truong, A.V.; Phung, T.A. A novel method based on adaptive cuckoo search for optimal network reconfiguration and distributed generation allocation in distribution network. *Int. J. Electron. Power Energy Syst.* **2016**, *78*, 801–815. [CrossRef]

33. Jiang, H.; Ding, F.; Zhang, Y. Short-term load forecasting based automatic distribution network reconfiguration. In Proceedings of the 2017 IEEE Power & Energy Society General Meeting, Chicago, IL, USA, 16–20 July 2017; pp. 1–5.

34. Gu, Y.; Jiang, H.; Zhang, J.J.; Zhang, Y.; Muljadi, E.; Solis, F. Load forecasting based distribution system network reconfiguration—A distributed data-driven approach. In Proceedings of the 2017 51st Asilomar Conference on Signals, Systems, and Computers, Pacific Grove, CA, USA, 29 October–1 November 2017; pp. 1358–1362.

35. Selvaraj, K.R.; Sundararaj, S.; Ravi, T. Artificial Neutral Network Based Load Forecasting and Economic Dispatch with Particle Swarm Optimization. *Int. J. Sci. Eng. Res.* **2013**, *4*, 139–145.

36. Arif, M.; Liu, Y.; Haq, I.U.; Ashfaq, A. Load forecasting using neural network integrated with economic dispatch problem. *Int. J. Electron. Comput. Eng.* **2018**, *12*, 900–905.

37. Wen, L.; Zhou, K.; Yang, S.; Lu, X. Optimal load dispatch of community microgrid with deep learning based solar power and load forecasting. *Energy* **2019**, *171*, 1053–1065. [CrossRef]

38. Wu, X.; Cao, W.-H.; Wang, D.; Ding, M. A Multi-Objective Optimization Dispatch Method for Microgrid Energy Management Considering the Power Loss of Converters. *Energies* **2019**, *12*, 2160. [CrossRef]

39. Gutiérrez-Alcaraz, G.; Galvan, E.; Cabrera, N.G.; Javadi, M. Renewable energy resources short-term scheduling and dynamic network reconfiguration for sustainable energy consumption. *Renew. Sustain. Energy Rev.* **2015**, *52*, 256–264. [CrossRef]

40. Xu, Y.; Xie, L.; Singh, C. Optimal scheduling and operation of load aggregator with electric energy storage in power markets. In Proceedings of the North American Power Symposium, Boston, MA, USA, 4–6 August 2011.

41. He, W. Load Forecasting via Deep Neural Networks. *Procedia Comput. Sci.* **2017**, *122*, 308–314. [CrossRef]

42. Zheng, J.; Xu, C.; Zhang, Z.; Li, X. Electric load forecasting in smart grids using Long-Short-Term-Memory based Recurrent Neural Network. In Proceedings of the 2017 51st Annual Conference on Information Sciences and Systems (CISS), Baltimore, MD, USA, 22–24 March 2017; pp. 1–6.

43. Bouktif, S.; Fiaz, A.; Ouni, A.; Serhani, M.A. Optimal Deep Learning LSTM Model for Electric Load Forecasting using Feature Selection and Genetic Algorithm: Comparison with Machine Learning Approaches †. *Energies* **2018**, *11*, 1636. [CrossRef]

44. Abdel-Nasser, M.; Mahmoud, K. Accurate photovoltaic power forecasting models using deep LSTM-RNN. *Neural Comput. Appl.* **2017**, *31*, 2727–2740. [CrossRef]

45. Chattopadhyay, D. Operational Planning of Power System: An Integrated Approach. *Energy Sources* **1994**, *16*, 59–73. [CrossRef]

46. Logenthiran, T.; Srinivasan, D. Short term generation scheduling of a Microgrid. In Proceedings of the TENCON 2009 - 2009 IEEE Region 10 Conference, Singapore, 23–26 November 2009; pp. 1–6.

47. Zhao, B.; Shi, Y.; Dong, X.; Luan, W.; Bornemann, J. Short-Term Operation Scheduling in Renewable-Powered Microgrids: A Duality-Based Approach. *IEEE Trans. Sustain. Energy* **2014**, *5*, 209–217. [CrossRef]

48. Yoon, A.-Y.; Moon, H.-J.; Moon, S.-I. Very short-term load forecasting based on a pattern ratio in an office building. *Int. J. Smart Grid Clean Energy* **2016**, *5*, 94–99. [CrossRef]

49. Jain, A.; Satish, B. Integrated approach for short term load forecasting using SVM and ANN. In Proceedings of the TENCON 2008—2008 IEEE Region 10 Conference, Hyderabad, India, 19–21 November 2008; pp. 1–6.

50. Hernandez, L.; Baladrón, C.; Aguiar, J.M.; Carro, B.; Sanchez-Esguevillas, A.J.; Lloret, J.; Massana, J.; Hernández-Callejo, L. A Survey on Electric Power Demand Forecasting: Future Trends in Smart Grids, Microgrids and Smart Buildings. *IEEE Commun. Surv. Tutor.* **2014**, *16*, 1460–1495. [CrossRef]

51. Hernández-Callejo, L.; Baladrón, C.; Aguiar, J.M.; Calavia, L.; Carro, B.; Sánchez-Esguevillas, A.; Sanjuan, J.; Gonzalez, A.; Lloret, J. Improved Short-Term Load Forecasting Based on Two-Stage Predictions with Artificial Neural Networks in a Microgrid Environment. *Energies* **2013**, *6*, 4489–4507. [CrossRef]

52. Chen, J.-F.; Wang, W.-M.; Huang, C.-M. Analysis of an adaptive time-series autoregressive moving-average (ARMA) model for short-term load forecasting. *Electron. Power Syst. Res.* **1995**, *34*, 187–196. [CrossRef]

53. Velasco, L.C.P.; Lou, D.; Paolo, G.; Bryan, M.B.f. Load Forecasting using Autoregressive Integrated Moving Average and Artificial Neural Network. *Int. J. Adv. Comput. Sci. Appl.* **2018**, *9*, 23–29. [CrossRef]

54. Saber, A.Y.; Alam, A.K.M.R. Short term load forecasting using multiple linear regression for big data. 2017 IEEE Symposium Series on Computational Intelligence (SSCI), Honolulu, HI, USA, 27 November–1 December 2017; pp. 1–6.

55. El-Hawary, M.; Mbamalu, G. Short-term power system load forecasting using the iteratively reweighted least squares algorithm. *Electron. Power Syst. Res.* **1990**, *19*, 11–22. [CrossRef]

56. Hernández-Callejo, L.; Baladrón, C.; Aguiar, J.M.; Calavia, L.; Carro, B.; Sánchez-Esguevillas, A.; Pérez, F.; Fernández, A.; Lloret, J. Artificial Neural Network for Short-Term Load Forecasting in Distribution Systems. *Energies* **2014**, *7*, 1576–1598. [CrossRef]

57. Moon, J.; Kim, Y.; Son, M.; Hwang, E. Hybrid Short-Term Load Forecasting Scheme Using Random Forest and Multilayer Perceptron. *Energies* **2018**, *11*, 3283. [CrossRef]

58. Chowdhury, D.; Sarkar, M.; Haider, M.Z.; Alam, T. Zone Wise Hourly Load Prediction Using Regression Decision Tree Model. In Proceedings of the 2018 International Conference on Innovation in Engineering and Technology (ICIET), Dhaka, Bangladesh, 25–27 October 2018; pp. 1–6.

59. Li, W.; Yang, X.; Li, H.; Su, L. Hybrid Forecasting Approach Based on GRNN Neural Network and SVR Machine for Electricity Demand Forecasting. *Energies* **2017**, *10*, 44. [CrossRef]

60. Mujeeb, S.; Javaid, N. Deep Long Short-Term Memory: A New Price and Load Forecasting Scheme for Big Data in Smart Cities. *Sustainability* **2019**, *11*, 987.

61. Gao, M.; Li, J.; Hong, F.; Long, D. Day-ahead power forecasting in a large-scale photovoltaic plant based on weather classification using LSTM. *Energy* **2019**, *187*, 115838. [CrossRef]

62. Wang, K.; Qi, X.; Liu, H. A comparison of day-ahead photovoltaic power forecasting models based on deep learning neural network. *Appl. Energy* **2019**, *251*, 113315. [CrossRef]

63. Gensler, A.; Henze, J.; Sick, B.; Raabe, N. Deep Learning for solar power forecasting—An approach using AutoEncoder and LSTM Neural Networks. In Proceedings of the 2016 IEEE International Conference on Systems, Man, and Cybernetics (SMC), Budapest, Hungary, 9–12 October 2016.

64. Alzahrani, A.; Shamsi, P.; Dagli, C.; Ferdowsi, M. Solar Irradiance Forecasting Using Deep Neural Networks. *Procedia Comput. Sci.* **2017**, *114*, 304–313. [CrossRef]

65. Dong, X.; Qian, L.; Huang, L. Short-term load forecasting in smart grid: A combined CNN and K-means clustering approach. In Proceedings of the 2017 IEEE International Conference on Big Data and Smart Computing (BigComp), Jeju Island, Korea, 13–16 February 2017; pp. 119–125.

66. Tian, C.; Ma, J.; Zhang, C.; Zhan, P. A Deep Neural Network Model for Short-Term Load Forecast Based on Long Short-Term Memory Network and Convolutional Neural Network. *Energies* **2018**, *11*, 3493. [CrossRef]

67. Zeng, P.; Li, H.; He, H.; Li, S. Dynamic Energy Management of a Microgrid Using Approximate Dynamic Programming and Deep Recurrent Neural Network Learning. *IEEE Trans. Smart Grid* **2019**, *10*, 4435–4445. [CrossRef]

68. Makridakis, S. Accuracy measures: theoretical and practical concerns. *Int. J. Forecast* **1993**, *9*, 527–529. [CrossRef]

69. Ben Hamida, I.; Salah, S.B.; Msahli, F.; Mimouni, M.F.; Msahli, F.; Mimouni, M.F. Optimal network reconfiguration and renewable DG integration considering time sequence variation in load and DGs. *Renew. Energy* **2018**, *121*, 66–80. [CrossRef]

70. Esmaeili, S.; Anvari-Moghaddam, A.; Jadid, S.; Guerrero, J.M. Optimal simultaneous day-ahead scheduling and hourly reconfiguration of distribution systems considering responsive loads. *Int. J. Electron. Power Energy Syst.* **2019**, *104*, 537–548. [CrossRef]

71. Hemmati, M.; Mohammadi-ivatloo, B.; Ghasemzadeh, S.; Reihani, E. Electrical power and energy systems risk-based optimal scheduling of reconfigurable smart renewable energy based microgrids. *Electron. Power Energy Syst.* **2018**, *101*, 415–428. [CrossRef]

72. Khan, S.; Javaid, N.; Chand, A.; Khan, A.B.M.; Rashid, F.; Afridi, I.U. Electricity Load Forecasting for Each Day of Week Using Deep CNN. *Adv. Intell. Syst. Comput.* **2019**, *927*, 1107–1119.
73. Ouyang, T.; He, Y.; Li, H.; Sun, Z.; Baek, S. Modeling and Forecasting Short-Term Power Load With Copula Model and Deep Belief Network. *IEEE Trans. Emerg. Top. Comput. Intell.* **2019**, *3*, 127–136. [CrossRef]
74. Golshannavaz, S.; Afsharnia, S.; Aminifar, F. Smart Distribution Grid: Optimal Day-Ahead Scheduling With Reconfigurable Topology. *IEEE Trans. Smart Grid* **2014**, *5*, 2402–2411. [CrossRef]

© 2020 by the authors. Licensee MDPI, Basel, Switzerland. This article is an open access article distributed under the terms and conditions of the Creative Commons Attribution (CC BY) license (http://creativecommons.org/licenses/by/4.0/).

Article

A Power Flow Control Strategy for Hybrid Control Architecture of DC Microgrid under Unreliable Grid Connection Considering Electricity Price Constraint

Faris Adnan Padhilah and Kyeong-Hwa Kim *

Department of Electrical and Information Engineering, Research Center for Electrical and Information
Technology, Seoul National University of Science and Technology, Seoul 01811, Korea; farisap.fa@gmail.com
* Correspondence: k2h1@seoultech.ac.kr; Tel.: +82-2-970-6406

Received: 17 August 2020; Accepted: 15 September 2020; Published: 16 September 2020

Abstract: This paper presents a power flow control strategy for a hybrid control architecture of the DC microgrid (DCMG) system under an unreliable grid connection considering the constraint of electricity price. To overcome the limitation of the existing schemes, a hybrid control architecture which combines the centralized control and distributed control is applied to control DCMG. By using the hybrid control approach, a more optimal and reliable DCMG system can be constructed even though a fault occurs in the grid or a central controller (CC). The power flow control strategy for the hybrid DCMG control architecture also takes the constraint of electricity price into account for the purpose of minimizing the electricity cost. In the proposed hybrid control, the high bandwidth communication (HBC) link is used in the centralized control to connect the CC with DCMG power agents. On the other hand, the low bandwidth communication (LBC) link is employed to constitute the distributed control. A small size of data is used to exchange the information fast between the agents and CC, or between each agent and its neighbors, which increases the reliability and robustness of the DCMG system in case of a fault in the communication link of the centralized control. A DCMG system with 400-V rated DC-link voltage which consists of a wind power agent, a battery agent, a grid agent, a load agent, and a CC is constructed in this study by using three power converters based on 32-bit floating point digital signal processor (DSP) TMS320F28335 controller. Various simulation and experimental results prove that the proposed scheme improves the system stability and robustness even in the presence of a fault in the communication link of the centralized control. In addition, the proposed scheme is capable of maintaining the DC-link voltage stably at the nominal value without severe transients both in the centralized control and distributed control, as well as both in the grid-connected case and islanded case. Finally, the scalability of the DCMG system is tested by adding and removing additional wind power agent and battery agent during a certain period.

Keywords: centralized control architecture; DC microgrid; distributed control architecture; electricity price constraint; hybrid control architecture; power flow control strategy

---

1. Introduction

Environmental concern and economic factors make microgrid (MG) research more intense lately. In general, an MG is a small power system which integrates loads, sources, and storage devices in an electric network by connecting them through power electronic converters [1]. MGs are supplemented by the distributed generation (DG) power from renewable energy sources such as photovoltaic (PV), wind turbine, and hydroelectric. The output from renewable energy sources often fluctuates from time to time as the renewable energy is highly dependent on natural conditions. To overcome this power fluctuation, an energy storage system (ESS) is usually added into the MG system [2]. In addition to the ESS, technology developments also make it possible to connect many power sources into MG. In [3], a supercapacitor is employed to alleviate the voltage transient during load changes. The research in [2] considers an electric vehicle with PV and ESS in a coordinated control to suppress the system frequency fluctuation in an MG system.

Depending on the bus voltage type, MGs can be classified as DC microgrids (DCMGs) and AC microgrids (ACMGs). Between them, recent interest is focused on DCMG because of its attractive feature [4]. DCMGs have several advantages over ACMGs since some constraints in ACMGs such as transformer inrush current, reactive power flow, frequency synchronization, current harmonics, and power quality can be avoided in DCMG systems [5]. Furthermore, as most DGs generate DC power or are inevitably converted to DC first, the integration process is much easier in the DCMG system, which leads to a construction of a more efficient MG system [6,7]. DCMG can also reduce the power conversion loss in the consumer sector because new electronic loads are dominated by DC loads [8]. Moreover, several AC loads like induction motors can be supplied by DCMGs using inverter-fed drive systems [9].

In order to guarantee that the MG system works effectively and efficiently, a coordination of power agents such as DGs, ESSs, and grid is needed in MG. Coordinated control of MG can be classified into three categories based on the MG architecture: the distributed control, decentralized control, and centralized control. In the distributed control approach, all the agents communicate with their respective neighbors and the operating conditions are decided based on the internal information of the agent and the external information of other agents. In the decentralized control scheme, all the agents determine their operating conditions by using only the internal information of the agent without any information from other agents. Meanwhile, in the centralized control method, all the agents send data to a central controller (CC), and then, CC decides the operating conditions of all the agents based on the acquired data.

All the coordination control architectures have their own advantages and disadvantages. By integrating the energy management system (EMS) to CC in the centralized control, it is possible to reach a global optimum solution while satisfying some constraints at the same time [10]. The centralized control approach also increases MG stability [11] and optimizes the power exchange [12] as well as the economic dispatch of DG units with high accuracy and controllability [13]. In spite of these advantages, there exist several disadvantages in the centralized control architecture. It is difficult to integrate many power units in this scheme. As the number of power units is increased, the computational burden of the CC is also increased [14]. The communication link between the CC and DG agents also becomes more fragile to communication failure [15]. Moreover, the centralized control mode is faced with a single point of failure issue because it highly relies on the CC to control the entire system of MG. In the centralized control, a fault detection delay caused by the fault in the communication link creates another problem in the MG system. With the aim of improving the system reliability of MG, the research in [16] proposes to install the local emergency mode in each DCMG agent in the centralized control architecture. In this local emergency mode, each agent determines its operating mode instead of the CC. However, this approach still has the limitations of the centralized control.

As for the decentralized control approach, the number of the power agents can be increased or decreased easily in the MG system. However, this scheme is lack of coordination to achieve an optimum solution [17] because the communication link does not exist.

On the other hand, the power units can be easily added or removed in the distributed control approach. This scheme has more robustness in terms of the communication link than the centralized control architecture. In addition, the existence of the communication links makes the MG system more probable to reach an optimum solution. In [18], an improved power management strategy is presented for multi-agent system (MAS)-based distributed control of DCMG. In this scheme, the DC-link voltage restoration algorithm is also developed to ensure the system power balance even under communication network problems. However, it is well known that the distributed control achieves only a sub-optimum solution since the objective of the distributed control is only to improve the reliability of MG [10].

Moreover, it is difficult to anticipate all the operating scenarios in a large MG system with many DGs [10].

To overcome such a limitation in each coordination control, a hybrid control architecture that combines two or more coordination control methods is introduced. The study in [19] proposes a hybrid control architecture to combine the centralized and distributed architectures for an islanded DCMG. In this work, the distributed control is used only as a backup plan when a fault occurs in the CC. However, excessive generated power from the DG units is not effectively utilized in this scheme when the DCMG operates in the distributed control mode. In addition to that, the DCMG system experiences a moderate transient period in the transition between the distributed control and centralized control. A combination of the centralized and decentralized architectures in a hierarchical control strategy is also presented in [20]. This study allows a robust and seamless transfer between the grid-connected and islanded cases in the ACMG operation.

The electricity generation by DG units and power flow control strategy mainly determine the power injected into or absorbed from the main grid. The power from the grid affects the electricity price which consumers should pay. The time-of-use pricing (TOU), real-time pricing (RTP), and stepwise power tariff (SPT) methods have been used to determine electricity bills worldwide [21]. The TOU and RTP tariffs have been used in optimal battery sizing in smart home systems [22,23]. The TOU pricing divides daily electricity bills into three different levels such as on-peak rate, mid-peak rate, and off-peak rate [22]. The TOU policy has been adopted in Ontario, Canada to calculate the electricity bill for in-home energy management systems which use the wind turbine and battery energy storage system (BESS) [24]. The RTP method has been implemented in the U.S. and Australia for years [25]. Compared to the TOU, the RTP method that depends on the real market cost of delivering electricity offers a higher variability of electricity bill [26]. The SPT policy divides monthly electricity bills based on consumer electricity consumption. In this method, higher electricity consumption leads to higher electricity price [27]. China is one of the countries to use the SPT method to determine electricity bills [28].

Considering the limitation of the existing coordination control architectures, this paper presents a hybrid control architecture of DCMG which combines the centralized control and distributed control. By using this hybrid control approach, a more optimal and reliable DCMG system can be constructed even though a fault occurs in the grid or the CC. The power flow control strategy of the hybrid DCMG control architecture is also developed by taking the constraint of electricity price into account. In the proposed hybrid control architecture, the high bandwidth communication (HBC) link is used in the centralized control to connect the CC with DCMG power agents. Based on the information from all the agents, the CC determines the operating modes of all the agents. On the contrary, the low bandwidth communication (LBC) link is employed to constitute the distributed control. To reduce the communication burden, only one-bit binary data is used to exchange the information between the agents via the LBC link. For fast data transfer between the agent and CC, or, between each agent, a small size of data is exchanged in both the HBC link and LBC link. A small size of exchange data in the HBC and LBC links facilitates the reliability of the DCMG system as is stated in [29]. DCMG system considered in this study consists of a wind power agent, battery agent, grid agent, and load agent. The operating modes of the agents are determined based on the wind power generation, battery state-of-charge (SOC) level, grid availability, and electricity price constraint. To verify the proposed hybrid control scheme, the simulation is conducted by using the Powersim (PSIM) software (9.1, Powersim, Rockville, MD, USA) under three different levels of SOC, two grid conditions, and two electricity price levels. The simulation results validate that the proposed scheme can stably and reliably maintain the DC-link voltage at the nominal value irrespective of the control mode transition between the centralized control and distributed control. In addition, the system scalability is tested to allow the addition and removal of additional DG source and battery during a certain period. Finally, experimental results are presented to validate the simulation results by using a prototype of the DCMG system. The contributions of this paper are as follows:

(i) A power flow control strategy for a hybrid DCMG control architecture is proposed by combining the centralized control and distributed control. The proposed hybrid DCMG control scheme stably maintains the DC-link voltage at the nominal value with high reliability and robustness against the communication link fault and the absence of utility grid.

(ii) The proposed hybrid power flow control provides a stable and smooth DCMG operation during the transition between the distributed control and the centralized control, which overcomes the limitations such as the common single point of failure in the centralized control as well as the limitations related with the lack of information exchange in the decentralized control.

(iii) The proposed scheme can be implemented by the exchange of only a small size of data through the HBC and LBC links, which reduces communication burden and helps the DCMG system reach the optimum solution.

(iv) This study also focuses on minimizing the electricity cost on the consumer side by introducing a power flow control for a hybrid DCMG system with the consideration of electricity price constraint.

This paper is organized as follows: Section 2 describes the configuration of DCMG and each power agent. Section 3 describes the proposed hybrid DCMG architecture and power flow control strategy under various conditions. Sections 4 and 5 present the simulation and experimental results. Finally, Section 6 concludes the paper.

2. Configuration of DCMG System

Figure 1 shows the configuration of the DCMG system considered in this study. The DCMG system consists of four agents, which are the grid agent, battery agent, wind power agent, and load agent. The grid agent and battery agent can export or import the power from the DC-link, while the wind power agent only provides the power to the DC-link and the load agent only absorbs the power from the DC-link. In this figure, P_G, P_B, P_W, and P_L denote the power from or into the grid agent, the power from or into the battery agent, the power from the wind power agent, and the power into the load agent, respectively. Figure 1 also shows the current direction of each power agent. For convenience, the reference direction of all the currents are taken as out of the DC-link. In order to connect the main grid source to the DCMG system, a transformer and a bidirectional AC-to-DC converter are employed. To supply the power from the wind turbine into the DCMG system, a permanent magnet synchronous generator (PMSG) and a unidirectional AC-to-DC converter are used. A battery is connected with a bidirectional DC-to-DC converter to exchange the power with the DCMG. In the load agent, load shedding or reconnection is achieved through electronic switches.

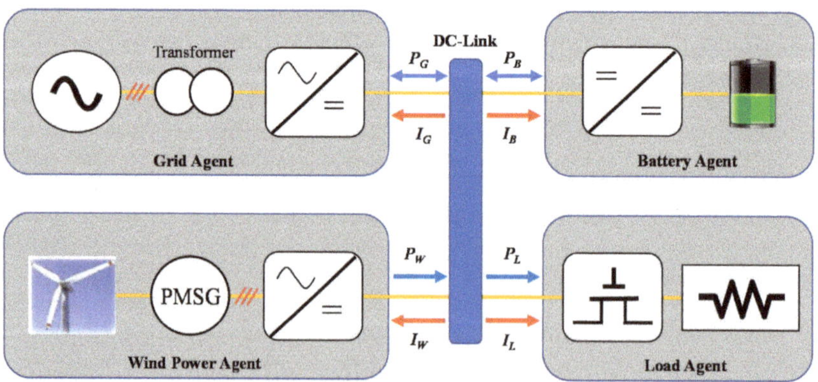

Figure 1. Configuration of DC microgrid (DCMG).

2.1. Grid Agent

Figure 2 shows the configuration of the grid agent. In order to damp the harmonic currents caused by the main grid source, an inductive-capacitive-inductive (LCL) filter is placed between the grid transformer and AC-to-DC converter. The parameters R_1, R_2, L_1, L_2, and C_f represent the filter resistances, filter inductances, and filter capacitance, respectively. The inverter-side current and grid-side current are denoted as i_1 and i_2, respectively. To regulate the currents in bidirectional way, an integral state feedback current controller with a full state observer is employed based on only the measurements of the grid currents and grid voltages [30]. The detailed control design process and observer implementation for the integral state feedback current controller in the grid-connected inverter is presented in [30–32].

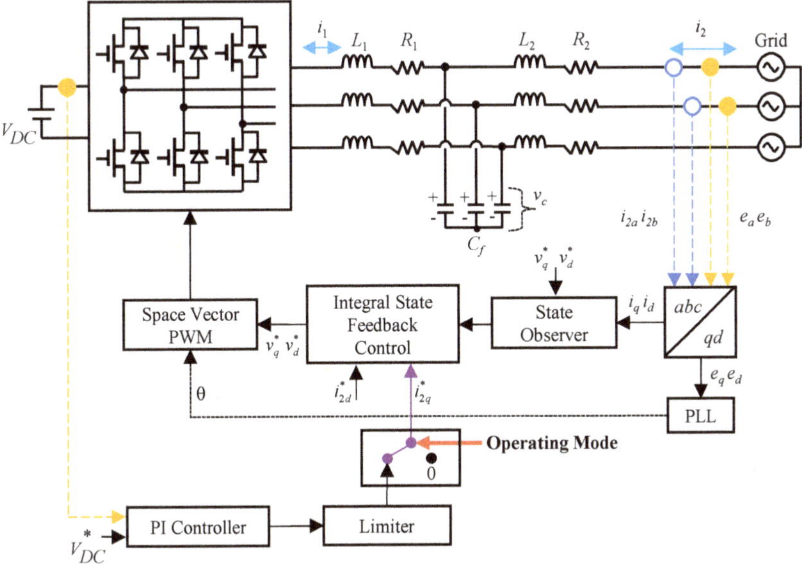

Figure 2. Configuration of grid agent.

The grid agent has three operating modes in the DCMG system, which are V_{DC} control converter (CON) mode, V_{DC} control inverter (INV) mode, and IDLE mode. The grid agent operates as V_{DC} control CON mode when the wind power cannot supply the load demand, or the wind power cannot supply both the load demand and battery power in the charging mode of battery. In this operating mode, the grid agent maintains the DC-link voltage at the nominal value by injecting the required power from the grid to the DCMG. The second operating mode, namely, V_{DC} control INV mode is used when the wind power is sufficient to supply the load demand. The grid agent also operates in this operating mode when the wind power can supply both the load demand and the maximum charging power of the battery ($P_{B,Max,chr}$). In V_{DC} control INV mode, the grid agent absorbs the surplus power from DCMG to inject it to the main grid, while regulating the DC-link voltage at the nominal value. The grid agent operates in IDLE mode instead of V_{DC} control CON mode when the CC operates under the condition of high electricity price. In IDLE mode, the grid agent avoids the use of additional power from the grid to reduce utility cost.

When the grid agent regulates the DC-link voltage both in V_{DC} control CON mode and V_{DC} control INV mode, the grid agent implements two cascaded control loops as shown in Figure 2. In the outer control loop, the DC-link voltage controller is designed by using the proportional-integral (PI) controller to generate the current reference. Otherwise, the current reference is also generated as zero according to the operating modes of the grid agent. The inner control loop is designed to control the grid-side currents by using the integral state feedback control and full state observer. The full state observer estimates the system states from the system model and the measurement of the grid currents and grid voltages in the synchronous reference frame (SRF). The voltage references from the inner control loop are applied through the space vector pulse width modulation (PWM) modulation to generate the converter drive signals.

2.2. Battery Agent

Figure 3 shows the configuration of the battery agent. The battery is connected to the DCMG system with an inductive (L) filter and interleaved bidirectional DC-to-DC converter for the purpose of reducing current ripples. The battery agent has four operating modes which are V_{DC} control by charging, V_{DC} control by discharging, charge with the maximum allowable current ($I_{B,Max}$), and IDLE mode. The battery agent operates in V_{DC} control mode by charging if a fault occurs in the grid, and the wind power generation P_W is higher than load demand P_L. This operating mode is also chosen when the battery SOC is lower than the minimum SOC level, SOC_{min}, and P_W is lower than P_L. Under the condition that P_W is lower than P_L, and the grid agent cannot operate in V_{DC} control CON mode, the battery agent operates in V_{DC} control mode by discharging. When the battery SOC is less than the maximum SOC level (SOC_{max}) and the grid agent is connected to the DCMG, the battery operates in charge with $I_{B,Max}$. This operating mode increases the battery SOC as fast as possible. The battery agent operates with IDLE mode when the battery SOC is greater than SOC_{max} and other power agents operate with V_{DC} control mode. This operating mode maintains the battery SOC level without exchanging the power with the DCMG.

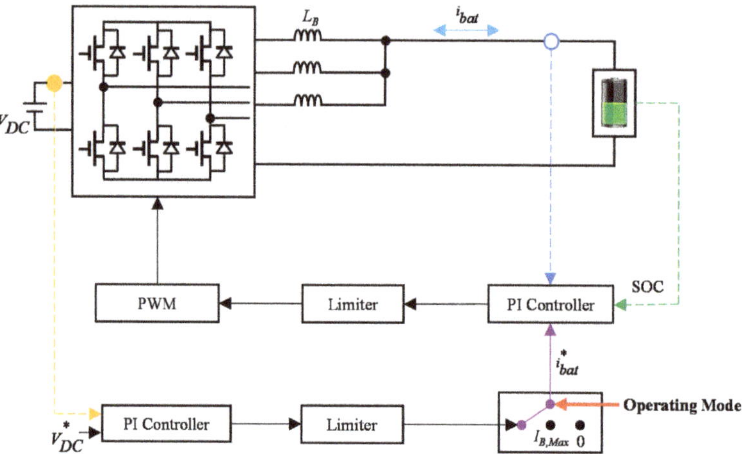

Figure 3. Configuration of battery agent.

To implement the power flow control, the battery agent uses two PI controllers in the outer control loop and inner control loop. Similar to the grid agent, the outer control loop regulates the DC-link voltage stably and produces the current reference of the battery agent. The inner control loop controls the battery charging/discharging currents in a bidirectional way. In addition to the battery current reference, the inner control loop uses the information on the battery current feedback, the battery SOC, and the operating modes of the battery agent. If the battery agent operates in V_{DC} control mode by charging or discharging, the battery current reference is determined from the outer PI control loop for the DC-link voltage regulation. The battery current reference is selected as $I_{B,Max}$ when the battery agent operates in charge with $I_{B,Max}$. Otherwise, the battery current reference is set to zero for IDLE mode operation of the battery agent.

2.3. Wind Power Agent

Figure 4 shows the configuration of the wind power agent. An L filter is interfaced between the PMSG and unidirectional AC-to-DC converter, and the output of AC-to-DC converter is connected to the DC-link. The wind power agent has two operating modes which are the maximum power point tracking (MPPT) mode and V_{DC} control mode. The MPPT mode in which the wind power agent operates mostly aims to draw the maximum power from the wind turbine. On the other hand, when the wind power is higher than the load demand and other power agents cannot absorb the surplus power from the wind power, the wind power agent initiates V_{DC} control mode to maintain the DC-link voltage of DCMG reliably. The V_{DC} control mode is implemented by two cascaded control loops, i.e., the outer loop PI controller for the DC-link voltage regulation and inner loop synchronous PI decoupling current controller. The MPPT mode is implemented by the MPPT algorithm, PI speed controller, and inner current control loop.

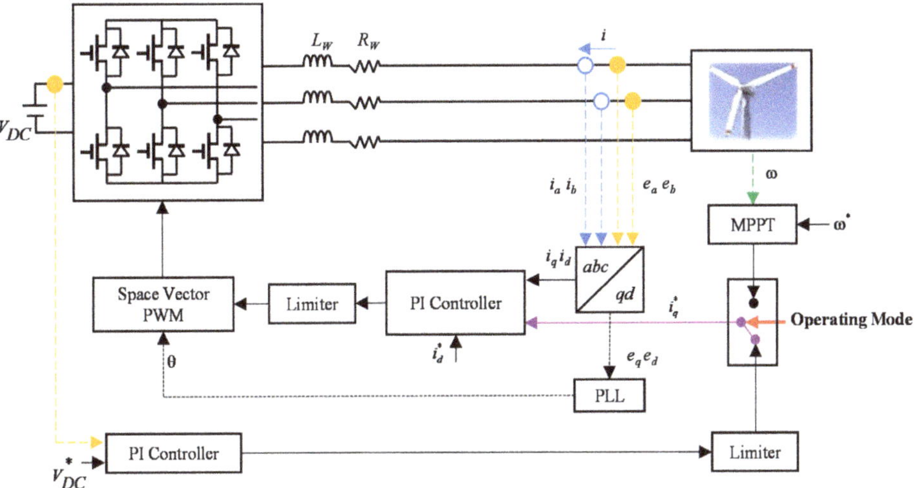

Figure 4. Configuration of wind power agent.

2.4. Load Agent

Figure 5 shows the configuration of the load agent. In a critical situation, the load agent can shed unnecessary load to prevent the DCMG from collapsing. After the termination of a critical situation, the load agent may reconnect the shedded load through the load reconnection algorithm. The load shedding algorithm is activated when all the agents cannot provide the demanded power of load. In this condition, the load agent disconnects load one by one from the least important load. This process lasts until the system power balance of DCMG is ensured by supplying the necessary load demand at some point. When the grid is reconnected to DCMG or the battery has a sufficient SOC level, the load reconnection algorithm is initiated to reconnect the shedded load [16].

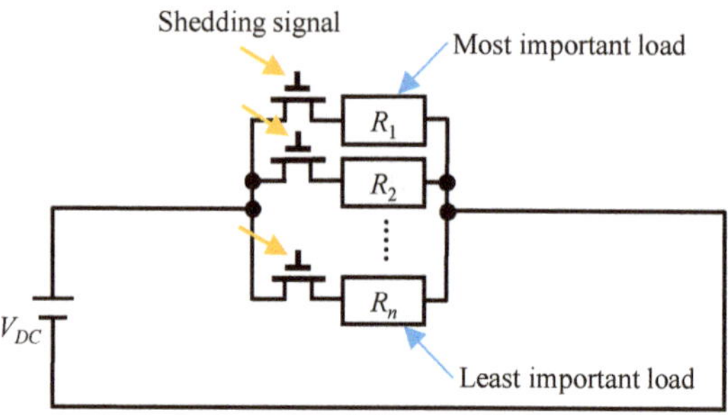

Figure 5. Configuration of load agent.

3. Proposed Hybrid DCMG Control Architecture and Power Flow Control Strategy

3.1. Hybrid DCMG Architecture

Figure 6 shows the concept of a hybrid control architecture in the DCMG system. In the hybrid control architecture, the centralized and distributed control schemes are combined. In this figure, the blue dashed line represents an HBC link for the centralized control architecture, while the black solid line represents an LBC link for the distributed control architecture. The exchange of information between the CC and power agents in the centralized control architecture is achieved through the HBC link. On the contrary, all the power agents exchange the information with adjacent neighbors in the distributed control architecture by using one-bit binary data format through the LBC link. In the centralized control of DCMG, the CC collects the information from all the power agents, processes the acquired data, and determines the operating modes of all the power agents to achieve a global optimum solution of the DCMG system. On the other hand, the distributed control of DCMG aims to improve the system reliability and robustness against a fault in the communication link.

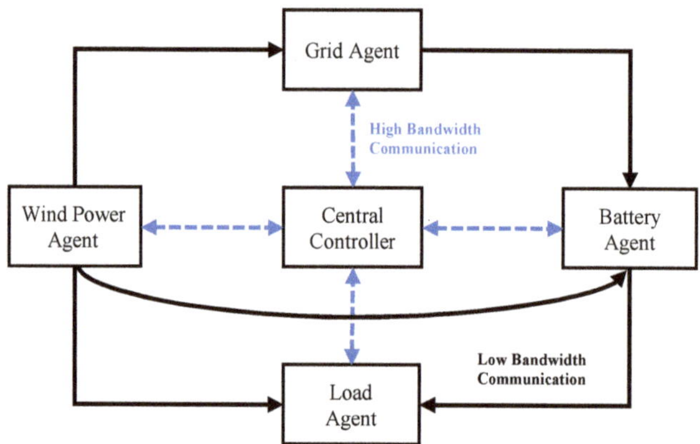

Figure 6. Concept of the hybrid DCMG control architecture.

In this study, the control mode transition between the centralized control and distributed control is determined based on the availability of the CC. Figure 7 describes the control mode transition in the hybrid DCMG architecture. During the normal condition without the CC fault, the CC operates the entire DCMG system by the centralized control architecture. Once a fault occurs in the CC, the distributed control takes control to operate the DCMG system.

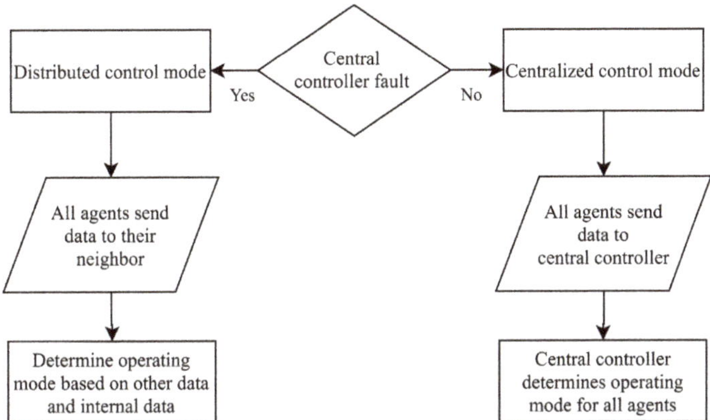

Figure 7. Control mode transition in the hybrid DCMG architecture.

In the centralized control mode, the data is exchanged between the CC and all power agents. All the power agents send specific data to the CC through the HBC link. At the same time, the CC also investigates the electricity price condition from the external sources like the internet. Based on the acquired information, the CC makes the best decision for the operating mode of all agents. Finally, the CC sends data which contains the agent operating mode along with a control signal (CS) to all the agents. Each power agent uses the CS value to determine the operation by the centralized control mode (CS signal is high) or distributed control mode (CS signal is low). When the CS signal is high, all the agents obviously use the operating mode given by the CC. On the other hand, when the CS signal is low, all the agents determine the operating modes based on the agent internal data and received information through the LBC link. In the distributed control mode, all the agents send specific data in a one-bit binary format to the adjacent neighbors. In this scheme, the data transfer by the HBC and LBC methods is accomplished every control period.

3.2. Power Flow Control Strategy

The power flow control strategy in the hybrid DCMG architecture is autonomously and reliably determined under both the centralized control and distributed control by using the relationship of supply-demand power. To determine the operating modes of power agents, several conditions and key parameters such as the availability of the CC and grid agent, battery SOC level, generated power from the wind power agent, and electricity price are primarily considered. In the proposed hybrid DCMG control scheme, the distributed control mode is composed of ten operating modes as the power flow control strategy. In the centralized control mode, an additional operating mode is added for the power flow control to consider explicitly the constraint of electricity price. Figure 8 shows the power flow control strategy in the distributed control mode. The detailed operating modes of each power agent for power flow control are listed in Table 1.

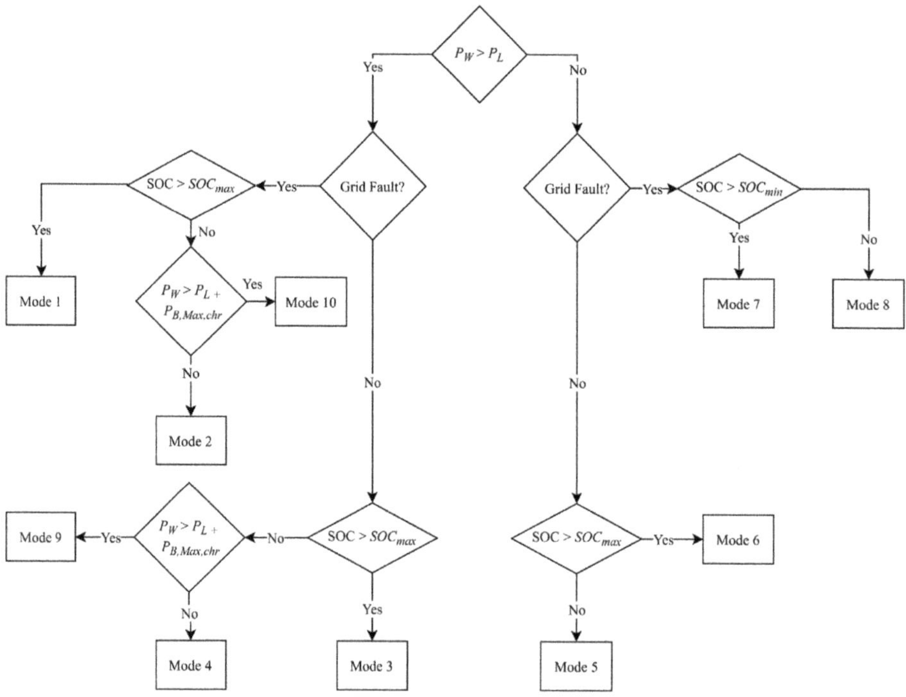

Figure 8. Power flow control strategy in the distributed control mode.

Table 1. Operating modes of each agent for power flow control strategy in the distributed control.

Mode	Wind Power Agent	Grid Agent	Battery Agent	Load Agent
1	V_{DC} control	Fault	IDLE	Normal/Reconnection
2	Maximum power point tracking (MPPT)	Fault	V_{DC} control by charging	Normal/Reconnection
3	MPPT	V_{DC} control (inverter (INV))	IDLE	Normal/Reconnection
4	MPPT	V_{DC} control (converter (CON))	Charge with $I_{B,Max}$	Normal/Reconnection
5	MPPT	V_{DC} control (CON)	Charge with $I_{B,Max}$	Normal/Reconnection
6	MPPT	V_{DC} control (CON)	IDLE	Normal/Reconnection
7	MPPT	Fault	V_{DC} control by discharging	Normal/Reconnection
8	V_{DC} control	Fault	IDLE	Shedding
9	MPPT	V_{DC} control (INV)	Charge with $I_{B,Max}$	Normal/Reconnection
10	V_{DC} control	Fault	Charge with $I_{B,Max}$	Normal/Reconnection

The power flow control strategy of DCMG in the distributed control mode is explained as follows.

- Operating mode 1: This operating mode occurs when the generated wind power P_W is higher than the required load power demand P_L and the battery SOC is higher than SOC_{max} under a fault in the grid agent. Because P_W cannot be absorbed by other agents, the operating mode of the wind power agent is changed into V_{DC} control mode while the battery agent is in IDLE mode.

- Operating mode 2: This operating mode is selected when the generated wind power P_W is higher than the required load demand P_L, and the battery is not fully charged (SOC < SOC_{max}) under the grid fault. This operating mode also requires an additional condition that P_W is less than the sum of load demand P_L and the maximum charging power of battery $P_{B,Max,chr}$. In this case, the wind power agent operates with the MPPT mode and the battery agent operates with the V_{DC} control mode by charging to regulate the DC-link voltage.

- Operating mode 3: This operating mode is used when the grid agent is normal, the generated wind power P_W is higher than the required load demand P_L, and the battery has been fully charged. In this case, the grid agent operates in V_{DC} control INV mode to absorb the surplus power from the wind power agent which operates in the MPPT mode, while the battery agent operates in IDLE mode.
- Operating mode 4: This operating mode requires the conditions that the grid agent is normal and the generated wind power P_W is higher than the required load demand P_L. Additional conditions to select this operating mode are the battery SOC is less than SOC_{max}, and P_W is less than the sum of load demand P_L and the maximum charging power of battery $P_{B,Max,chr}$. In this mode, the wind power agent operates in the MPPT mode, the battery agent charges the battery with the maximum current $I_{B,Max}$, and the grid agent provides the insufficient power from the grid by operating in V_{DC} control CON mode.
- Operating mode 5: This operating mode requires the conditions that the generated wind power P_W is less than the required load demand P_L and the battery SOC is less than SOC_{max} without a fault in the grid agent. The wind power agent draws the maximum power by operating in the MPPT mode, the battery agent charges the battery with the maximum current $I_{B,Max}$, and the grid agent provides the insufficient power from the grid by operating in V_{DC} control CON mode.
- Operating mode 6: Operating mode 6 follows the same flow with the operating mode 5. The only difference is that this mode is used with the condition of $SOC > SOC_{max}$. In this operating mode, while both the wind power agent and grid agent maintain their operations with the same as in the operating mode 5, the battery agent changes its operation into IDLE mode.
- Operating mode 7: When the generated wind power P_W is less than the required load demand P_L and the battery SOC is higher than SOC_{min} under a fault in the grid agent, the operating mode 7 is chosen. In this mode, the wind power agent operates in the MPPT mode and the battery agent regulates the DC-link voltage by operating in V_{DC} control by discharging.
- Operating mode 8: Operating mode 8 follows the same flow with the operating mode 7. The only difference is that this mode is used with the condition of $SOC < SOC_{min}$. When the DCMG system operates in operating mode 7 for a long period, the battery SOC is gradually decreased and reaches the threshold level of SOC_{min}. In this case, the battery agent can not supply the power any more. To avoid the DCMG system collapse under this critical condition, the load shedding algorithm is activated by the load agent. The load agent removes less important loads, the battery agent is in IDLE mode, and the wind power agent changes its operating mode into V_{DC} control mode to regulate the DC-link voltage.
- Operating mode 9: Operating mode 9 is selected by following the same flow with the operating mode 4. The difference is that this mode is determined when the generated wind power P_W is greater than the sum of load demand P_L and the maximum charging power of battery $P_{B,Max,chr}$. In this mode, the wind power agent still operates in the MPPT mode, the battery agent charges the battery with the maximum current $I_{B,Max}$. However, contrary to operating mode 4, the grid agent provides excessive power into the grid by operating in V_{DC} control INV mode.
- Operating mode 10: Operating mode 10 follows the same flow with the operating mode 2. The only difference is that this mode is chosen with the condition of $P_W > P_L + P_{B,Max,chr}$. In this condition, the generated wind power P_W is higher than the required load demand P_L, and the battery is not fully charged ($SOC < SOC_{max}$) under the grid fault. In addition, since P_W can supply additional charging power of battery, the battery agent charges the battery with the maximum current $I_{B,Max}$, and the wind power agent changes its operating mode into V_{DC} control mode to regulate the DC-link voltage.

Figure 9 shows the power flow control strategy in the centralized control mode. The detailed operating modes of each power agent for power flow control are listed in Table 2. In the centralized control, all the operating modes are the same as those of the distributed control except for the operating mode 11. The study in [33] shows that 40% of the world wide energy usage comes from the energy consumption in building area. As the number of building is increased, many researchers have been interested in the optimization of energy savings by considering electricity price [21–26]. The research in [34] reports that an effective algorithm on EMS can reduce electricity price with the average of 23.1%. Another study in [10] shows that the use of EMS in the centralized control mode leads to cheaper electricity price than in the distributed control mode. Generally, the electricity cost saving is influenced by many factors such as the load profile, distributed power generation, pricing method, optimization algorithm, and total usage of grid power during high electricity cost condition. In spite of such a difficulty, a particular study in [35] shows the electricity cost saving up to 28% by restraining the electricity injection from the grid during high electricity cost period.

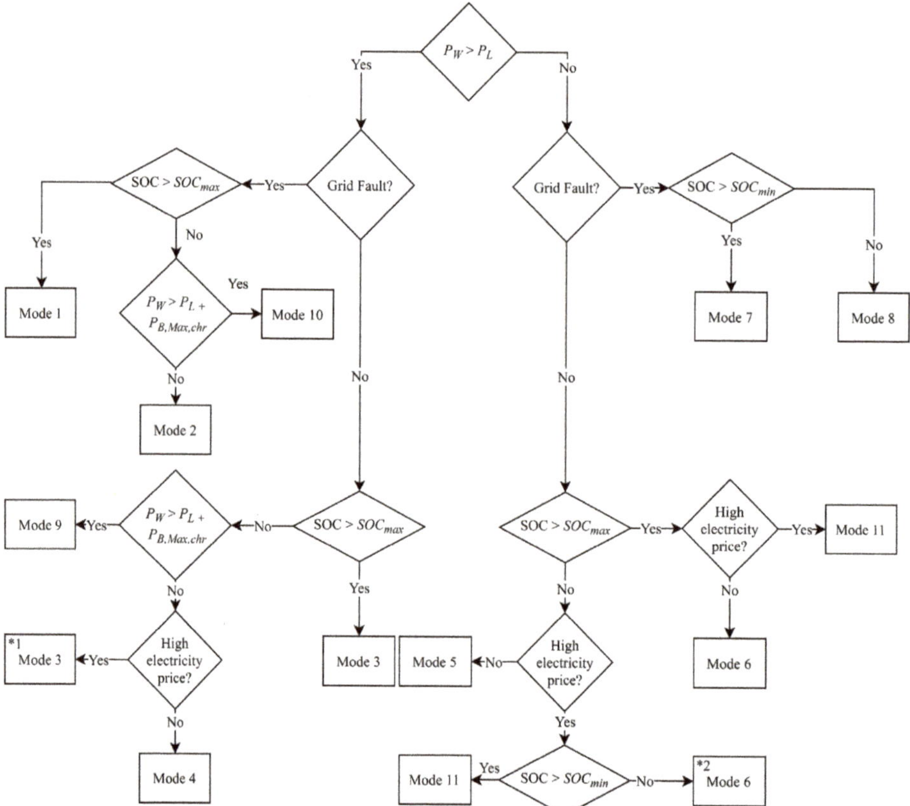

Figure 9. Power flow control strategy in the centralized control mode.

Table 2. Operating modes of each agent for power flow control strategy in the centralized control.

Mode	Wind Power Agent	Grid Agent	Battery Agent	Load Agent
1	V_{DC} control	Fault	IDLE	Normal/Reconnection
2	MPPT	Fault	V_{DC} control by charging	Normal/Reconnection
3	MPPT	V_{DC} control (INV)	IDLE	Normal/Reconnection
4	MPPT	V_{DC} control (CON)	Charge with $I_{B,Max}$	Normal/Reconnection
5	MPPT	V_{DC} control (CON)	Charge with $I_{B,Max}$	Normal/Reconnection
6	MPPT	V_{DC} control (CON)	IDLE	Normal/Reconnection
7	MPPT	Fault	V_{DC} control by discharging	Normal/Reconnection
8	V_{DC} control	Fault	IDLE	Shedding
9	MPPT	V_{DC} control (INV)	Charge with $I_{B,Max}$	Normal/Reconnection
10	V_{DC} control	Fault	Charge with $I_{B,Max}$	Normal/Reconnection
11	MPPT	IDLE	V_{DC} control by discharging	Normal/Reconnection

For the purpose of minimizing the customer electricity cost in the DCMG system, the electricity price constraint is also considered in this work. In the proposed scheme, the DCMG system avoids the power supplied from the grid during periods of high electricity price, absorbing the least power only in the critical condition to prevent the system from collapsing. In the presented power flow control strategy, contrary to the distributed control, there is additional operating mode 3 (denoted by *1) in the centralized control mode by considering the constraint of electricity price. As shown in the left part of Figure 9 for the case $P_W > P_L$, the high electricity price condition in the centralized control changes the DCMG operation from operating mode 4 to operating mode 3. Whereas the operating mode 4 is selected for the DCMG operation in the normal electricity price condition, the DCMG system changes its operation into the operating mode 3 under high electricity price condition to avoid additional cost. Instead of operating converter mode to supply the power from the grid, the grid agent operates in inverter mode in operating mode 3 to avoid the power supply from the grid.

Based on the constraint of electricity price, the operating mode 11 is considered for the DCMG operation. The first condition to choose the operating mode 11 follows the same flow with the operating mode 6. The utility condition having high electricity price divides the DCMG operation into the operating mode 6 and operating mode 11. Instead of operating converter mode to supply the power from the grid, the grid agent operates in IDLE mode to avoid the use of power from the grid. The second condition to choose the operating mode 11 follows the same flow with the operating mode 5. Because of high electricity price condition, the DCMG selects the operating mode 11 rather than operating mode 5 to operate the grid agent in IDLE mode instead of converter mode. Similarly, there is additional operating mode 6 (denoted by *2) by considering the constraint of electricity price as shown in Figure 9 for the case $P_W < P_L$. In spite of high electricity price, the grid agent supplies the power into the DCMG as converter mode only when the battery has an extremely low SOC level in this additional operating mode 6 (denoted by *2).

4. Simulation Results

To verify the effectiveness of the power flow strategy in the proposed hybrid DCMG control scheme under the CC fault, unreliable grid connection, and uncertainty of electricity price, the simulation has been done using the PSIM software. Table 3 lists the system parameters used for simulation. Simulations are carried out for four cases which are low battery SOC case, safe regional battery SOC case, high battery SOC case, and high electricity price case. In addition to that, the DCMG system scalability is tested by adding and removing additional power agents during a certain period.

Table 3. System parameters of DCMG.

Unit	Parameters	Symbol	Value
Grid agent	Grid voltage	V_G^{rms}	220 V
	Grid frequency	F_G	60 Hz
	Transformer Y/Δ	T	380/220 V
	Inverter-side inductance of LCL filter	L_1	1.7 mH
	Inverter-side resistance of LCL filter	R_1	0.5 Ω
	Grid-side inductance of LCL filter	L_2	1.7 mH
	Grid-side resistance of LCL filter	R_2	0.5 Ω
	Filter capacitance of LCL filter	C_f	4.5 μF
Wind power agent	PMSG stator resistance	R_S	0.64 Ω
	PMSG dq-axis inductance	L_{dq}	0.82 mH
	PMSG number of poles	P	6
	PMSG inertia	J	0.111 kgm^2
	PMSG flux linkage	ψ	0.18 Wb
	Converter filter inductance	L_W	7 mH
	Converter filter resistance	R_W	0.2 Ω
Battery agent	Maximum allowable current	$I_{B,Max}$	3 A
	Maximum SOC	SOC_{max}	90%
	Minimum SOC	SOC_{min}	20%
	Rated capacity	C	30 Ah
	Maximum voltage	V_B^{max}	265 V
	Converter filter inductance L	L_B	7 mH
Load agent	Power of load 1	P_{L1}	200 W
	Power of load 2	P_{L2}	200 W
	Priority level: load 1 > load 2	-	-
DC-link	Nominal voltage	V_{DC}^{nom}	400 V
	Capacitance	C_{DC}	4 mF

In the simulations for three different battery SOC cases, the grid has a fault from $t = 0.5$ s to $t = 1.5$ s. It is also assumed that the CC fault occurs from $t = 0$ to $t = 0.25$ s, from $t = 0.75$ s to $t = 1.25$ s, and from $t = 1.75$ s to $t = 2$s. Initially, the generated wind power P_W is less than the required load demand P_L until $t = 1$ s. After that, P_W is increased and greater than P_L until $t = 2$ s. In these simulations, the battery agent current, the grid agent current, the wind power agent current, and the load agent current are represented as I_B, I_G, I_W, and I_L with the reference direction as in Figure 1, in which the positive current denotes the current into the agent out of the DC-link and the negative current denotes the current from the agent into the DC-link.

4.1. Case of Low Battery SOC

Figure 10 shows the simulation results of the power flow strategy in the proposed hybrid DCMG control scheme for the case of low battery SOC level. Initially, the battery SOC is less than SOC_{min} and all the agents start the control operation at $t = 0.05$ s. Since the generated wind power P_W is less than the load demand P_L and the battery SOC is very small without a grid fault, the DCMG system initially starts the operation with the operating mode 5, which lasts until a fault occurs in the grid at $t = 0.5$ s.

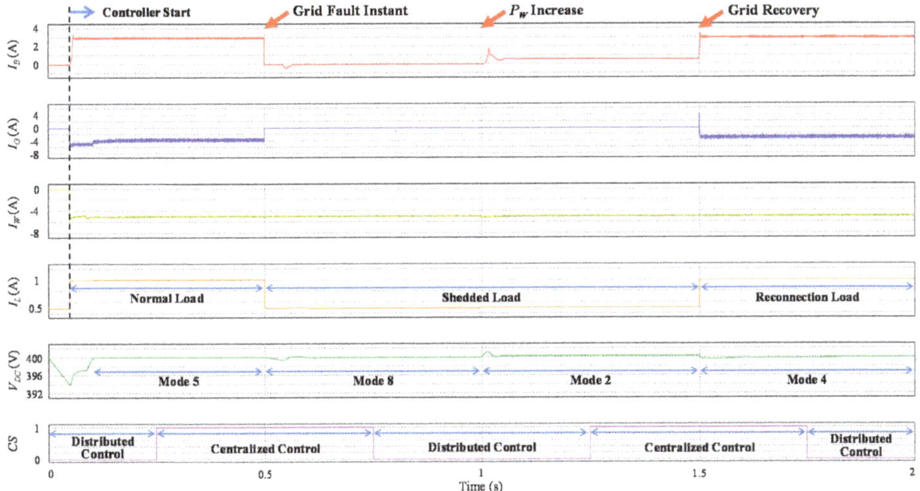

Figure 10. Simulation results for the case of battery state-of-charge (SOC) level lower than the minimum value.

From this instant, the DCMG operation is changed into the operating mode 8 because both the wind power agent and battery agent cannot provide the required load demand P_L. In this case, the load shedding algorithm is activated by the load agent to remove less important load while the battery agent changes the operation into IDLE mode, and the wind power agent is changed into V_{DC} control mode to regulate the DC-link voltage.

When P_W is increased higher than P_L at t = 1 s, the DCMG changes its operation into the operating mode 2. In this mode, the wind power agent operates in the MPPT mode, and the battery agent regulates the DC-link voltage by operating with the V_{DC} control mode by charging. Because the battery SOC level is still lower than SOC_{min}, the load reconnection algorithm cannot be activated by the load agent.

As the grid is recovered from fault at $t = 1.5$ s, the DCMG operation is changed into the operating mode 4 and the load reconnection is activated by the load agent at the same time. Consequently, the wind power agent maintains its operation in the MPPT mode, the battery agent charges the battery with the maximum current $I_{B,Max}$, and the grid agent supplies the inadequate power by operating in V_{DC} control CON mode.

As stated earlier, it is assumed that the CC has a fault from $t = 0$ to $t = 0.25$ s, from $t = 0.75$ s to $t = 1.25$ s, and from $t = 1.75$ s to $t = 2$s in this simulation. During the CC fault, the distributed control scheme takes the role to operate the DCMG. As shown in this result, in spite of the transition between the distributed control mode and centralized control mode, all the agents operate stably within the DCMG system without severe transient, which validates that the proposed scheme contributes to improve the system reliability and robustness against a fault in the communication link of the centralized control.

4.2. Case of Safe Regional Battery SOC

Figure 11 shows the simulation results of the power flow strategy in the proposed hybrid DCMG control scheme for the case of safe regional SOC level between SOC_{min} and SOC_{max}. Similarly, all the agents start the operation at $t = 0.05$ s and the DCMG initially starts with the operating mode 5. This simulation test considers two cases of P_W. Figure 11a represents the case for $P_W < P_L + P_{B,Max,chr}$, and Figure 11b represents the case for $P_W > P_L + P_{B,Max,chr}$.

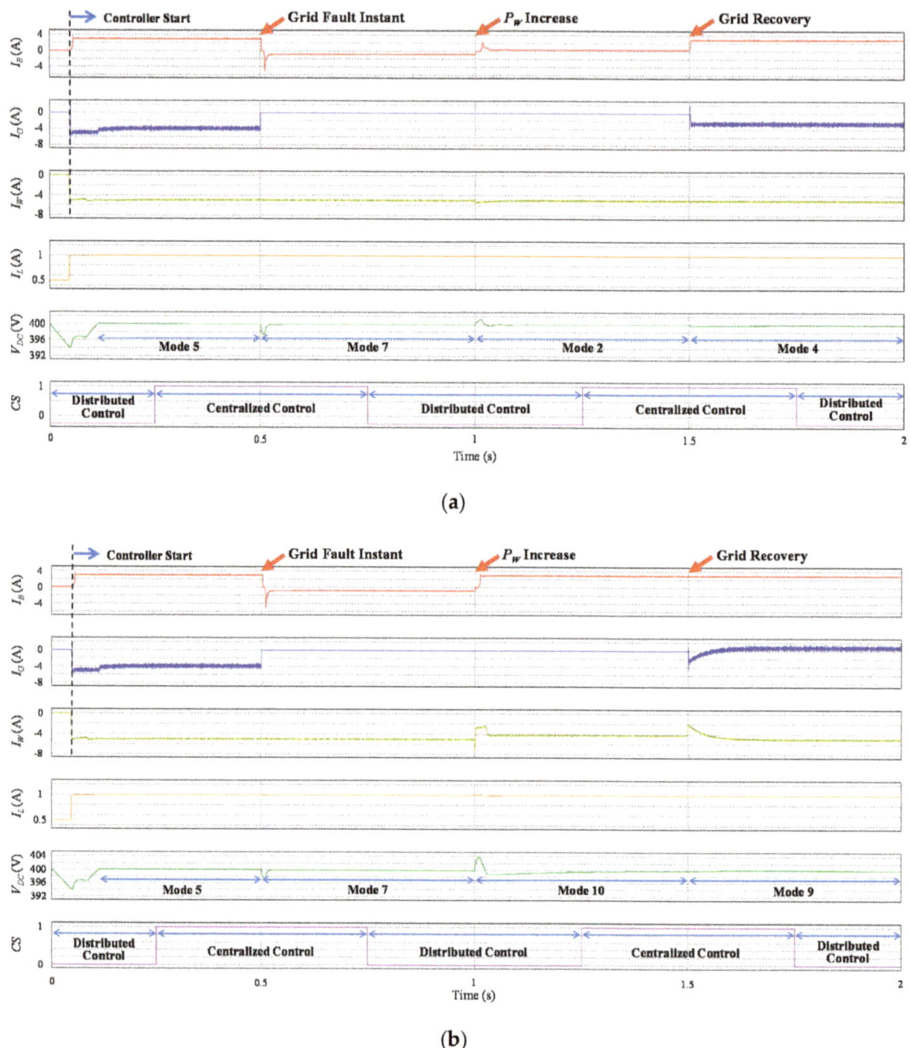

Figure 11. Simulation results for the case of safe regional battery SOC level: (**a**) Case of $P_W > P_L$ and $P_W < P_L + P_{B,Max,chr}$; (**b**) Case of $P_W > P_L$ and $P_W > P_{B,Max,chr}$.

As soon as a fault occurs in the grid at $t = 0.5$ s, the DCMG system operation is changed into the operating mode 7 in both cases. In this mode, the wind power agent still operates in the MPPT mode and the battery agent changes its operating mode to V_{DC} control by discharging to regulate the DC-link voltage.

As the generated wind power P_W is increased at $t = 1$ s, the DCMG operation shows different behavior in Figure 11a,b. Under the condition of $P_W < P_L + P_{B,Max,chr}$, the operating mode 2 is used before the grid is recovered. In this mode, the wind power agent operates in the MPPT mode and the excessive power from P_W is absorbed by the battery agent which operates in V_{DC} control mode by charging. On the other hand, under the condition of $P_W > P_L + P_{B,Max,chr}$, the operating mode 10 is used, in which the battery agent charges the battery with the maximum current $I_{B,Max}$ and the wind power agent regulates the DC-link voltage by operating in V_{DC} control mode.

When the grid is recovered $t = 1.5$ s, the DCMG operation is changed into the operating mode 4 in Figure 11a, and the operating mode 9 in Figure 11b. In the operating mode 4, the wind power agent maintains its operating mode, the battery agent still charges the battery with the maximum current $I_{B,Max}$, and the grid agent operates in V_{DC} control CON mode to provide the deficient power in Figure 11a. In the operating mode 9, the wind power agent operates in the MPPT mode, the battery agent maintains its operating mode, the grid agent regulates the DC-link voltage by operating in V_{DC} control INV mode.

A smooth and reliable transition between the centralized control and distributed control caused by the CC fault is also confirmed from this simulation without affecting the overall performance.

4.3. Case of High Battery SOC

Figure 12 shows the simulation results of the proposed hybrid DCMG control scheme for the case of high battery SOC level. Because the battery has been fully charged from the beginning, the DCMG starts with the operating mode 6 and all the agents start the control $t = 0.05$ s. In the operating mode 6, the wind power agent operates in the MPPT mode, the battery agent is in IDLE mode, and the grid agent operates in V_{DC} control CON mode to regulate the DC-link voltage.

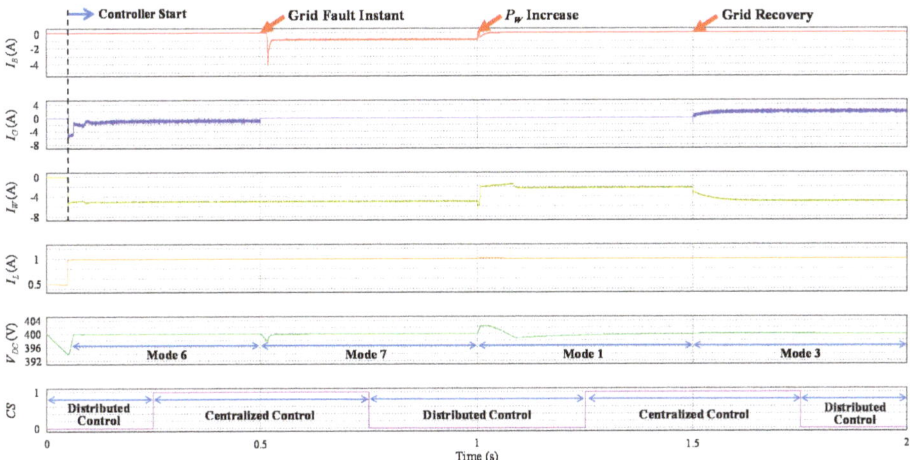

Figure 12. Simulation results for the case of battery SOC level higher than the maximum value.

In the presence of the grid fault at $t = 0.5$ s, while the wind power agent remains in the MPPT mode, the battery agent changes its operation into V_{DC} control by discharging to regulate the DC-link voltage, which realizes the operating mode 7. As the generated wind power P_W is increased higher than P_L at $t = 1$ s and the battery agent is still fully charged, the wind power agent executes V_{DC} control to regulate the DC-link voltage.

As the grid is recovered from fault at $t = 1.5$ s, the surplus wind power can be injected into the grid by the grid agent. At this instant, the DCMG system is controlled by the operating mode 3, in which the wind power agent changes the operation into the MPPT mode, the battery agent is still in IDLE mode, and the grid agent uses V_{DC} control INV mode to control the DC-link voltage.

It is also evident in this simulation that the transition between the centralized control and distributed control in a hybrid DCMG structure is very smooth and stable even in the presence of a sudden fault in the CC.

4.4. Case of Electricity Price Constraint

Figure 13 shows the simulation results with the constraint of high electricity price with the assumption that the CC has the information on the time-varying price of utility. This simulation test considers two cases of P_W. Figure 13a represents the case for $P_W > P_L$, and Figure 13b represents the case for $P_W < P_L$. Similar to the previous cases, all agents start the operation at $t = 0.05$ s in both cases.

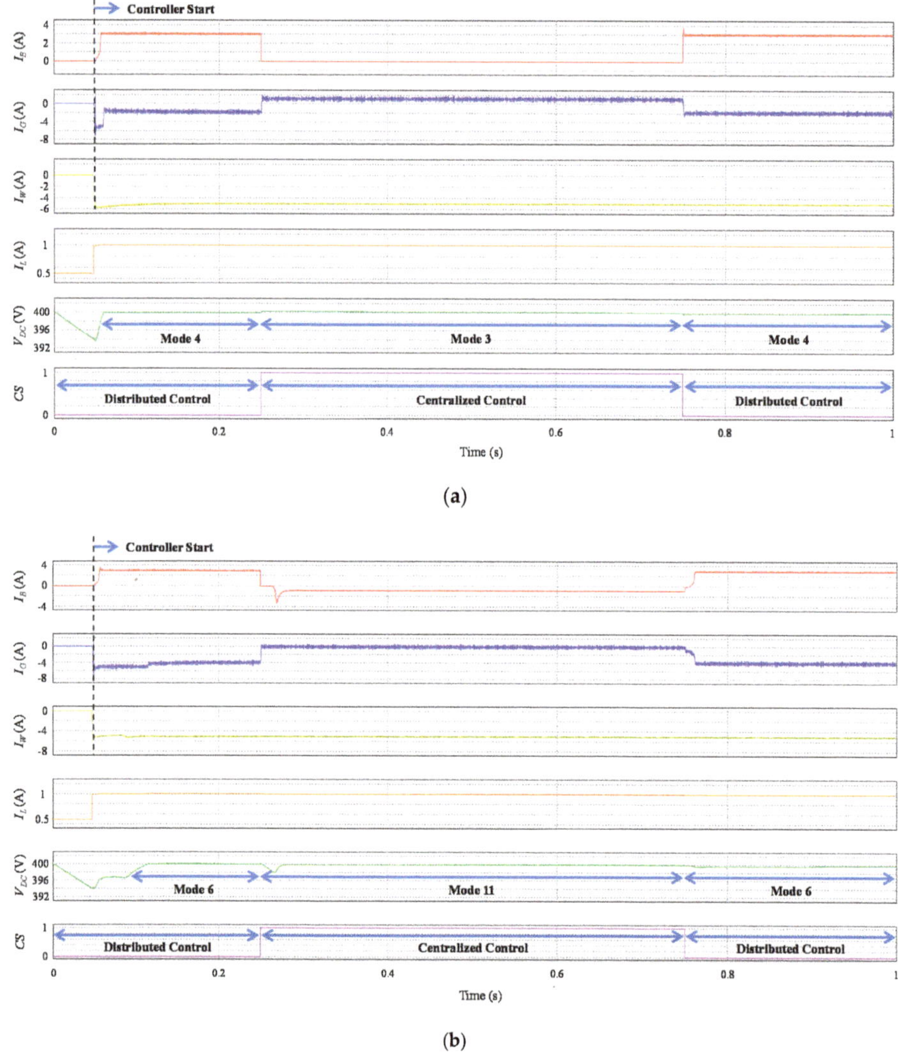

Figure 13. Simulation results with the constraint of high electricity price. (**a**) Operating mode transition for $P_W > P_L$; (**b**) Operating mode transition for $P_W < P_L$.

When the generated wind power P_W has the condition of $P_W > P_L + P_{B,Max,chr}$, the DCMG starts the operation with the operating mode 4, in which the grid agent operates in V_{DC} control CON mode. However, as soon as the distributed control is changed into the centralized control as a result of the recovery of the CC fault at $t = 0.25$ s, the DCMG operation is changed into the operating mode 3 as shown in Figure 13a. In the operating mode 3 of the centralized control, the battery agent changes the operation into IDLE mode to avoid the use of additional power from the grid, and the surplus power from P_W is injected into the grid by operating in V_{DC} control INV mode. If the fault occurs in the CC at $t = 0.75$ s, the DCMG system operation is returned to the previous operation.

With the condition of $P_W < P_L$, the DCMG starts with the operating mode 6, in which the battery agent is in IDLE mode and the grid agent operates in V_{DC} control CON mode. As soon as the distributed control is changed into the centralized control as a result of the recovery of the CC fault at $t = 0.25$ s, the DCMG operation is changed into the operating mode 11 as shown in Figure 13b. As a result, the grid agent stops the converter mode operation to avoid the use of power from the grid. Instead, the deficient power in the DCMG is supplied by battery with V_{DC} control mode by discharging. Similarly, the DCMG system operation is returned to the previous operation in case of the CC fault.

4.5. Scalability Test

The scalability of DCMG has been one of the important issues in DCMG structure. The scalability test is performed in this study based on the method in [7] by including an additional converter to the DCMG system at steady-state. Figure 14 shows the simulation result of the proposed scheme for the scalability test. In this simulation, the second battery agent is added at $t = 0.3$ s, and removed at $t = 0.9$ s. The second wind power agent is added at $t = 0.4$ s, and removed at $t = 0.7$ s. The second wind turbine operates in the MPPT mode and the second battery agent operates in charge with $I_{B,Max}$ mode. In addition to that, a fault occurs in the grid at $t = 0.5$ s, which forces the DCMG system to operate in the islanded mode. It is confirmed from this simulation that the DCMG can still maintain the DC-link voltage at the nominal value stably in spite of the addition as well as the removal of additional power agents even under the grid-connected and islanded cases.

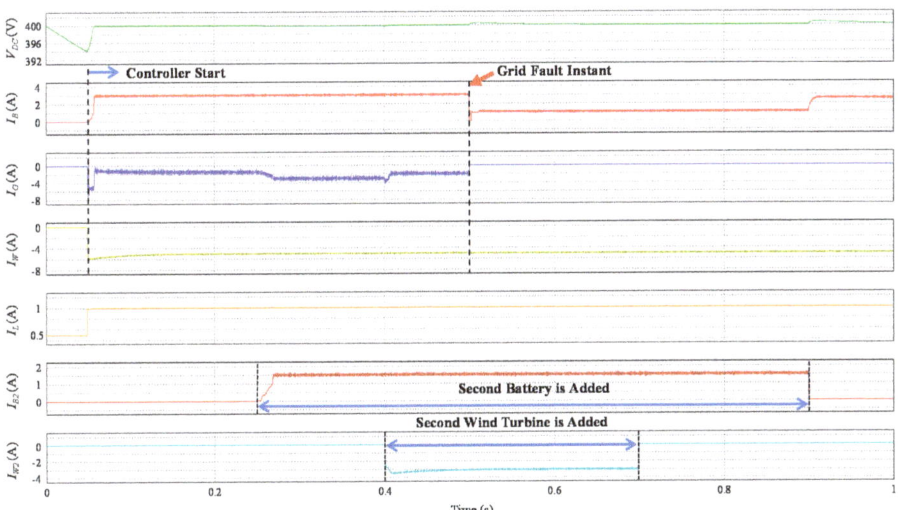

Figure 14. Simulation result for scalability when other agents are added into DCMG.

In all the simulation tests, the maximum deviation of the DC-link voltage from the nominal value of 400 V is measured as 1.5325% in Figures 10–13b. In Figures 13a and 14, it is 1.6275% and 1.5225%, respectively. The maximum deviation of the DC-link voltage from the nominal value is quite unnoticeable.

5. Experimental Results

In order to verify the effectiveness of the power flow strategy in the proposed hybrid DCMG architecture, the experiments were conducted by using a laboratory DCMG testbed system. Figure 15 shows the configuration of the experimental DCMG system. The entire system consists of a dc load and three power converters to connect the main grid source, lithium-ion battery, and wind turbine emulator to the DC-link. The wind turbine emulator is constructed by a PMSG and an induction machine drive system. DC load is composed of two 800 Ω resistors in parallel connection, and the magnetic contactor works as a switch to shed and reconnect the load. A 32-bit floating-point digital signal processor (DSP) TMS320F28335 controller is employed to implement the proposed power flow control strategy as well as to control the power converters in each agent.

Figure 15. Configuration of the experimental DCMG system.

5.1. Case of Low Battery SOC

Figure 16 shows the experimental results of the power flow strategy in the proposed hybrid DCMG control scheme when the battery SOC level is lower than SOC_{min} and the grid fault occurs. Figure 16a shows the operating mode transition from 5 to 8 caused by the grid fault. Figure 16b,c show the steady-state responses at the operating mode 5 and 8, respectively. Initially, P_W is less than P_L and the battery SOC level is lower than SOC_{min}. In the operating mode 5, the wind power agent is in the MPPT mode, the battery agent is in charge with $I_{B,Max}$, and the grid agent is in V_{DC} control CON mode for the purpose of providing the insufficient power from the grid by regulating the DC-link voltage. When a fault occurs in the grid, it is impossible for the battery agent to supply the required power to the load agent, which makes the operating mode transition from 5 to 8. As a consequence, the load agent sheds unnecessary load, the battery agent is in IDLE, and the wind power agent is in V_{DC} control mode to maintain V_{DC} at the nominal value. The activation of load shedding algorithm by the load agent causes 15 ms of time delay due to the coil in the magnetic contactor as is seen in Figure 16a. A small delay (3 ms) is also caused by the data transfer and observed in the experimental results. However, as can be shown in the transient response in Figure 16a, the operating mode transition is quite smooth and the DC-link voltage regulation is very stable. The steady-state responses in Figure 16b,c also validate a reliable operation of the DCMG system.

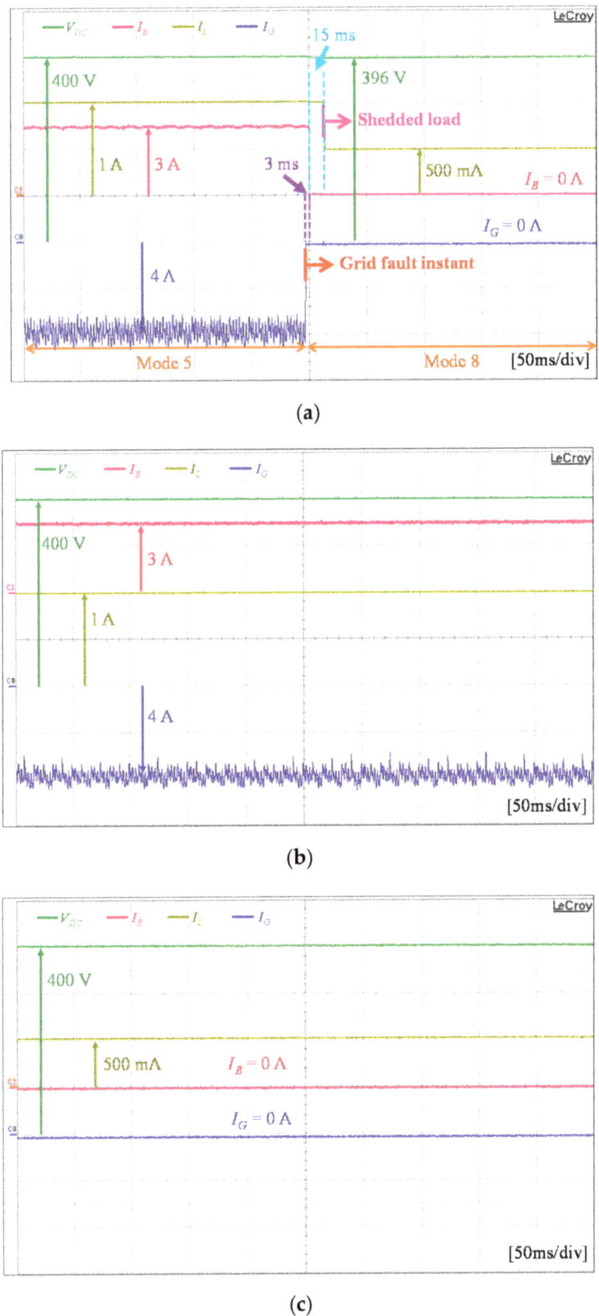

Figure 16. Experimental results when the battery SOC level is lower than SOC_{min} and the grid fault occurs. (**a**) Operating mode transition from 5 to 8; (**b**) Steady-state response at operating mode 5; (**c**) Steady-state response at operating mode 8.

5.2. Case of Safe Regional Battery SOC

Figure 17 shows the experimental result of the power flow strategy in the proposed hybrid DCMG control scheme for the case of safe regional SOC level. Initially, the generated wind power P_W is less than the load demand P_L, and the battery SOC level in the middle of SOC_{min} and SOC_{max}. Without the grid fault, the DCMG starts the operation with the operating mode 5. In this mode, the wind power agent is in the MPPT mode, the battery agent is in charge with $I_{B,Max}$, and the grid agent is in V_{DC} control CON mode. When a fault occurs in the grid, the DCMG changes the operation into the operating mode 7. Figure 17a shows the operating mode transition from 5 to 7 caused by the grid fault. Figure 17b,c show the steady-state responses at the operating mode 5 and 7, respectively. As a result of the operating mode transition into 7, the battery agent executes V_{DC} control mode by discharging to provide the load remand. These waveforms also confirm a smooth operating mode transition and stable operation of the proposed hybrid DCMG control scheme.

5.3. Case of High Battery SOC

Figure 18 shows the experimental result of the proposed hybrid DCMG control scheme for the case of high battery SOC level. Initially, the generated wind power P_W is less than the load demand P_L, the battery SOC is higher than SOC_{max}, and the grid is in the normal condition, which results in the operating mode 6. In this mode, the wind power agent is in the MPPT mode, the battery agent is in IDLE mode, and the grid agent is in V_{DC} control CON mode. As soon as a fault occurs in the grid, the DCMG operation is changed into the operating mode 7. Figure 18a shows the operating mode transition from 6 to 7 caused by the grid fault. Figure 18b,c show the steady-state responses at the operating mode 6 and 7, respectively. As soon as the operating mode is changed, the battery agent takes the role to regulate the DC-link voltage by operating in the V_{DC} control mode by discharging. Similarly, the usefulness of the proposed hybrid DCMG control scheme can be observed.

5.4. Case of Electricity Price Constraint

Figure 19 shows the experimental results of the proposed hybrid DCMG control scheme with the constraint of high electricity price. With the condition that the generated wind power P_W is less than the load demand P_L, and the battery is fully charged without the grid fault, the DCMG starts with the operating mode 6 at first. If the CC detects high electricity price condition at this instant, the grid agent operation is changed from V_{DC} control CON mode into the IDLE mode to avoid the increase of utility cost. Instead, the battery is discharged to compensate the deficient power in the DCMG system while regulating the DC-link voltage with V_{DC} control mode by discharging. In this case, the resultant DCMG operation is the operating mode 11. Obviously, if the electricity price is returned to the normal condition, the operation of the DCMG system is also restored into the operating mode 6. Figure 19a shows the experimental result of the DCMG operating mode transition when the CC detects high electricity price condition. Figure 19b,c show the steady-state responses at the operating mode 6 and 11, respectively.

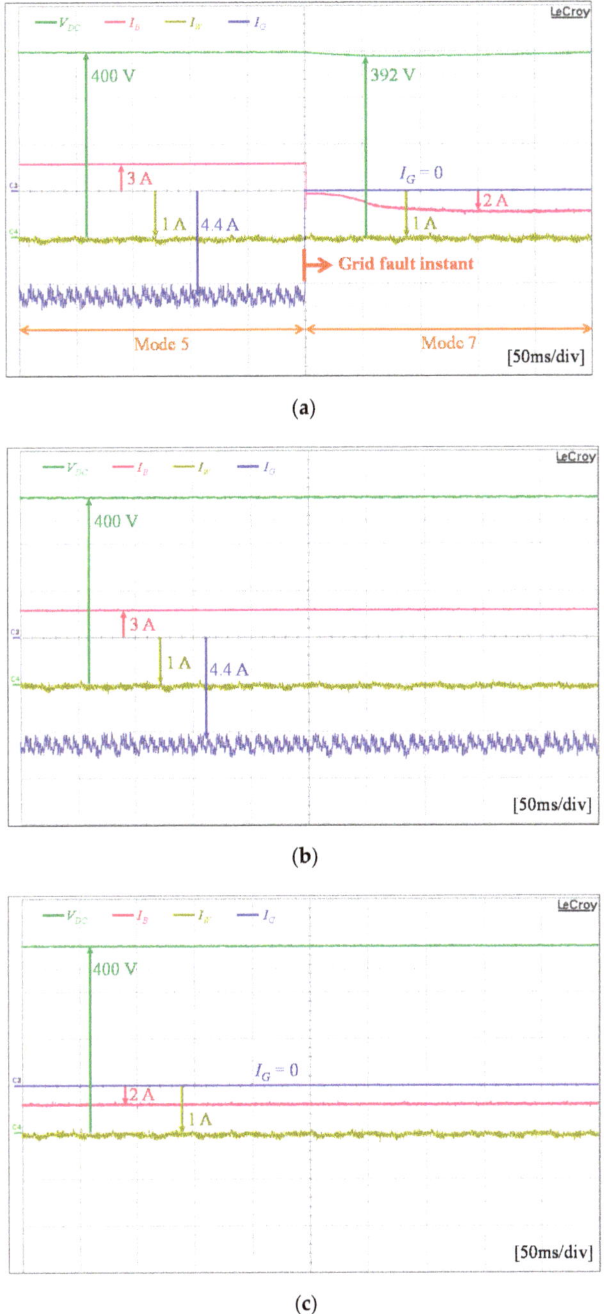

(**a**)

(**b**)

(**c**)

Figure 17. Experimental results when DCMG operates in the normal SOC level and the grid fault occurs. (**a**) Operating mode transition from 5 to 7; (**b**) Steady-state response at operating mode 5; (**c**) Steady-state response at operating mode 7.

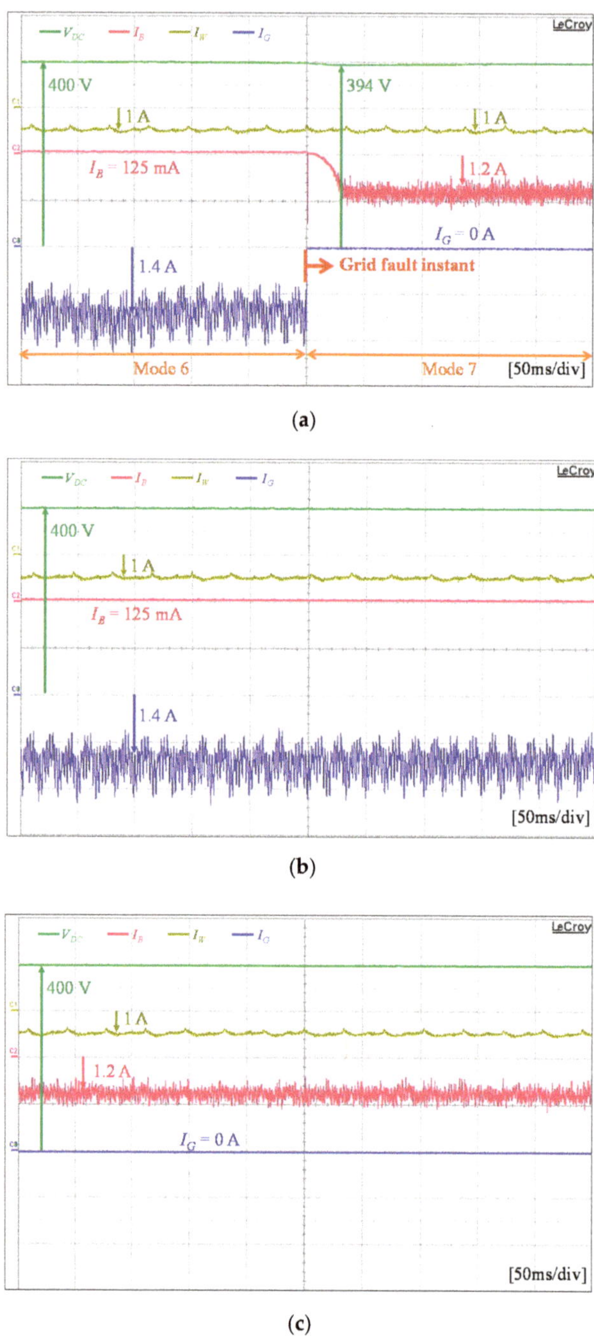

Figure 18. Experimental results when the battery SOC level is higher than SOC_{max} and grid fault occurs. (**a**) Operating mode transition from 6 to 7; (**b**) Steady-state response at operating mode 6; (**c**) Steady-state response at operating mode 7.

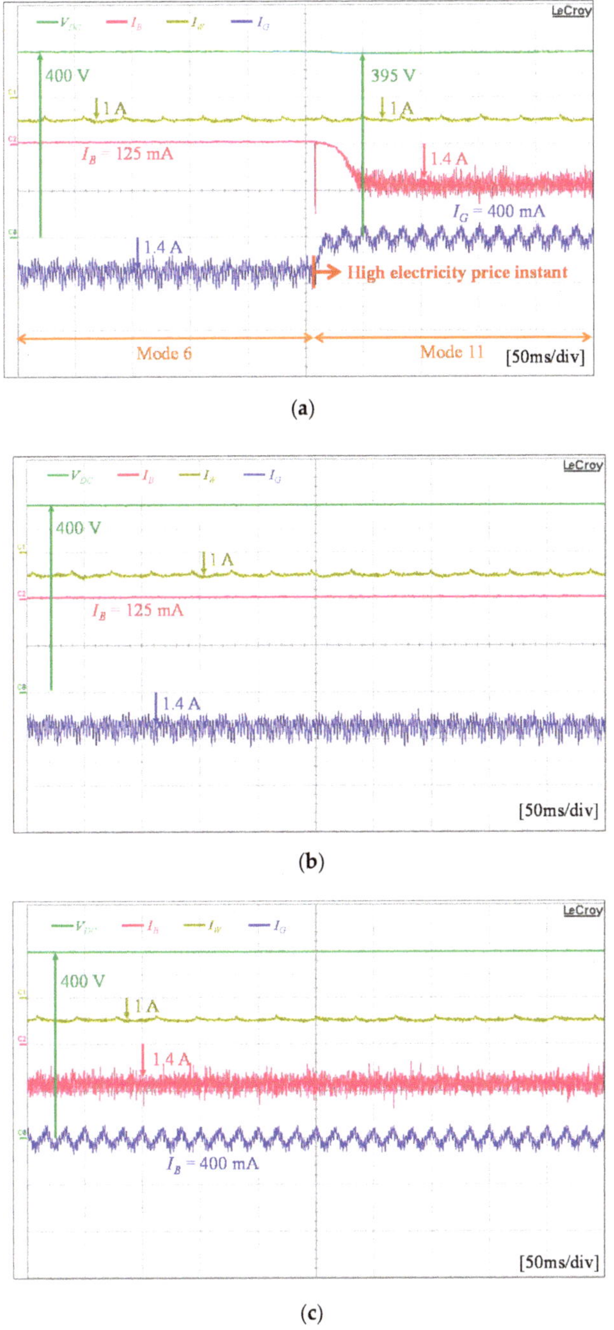

Figure 19. Experimental results with the constraint of high electricity price. (**a**) Operating mode transition from 6 to 11; (**b**) Steady-state response at operating mode 6; (**c**) Steady-state response at operating mode 11.

6. Conclusions

This paper has presented an effective power flow control strategy for a hybrid control architecture of DCMG under an unreliable grid connection considering the constraint of electricity price. To overcome the limitation of the existing coordination control architectures, a hybrid control method which combines the centralized control and distributed control is applied to control DCMG. By using the hybrid control approach, a more optimal and reliable DCMG system can be constructed even though a fault occurs in the grid or the CC. The power flow control strategy is developed for the hybrid DCMG control architecture by taking the constraint of electricity price into account. In the proposed architecture, the HBC link is used in the centralized control to connect the CC with DCMG power agents. Based on the information from all the agents, the CC determines the operating modes of all the agents. On the contrary, the LBC link is employed to constitute the distributed control. To reduce the communication burden, only one-bit binary data is used to exchange the information between the agents via the LBC link. Main contributions of this paper can be summarized as follows:

(i)   A power flow control strategy for a hybrid DCMG control scheme which combines the centralized control and distributed control is proposed. The proposed scheme improves the system stability and robustness even in the presence of a fault in the communication link of the centralized control. All the power agents operate stably within the DCMG system without severe transients in spite of the transition between the distributed control and the centralized control. The DC-link voltage can be also regulated stably with the proposed scheme in both the centralized control and distributed control even during transition periods.

(ii)   By using the proposed scheme, the common single point of failure as in the centralized control can be avoided. Even when the distributed control takes control of the DCMG system because of a fault in the CC, the DCMG system stably works. Furthermore, the problems related to the lack of information exchange as in the decentralized control can be solved by means of a communication network.

(iii)   The proposed scheme proves that the hybrid DCMG control scheme can be implemented by the exchange of only a small size of data through the HBC and LBC links to exchange the information with reduced communication burden.

(iv)   In the proposed hybrid DCMG control scheme, the constraint of electricity price is also considered to develop the power flow control with the aim to minimize the consumer electricity cost.

To verify the practical usefulness of the power flow control strategy of the proposed hybrid control architecture, a prototype DCMG system has been constructed by using 32-bit floating-point DSP TMS320F28335 controller. The DCMG consists of bidirectional AC-to-DC converter to connect the grid source, unidirectional AC-to-DC converter to connect the wind turbine emulator, and bidirectional DC-to-DC converter to connect the lithium-ion battery. Comprehensive simulation and experimental results have been presented to prove the effectiveness of the proposed power flow control strategy of the hybrid control architecture, which well confirms the stable operation and overall performance of the proposed scheme.

Author Contributions: F.A.P. and K.-H.K. conceived the main concept of the microgrid control structure and developed the entire system. F.A.P. carried out the research and analyzed the numerical data with guidance from K.-H.K., F.A.P. and K.-H.K. collaborated in the preparation of the manuscript. All authors have read and agreed to the published version of the manuscript.

Funding: This research received no external funding.

Acknowledgments: This research was supported by Basic Science Research Program through the National Research Foundation of Korea (NRF) funded by the Ministry of Education (NRF-2019R1A6A1A03032119). This work was supported by the National Research Foundation of Korea (NRF) grant funded by the Korea government (MSIT) (NRF-2020R1F1A1048262).

Conflicts of Interest: The authors declare no conflict of interest.

References

1. Rosero, C.X.; Velasco, M.; Marti, P.; Camacho, A.; Miret, J.; Castilla, M.A. Active Power Sharing and Frequency Regulation in Droop-Free Control for Islanded Microgrids Under Electrical and Communication Failures. *IEEE Trans. Ind. Electron.* **2020**, *67*, 6461–6472. [CrossRef]
2. Pahasa, J.; Ngamroo, I. Coordinated PHEV, PV, and ESS for Microgrid Frequency Regulation Using Centralized Model Predictive Control Considering Variation of PHEV Number. *IEEE Access* **2018**, *6*, 69151–69161. [CrossRef]
3. Mathew, P.; Madichetty, S.; Mishra, S. A Multi-Level Control and Optimization Scheme for Islanded PV Based Microgrid: A Control Frame Work. *IEEE J. Photovolt.* **2019**, *9*, 822–831. [CrossRef]
4. Espín-Sarzosa, D.; Palma-Behnke, R.; Núñez-Mata, O. Energy Management Systems for Microgrids: Main Existing Trends in Centralized Control Architectures. *Energies* **2020**, *13*, 547. [CrossRef]
5. Balog, R.S.; Weaver, W.W.; Krein, P.T. The Load as an Energy Asset in a Distributed DC SmartGrid Architecture. *IEEE Trans. Smart Grid* **2012**, *3*, 253–260. [CrossRef]
6. Khorsandi, A.; Ashourloo, M.; Mokhtari, H. A Decentralized Control Method for a Low-Voltage DC Microgrid. *IEEE Trans. Energy Convers.* **2014**, *9*, 793–801. [CrossRef]
7. Cook, M.D.; Parker, G.G.; Robinett, R.D.; Weaver, W.W. Decentralized Mode-Adaptive Guidance and Control for DC Microgrid. *IEEE Trans. Power Deliv.* **2017**, *32*, 263–271.
8. Nasirian, V.; Moayedi, S.; Davoudi, A.; Lewis, F.L. Distributed Cooperative Control of DC Microgrids. *IEEE Trans. Power Electron.* **2015**, *30*, 2288–2303. [CrossRef]
9. Kwasinski, A.; Onwuchekwa, C.N. Dynamic Behavior and Stabilization of DC Microgrids with Instantaneous Constant-Power Loads. *IEEE Trans. Power Electron.* **2011**, *26*, 822–834. [CrossRef]
10. Feng, X.; Shekhar, A.; Yang, F.; Hebner, R.E.; Bauer, P. Comparison of Hierarchical Control and Distributed Control for Microgrid. *Elect. Power Compon. Syst.* **2017**, *45*, 1043–1056. [CrossRef]
11. Diaz, N.L.; Luna, A.C.; Vasquez, J.C.; Guerrero, J.M. Centralized Control Architecture for Coordination of Distributed Renewable Generation and Energy Storage in Islanded AC Microgrids. *IEEE Trans. Power Electron.* **2017**, *32*, 5202–5213.
12. Katiraei, F.; Iravani, R.; Hatziargyriou, N.; Dimeas, A. Microgrids management. *IEEE Power Energy Mag.* **2008**, *6*, 54–65.
13. Xu, Q.; Xiao, J.; Wang, P.; Wen, C. A Decentralized Control Strategy for Economic Operation of Autonomous AC, DC, and Hybrid AC/DC Microgrids. *IEEE Trans. Energy Convers.* **2017**, *32*, 1345–1355.
14. Liu, W.; Gu, W.; Sheng, W.; Meng, X.; Wu, Z.; Chen, W. Decentralized Multi-Agent System-Based Cooperative Frequency Control for Autonomous Microgrids with Communication Constraints. *IEEE Trans. Sustain. Energy* **2014**, *5*, 446–456.
15. Karimi, Y.; Oraee, H.; Golsorkhi, M.S.; Guerrero, J.M. Decentralized Method for Load Sharing and Power Management in a PV/Battery Hybrid Source Islanded Microgrid. *IEEE Trans. Power Electron.* **2017**, *32*, 3525–3535.
16. Nguyen, T.V.; Kim, K.H. Power Flow Control Strategy and Reliable DC-Link Voltage Restoration for DC Microgrid under Grid Fault Conditions. *Sustainability* **2019**, *11*, 3781.
17. Mehdi, M.; Kim, C.H.; Saad, M. Robust Centralized Control for DC Islanded Microgrid Considering Communication Network Delay. *IEEE Access* **2020**, *8*, 77765–77778.
18. Nguyen, T.V.; Kim, K.-H. An Improved Power Management Strategy for MAS-Based Distributed Control of DC Microgrid under Communication Network Problems. *Sustainability* **2020**, *12*, 122.
19. Mathew, P.; Madichetty, S.; Mishra, S. A Multilevel Distributed Hybrid Control Scheme for Islanded DC Microgrids. *IEEE Syst. J.* **2019**, *13*, 4200–4207.
20. Mohamed, Y.A.I.; Radwan, A.A. Hierarchical Control System for Robust Microgrid Operation and Seamless Mode Transfer in Active Distribution Systems. *IEEE Trans. Smart Grid* **2011**, *2*, 352–362.
21. Zhou, L.; Zhang, Y.; Lin, X.; Li, C.; Cai, Z.; Yang, P. Optimal Sizing of PV and BESS for a Smart Household Considering Different Price Mechanisms. *IEEE Access* **2018**, *6*, 41050–41059.
22. Wu, X.; Hu, X.; Yin, X.; Zhang, C.; Qian, S. Optimal Battery Sizing of Smart Home via Convex Programming. *Energy* **2017**, *45*, 444–453.
23. Hemmati, R.; Saboori, H. Stochastic Optimal Battery Storage Sizing and Scheduling in Home Energy Management Systems Equipped with Solar Photovoltaic Panels. *Energy Build.* **2017**, *152*, 290–300.

24. Hemmati, R. Technical and Economic Analysis of Home Energy Management System Incorporating Small-Scale Wind Turbine and Battery Energy Storage System. *J. Clean. Prod.* **2017**, *159*, 106–118.
25. Wang, H.; Fang, H.; Yu, X.; Liang, S. How Real Time Pricing Modifies Chinese Households' Electricity Consumption. *J. Clean. Prod.* **2018**, *178*, 776–790.
26. Newsham, G.R.; Bowker, B.G. The Effect of Utility Time-Varying Pricing and Load Control Strategies on Residential Summer Peak Electricity Use: A Review. *Energy Policy* **2010**, *38*, 3289–3296.
27. Zhou, B.; Yang, R.; Li, C.; Wang, Q.; Liu, J. Multiobjective Model of Time-of-Use and Stepwise Power Tariff for Residential Consumers in Regulated Power Markets. *IEEE Syst. J.* **2018**, *12*, 2676–2687.
28. Li, C.; Tang, S.; Cao, Y.; Xu, Y.; Li, Y.; Li, J.; Zhang, R. A New Stepwise Power Tariff Model and Its Application for Residential Consumers in Regulated Electricity Markets. *IEEE Trans. Power Syst.* **2013**, *28*, 300–308.
29. Lu, X.; Guerrero, J.M.; Sun, K.; Vasquez, J.C. An Improved Droop Control Method for DC Microgrids Based on Low Bandwidth Communication with DC Bus Voltage Restoration and Enhanced Current Sharing Accuracy. *IEEE Trans. Power Electron.* **2014**, *29*, 1800–1812.
30. Yoon, S.J.; Lai, N.B.; Kim, K.H. A Systematic Controller Design for a Grid-connected Inverter with LCL Filter Using a Discrete-time Integral State Feedback Control and State Observer. *Energies* **2018**, *11*, 437.
31. Tran, T.V.; Yoon, S.-J.; Kim, K.-H. An LQR-Based Controller Design for an LCL-Filtered Grid-Connected Inverter in Discrete-Time State-Space under Distorted Grid Environment. *Energies* **2018**, *11*, 2062. [CrossRef]
32. Bimarta, R.; Tran, T.V.; Kim, K.-H. Frequency-Adaptive Current Controller Design Based on LQR State Feedback Control for a Grid-Connected Inverter under Distorted Grid. *Energies* **2018**, *11*, 2674. [CrossRef]
33. Kolokotsa, D.; Rovas, D.; Kosmatopoulos, E.; Kalaitzakis, K. A Roadmap towards Intelligent Net Zero- and Positive-Energy Buildings. *Sol. Energy* **2011**, *85*, 3067–3084. [CrossRef]
34. Beaudin, M.; Zareipour, H. Home Energy Management Systems: A Review of Modelling and Complexity. *Renew. Sustain. Energy Rev.* **2015**, *45*, 318–335.
35. Boynuegri, A.R.; Yagcitekin, B.; Baysal, M.; Karakas, A.; Uzunoglu, M. Energy management algorithm for smart home with renewable energy sources. In Proceedings of the 4th International Conference on Power Engineering, Energy and Electrical Drives, Istanbul, Turkey, 13–17 May 2013.

© 2020 by the authors. Licensee MDPI, Basel, Switzerland. This article is an open access article distributed under the terms and conditions of the Creative Commons Attribution (CC BY) license (http://creativecommons.org/licenses/by/4.0/).

 *sustainability*

Article

A Novel VSG-Based Accurate Voltage Control and Reactive Power Sharing Method for Islanded Microgrids

Bowen Zhou [1],*, Lei Meng [2],*, Dongsheng Yang [1], Zhanchao Ma [1] and Guoyi Xu [3]

[1] College of Information Science and Engineering, Northeastern University, Shenyang 110819, China; yangdongsheng@mail.neu.edu.cn (D.Y.); mazhanchao@stumail.neu.edu.cn (Z.M.)
[2] Department of Electrical Engineering, the Hong Kong Polytechnic University, Kowloon, Hong Kong, China
[3] The State Key Laboratory of Alternate Electrical Power System with Renewable Energy Sources, North China Electric Power University, Beijing 102206, China; xu_gy@ncepu.edu.cn
* Correspondence: zhoubowen@ise.neu.edu.cn (B.Z.); mengleischolar@gmail.com (L.M.)

Received: 4 October 2019; Accepted: 19 November 2019; Published: 25 November 2019

Abstract: Islanded microgrids (IMGs) are more likely to be perturbed by renewable generation and load demand fluctuation, thus leading to system instability. The virtual synchronous generator (VSG) control has become a promising method in the microgrids stability control area for its inertia-support capability. However, the improper power sharing and inaccurate voltage control problems of the distributed generations (DGs) in microgrids still has not been solved with a unified method. This paper proposes a novel VSG equivalent control method named Imitation Excitation Control (IEC). In this method, a multi-objective control strategy for voltage and reactive power in a low voltage grid that considers a non-negligible resistance to reactance ratio (R/X) is proposed. With the IEC method, the voltage drop across feeders is compensated, thus the terminal voltage of each inverter will be regulated, which will effectively stabilize the PCC (point of common coupling) voltage and inhibit the circular current. Meanwhile, this method can realize accurate reactive power tracking the reference value, making it accessible for reactive power scheduling. What is more, the reasonability of the IEC model, namely the equivalent mechanical characteristic and transient process inertia support between VSGs and conventional synchronous generators (SG), is illustrated in this paper. Moreover, steady-state stability is proved by the small-signal modeling method, and the energy required by inertia support is given. Finally, the simulation result validates the effectiveness of the proposed method, and it is also demonstrated that the proposed method outperforms the conventional droop control method.

Keywords: VSG; power sharing; inertia support; energy support; small signal stability

1. Introduction

In recent years, to solve the increasing environmental and energy issues, distributed generation (DG) with renewable energy sources such as solar energy and wind energy has been greatly developed. To facilitate DG integration and to relieve DG's direct strike to the main grid, microgrids are developed. Microgrids (MG), also named minigrids, are becoming popular for their flexibility and convenience in distributed control system [1,2].

However, when microgrids operate in islanded mode, due to the loss of stable voltage-support from the main grid, they are more likely to be affected by the fluctuation of loads and renewable energy sources. What is more, the high penetration of electronic devices like inverters/converters lower the system inertia, making the MGs much weaker than the traditional high voltage power system. Unstable conditions like harmonics [3], resonances [4], frequency and voltage instability, power fluctuations and inaccurate power sharing, and even severe failures are more likely to happen accordingly.

This paper focuses on issues of islanded MGs. We summarize the problems in the islanded MGs into two aspects. The first one is the stability problem. Since a large number of power electronic devices like inverters and converters are integrated in the microgrids, they have a very fast response speed and lack inertia support. Thus, they are very likely to have step responses to the disturbances [5,6]. Moreover, because of the lack of damping, it is easy to generate excessive overshoots during the transient response process, causing damage to the electronic devices and the load. Another major issue is the accuracy problem. Since the renewable energy generating microgrids mostly lie in the promote areas, the distances between DGs tend to be relatively large, so that it is easy to generate a large voltage drop in the transmission lines, making the voltage control inaccurate [7]. In addition, for low-voltage networks, the resistances in transmission lines are often nonnegligible. Thus, the high R/X ratio will produce active and reactive coupling [7–9]. If the lines are still taken as purely inductive, like in the traditional high-voltage power system, the problem of uneven power sharing will occur.

A lot of meaningful work has been carried out on the above two issues of the microgrid [6,10–21]. On the one hand, for the stability problem of microgrids, a new control method virtual synchronous generator (VSG) is proposed [11–16]. In the VSG method, the system inertia is generated by mimicking the mechanical characteristic of the conventional synchronous generator (SG). Therefore, the frequency stability and dynamic performance of the system can be modified simultaneously [13,14]. What is more, to mimic conventional synchronous generators (SGs), inertia support and energy support in real systems must be considered. However, most research mainly focusses on modifying VSGs' control loops to improve their dynamic response characteristics [15–17], while ignoring their fundamental theoretical issues. For instance, the inherent relations between VSG and SG, the equivalent physical mechanism of VSG, the feasibility analysis of transforming the SG control method to VSG, etc., are all important to the research and application of VSGs.

In [22], the authors equate the DC-side capacitor of the inverter with the rotor of SG . And the reasonability of equivalency is demonstrated from the perspective of their physical structures, characteristic parameters and mathematical models. However, the critical issue of VSG, that is the meaning of inertia support and energy transformation, is not discussed in this paper. In [23], the electromagnetic power that is generated by the variation of SG rotor kinetic energy is deduced. Additionally, the request of SG primary frequency regulation characteristics is analyzed as well. However, a similar analysis for VSG is not given. In this paper, the SG rotor and VSG DC-side capacitor inertia support during the system transient processes are analyzed comparatively, and are of great significance to VSG modeling and parameter configuration.

On the other hand, for the problem of accurate voltage control and power sharing, a lot of methods have been proposed. Most of the methods are based on the reactive power-voltage (Q-V) droop relation [7,18–21,24–28]. In [24–28], the power sharing errors caused by unmatched line impedance are inhibited by increasing the droop slope. However, this method would result in severe point of common coupling (PCC) voltage drop and therefore influence the system stability. In [7,18–21], the authors adopt virtual impedance to revise the Q-V droop relation of the DGs in microgrids, making it reflect the reactive power and voltage relation accurately. In [7,18,19], the reactive power sharing is implemented by employing virtual impedance controllers, which are based on the state estimator and local load measurement. In [20], the output impedance of the inverter is modified by a virtual impedance loop, which is predominantly inductive. Thus the decoupling of the power flows will be effectively realized even for the high R/X ratio feeders. The authors of [18] use an adaptive virtual impedance method to compensate the voltage drops on feeders.

However, all of the above distributed control methods are based on a Q-V droop relationship that cannot completely solve the problem of uneven power sharing. Because the reactive power and the voltage of each DG meet a droop-relation, which is determined by the DG's characteristics. When the output voltage of the DG is maintained at the rated value, the value of the output reactive power is determined, thus unable to follow its reference value. As a result, optimal reactive power generation in the microgrid system cannot be realized. Nevertheless, we believe that optimizing the reactive power is of great economic significance because it can reduce the investment of reactive-load compensation equipment. The core idea of our method regards the voltage stability and accurate reactive power tracking as a multi-object control issue. To realize this, an imitation excitation control (IEC) strategy to realize the accurate voltage controlling and reactive sharing of DG units in islanded microgrids is first proposed. The novel method employs a central controller and communication bus to generate and dispatch the control commands. The voltage phasor diagram of VSGs is utilized to calculate and adjust the output voltage of the inverter, while the voltage drop on feeder impedance is taken into consideration and compensated. Meanwhile, the power can be output according to the demands and thus the circulating current is eliminated simultaneously.

This paper focuses on the voltage control and reactive power sharing of the islanded microgrid. The main contributions of this paper are:

- Adoption of a VSG based control method, which can maintain the frequency stability of the system. On the basis of [22], we prove the equivalence of SG and VSG from the aspect of equivalent mechanical characteristics and inertia support in transient processes.
- Proposing a new method for voltage control and reactive power sharing by mimicking the excitation control of the traditional SG on the voltage control part. It can compensate a problem of the droop relation-based method, that the reactive power cannot follow the reference value.
- Instead of treating the transmission lines as pure inductive, this paper takes the R/X ratio which leads to an inaccurate voltage value and power sharing into consideration. This problem is also solved by this IEC method.

The rest of this paper is organized as follows. In Section 1, the basic microgrid configuration and essential assumptions are given. In Section 2, the equivalent mechanical characteristics, inertia-support and the transient adjusting process of VSGs are discussed. In Section 3, the principle and the mechanism of IEC for accurate voltage control and reactive power tracking method is illustrated, and the equivalence of the inverter with an inductance-capacitor-inductane (LCL) type filter and the salient-pole SG is proved. In Section 4, power regulation and voltage adjustment is illustrated. The system steady-state stability analysis under the IEC method is conducted by small-signal analysis. VSG transient energy analysis is also organized in this section, which gives the capacity requirement of the DC-capacitor. In Section 4, the MG simulation environment with two paralleled inverters is set up to verify the validity of the proposed method in this paper. Small signal stability of the studied system is also analyzed. Finally, there is the conclusion, pointing out the defections and prospective research of the novel method.

2. Configuration and Assumptions of the Studied Islanded Microgrid

The microgrids are usually comprised of distributed resources like photovoltaic (PV) panels and wind turbines, energy storage devices like battery and super-capacitors and local loads which include the critical loads and normal loads. Microgrids can operate in interconnected mode or islanded mode; in this paper, only the islanded mode system is discussed.

In microgrids, most of the primary sources of DG units are DC sources, like the PV generators and battery storage devices; that means the power electronic interfaces for DC–AC transformation are required; therefore, the inverter control is of great significance. In this paper, the VSG method is used for voltage, frequency and power control of each DG in microgrids.

The islanded microgrid configuration discussed in this paper is shown in Figure 1.

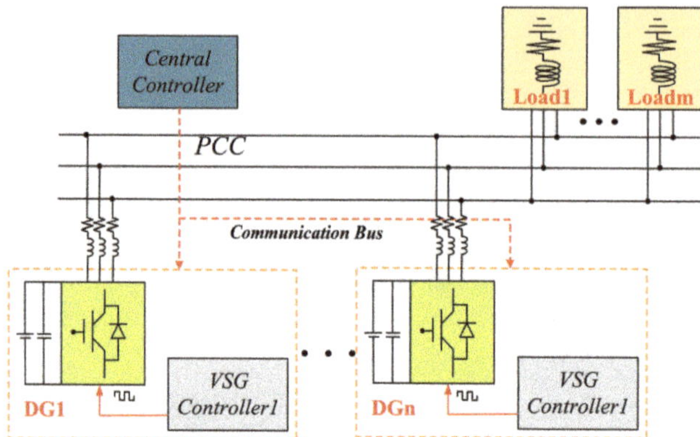

Figure 1. The configuration of microgrids in this paper.

In order to implement the VSG control method that is proposed in this paper, some conditions and hypotheses about the microgrids need to be made. The voltage class of three-phase common bus in this microgrid is 380 V, and each DG unit is connected to the common bus through a feeder, whose impedance is acquirable and is supposed to be $Z_{li} = R_{li} + jX_{li}$, where i refers to the ith DG. For the simplicity of the output power programme for DGs, all of the loads in the microgrid are lumped together as one; that means the local loads of each DG are not taken into consideration.

A central controller (CC) is equipped at the PCC; it is works for optimal power scheduling and can dispense reference active and reactive power for each DG. The voltage and current signals of the PCC bus are collected and sent to the CC, and the operation conditions of each DG are also sent to the CC through a low bandwidth communication bus. What should be illustrated is that the communication is only needed between the CC and DG units; the operating status of each load does not need to be sent to the CC. What is more, as a low voltage grid, the feeder resistance of the MG cannot be neglected, and the impedance of each feeder can be calculated by feeder specification parameters and distance.

3. VSG Equivalent Mechanical Characteristics Illustration

The inertia support of VSGs and their frequency stability will be analyzed in this part. Droop control is a conventional method for islanded mode microgrids. However, as analyzed in [5], when comparing with VSG control, the droop control has a poor performance in both frequency response and small signal oscillation rejection. The droop control always suffers step frequency-change when it is disturbed by a small signal, whereas the frequency in VSG control can change slowly. That is because the VSG control model has a virtual inertia link in the equivalent swing equation, while the droop control has no inertia. As compared in [22], the physical mechanism and characteristic parameters of SG and VSG share great correlations. In this part, the correspondence of an SG rotor and a VSG DC-side capacitor is analyzed, and a simpler equivalent method of capacitor voltage u_C and rotor angular speed ω is proposed.

The three-phase inverter configuration with a VSG based control diagram is shown in Figure 2. The three phase inverter legs operate under pulse width a modulation (PWM) command, which is determined by a VSG control loop. $e_i(i = a, b, c)$ is the terminal electromotive force (EMF) of the VSG, and is equivalent to the field EMF in SG; the voltage $v_i(i = a, b, c)$ after the LCL filter is the output voltage of the inverter, while $i_i(i = a, b, c)$ is the line current. $Z_{line-i} = R_{line-i} + jX_{line-i}$ is the impedance of the transmission line between the inverter and the PCC. Proper control methodologies (like MPPT for photovoltaic and the wind turbine) are able to maintain DC voltage stability of each DG, and will not be discussed in this paper.

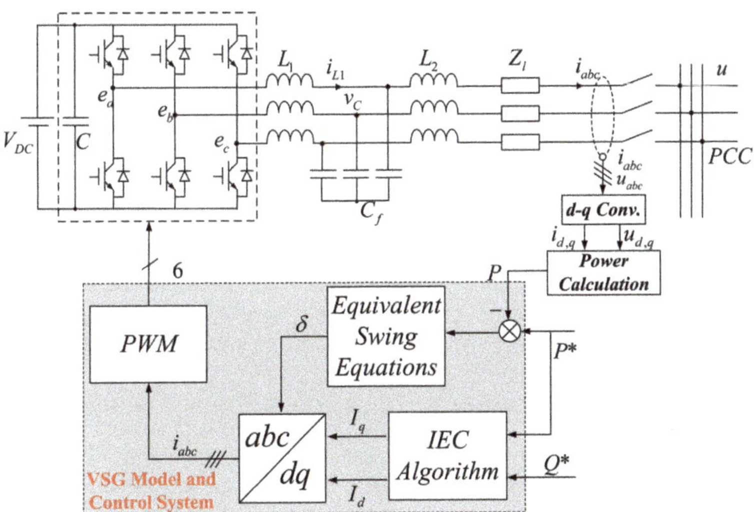

Figure 2. VSG diagram using the proposed control algorithm.

The DC-side capacitor of VSG and the rotor of SG have similar characteristics in energy transmission and dynamic character. For energy transmission, the SG transforms the mechanical energy of a prime motor to electrical energy through the rotor, which stores the kinetic energy in synchronous rotation; a similar energy transfer process of VSGs is realized by the charging and discharging process of a DC capacitor. The stored kinetic energy of an SG rotor at the synchronous angular speed W_k and the stored electrical energy of a VSG capacitor E_k are expressed as Equations (1) and (2), respectively:

$$W_k = 1/2J\Omega_s^2 \tag{1}$$

$$E_k = 1/2CV_{dc}^2 \tag{2}$$

where J is the rotor's moment of inertia in SG and Ω_s is the synchronous mechanical angular velocity (rad/s) of the SG rotor. Assume that the number of pole-pair of the SG rotor is one, and there is one pair of poles on each phase of the stator. So that Ω_s equals the electrical angular velocity ω. C is the capacitance of the capacitor, and V_{dc} is the rated voltage of it.

For synchronous generators, the mechanical part of the machine is governed by:

$$J\frac{d^2\delta}{dt} = T_m - T_e - D\frac{d\delta}{dt} \tag{3}$$

where T_m is the mechanical torque, T_e is the SG electromagnetic torque and D is the damping coefficient.

For the rotor, the angular velocity and the rotating angle meet:

$$\omega = \frac{\delta}{dt} \tag{4}$$

Multiplying ω_m on both sides of Equation (3), and taking $T_J = 2W_K / P_m^*$ as the inertia time constant of SG, P_m^* is the rated mechanical power of SG; the rotor kinematic equation, which is also called the Swing equation, can be derived as:

$$\begin{cases} \dfrac{d\delta}{dt} = \omega - \omega_s \\ T_J \dfrac{d\Delta\omega}{dt} = P_m - P_e - D\omega \end{cases} \tag{5}$$

where $P_m = T_m\omega$ and $P_e = T_e\omega$ are the mechanical power and electromagnetic power, respectively. The damping power $D\omega$, which is proportional to electrical angular velocity ω, is generated by damping winding. The main functions of damping winding are damping the system disturbances and eliminating the synchronous rotation speed difference of the SG rotor. The mechanical and electromagnetic torques meet:

$$\Delta T = T_m - T_e = J\dfrac{d\omega}{dt} \tag{6}$$

Equations (5) and (6) describe the mechanical characteristics of SG, while similar characteristics also exist in the VSG. For the DC-side capacitor in Figure 2, the capacitor current meets:

$$i_C = C\dfrac{du_{DC}}{dt} = i_{DC} - i_S - i_D \tag{7}$$

where i_C, i_{DC} and i_S are the capacitor current, DC input current and DC output current, respectively. i_D is the chopper circuit current, which is utilized to restrict the DC voltage to a given region. Therefore, the chopper in the VSG is equivalent to damping winding in the SG. Comparing Equations (6) and (7), it can be learnt that ΔT in the SG and capacitor current i_C in the VSG, and ω in the SG and u_{DC} in the VSG are counterparts, respectively, because each pair of them share the same dynamic character.

We define the equivalent inertia time constant of the VSG as:

$$T_C = 2E_K / P_{DC} \tag{8}$$

(7) can be rewritten as:

$$T_C \dfrac{du_{DC}*}{dt} = i_{DC}* - i_g* - i_D*$$
$$= P_{DC}* - P_g* - P_D* \tag{9}$$

where P_{DC} is the rated DC power, P_g is the inverter output power and P_D is the power consumed on discharge resistance R_D. The chopper circuit is activated when $|\Delta u_{DC}|$ exceeds the pre-defined bounds. $\{\cdot\}*$ represents the per unit value of a specific variable. Therefore, the VSG equivalent swing equation can be constructed as:

$$T_C \dfrac{du_{DC}}{dt} = P_{DC} - P_g - D_p\omega \tag{10}$$

Under the circumstance of DC voltage variation, the standard dynamical equation of the VSG capacitor can be obtained by linearizing P_{DC}:

$$T_C \dfrac{d\Delta u_{DC}}{dt} = \Delta P_{DC} - D_p \cdot \Delta u_{DC} \tag{11}$$

where $D_p = 2 \cdot U_{d0} / R_d$ is an equivalent damping coefficient, which adds inertia to a system. What should be noticed is that u_{DC} cannot be equivalent to ω directly, to compromise this, we multiply a coefficient to Δu_{DC}:

$$\Delta\omega = K\Delta u_{DC} \tag{12}$$

Therefore, the following holds:

$$\frac{d\delta}{dt} = K \cdot (u_{DC} - u_s) \tag{13}$$

Therefore, Equations (10) and (13) together build the equivalent swing equation of VSGs.

The equivalent mathematical model of an inverter's DC-side capacitor is constructed to provide inertia and energy support to a VSG system. The process of operating status adjustment of a SG and VSG are shown in Figure 3a,b, respectively. The mechanical characteristic curve of an SG can be approximated to a droop line of rotational speed and mechanical torque. Firstly, assume that, at the very beginning, the SG is operating at the static stable point a_1 with the load torque T_{L1}, while at time t there is a drastic load variation from T_{L1} to T_{L2}, which triggers a power rescheduling; thus, the SG starts to operate along the new mechanical characteristic curve 2. Due to the effect of inertia, the rotational speed of the rotor cannot change instantaneously. Thereby, the operating status of the SG would jump to point b at time t and then increase its rotation speed along line 2. The transient process would not stop until the SG comes to another static stable point c_1. Similar transient process would occurs in VSG with the capacitor at DC-side works as energy storage devices. When the load of the DG changes, the appropriate adjustment of the operation status is made. Since the step change cannot be made to capacitor voltage, the energy of the inverter would not be released or increased instantaneously; therefore, the strike to the system can be relieved effectively.

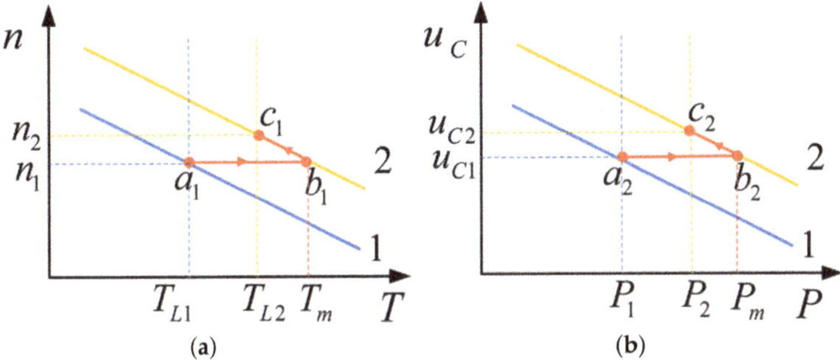

(a)   (b)

Figure 3. (a) Operating status adjustment process of conventional synchronous generators (SGs); (b) Operating status adjustment process of virtual synchronous generators (VSGs).

The dynamic model that is constructed in Equations (10) and (12) is shown in Figure 4.

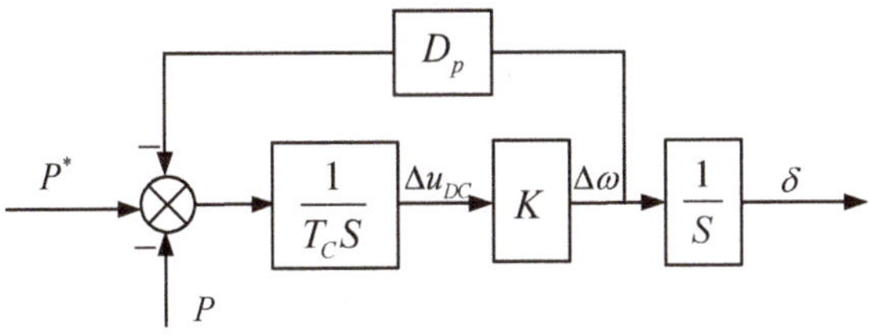

Figure 4. The frequency control diagram of VSG.

This model is a second-order system, and under the control of this system the frequency can be stabilized in finite time. The specific active power-frequency stability and energy issues of the DC-side capacitor will be analyzed in Sections 5.1 and 5.2, respectively.

4. IEC Method for Voltage Stabilization and Accurate Reactive Power Tracking

In this section, a comprehensive equivalent strategy for stabilizing VSG voltage and achieving accurate reactive power tracking is proposed. The capacitor on the DC side of the inverter plays a critical role in the dynamic stability of the VSG. This DC capacitor is equivalent to the SG rotor, thus the equivalent swing equations that relate to the the frequency stability are derived from its dynamical characteristics. Moreover, later it will be proved that, when considering the R/X ratio of transmission line, both the active and reactive power will contribute to the VSG terminal voltage. Thus a comprehensive control strategy is proposed in this section, which is an imitation of the SG's excitation control. The output active and reactive power of each DG will accurately track the set values, which are given by the central controller. Meanwhile, the PCC voltage can always remain at the rated value. Since the variation of load demands and the fluctuation of distributed generating resources happens frequently, it is necessary for DG units to adjust their output power to stabilize the PCC voltage of the islanded MG. In this part the imitation excitation control (IEC) method is first proposed, which focuses on outputting specific VSG power and maintaining the PCC voltage U constant.

First of all, the equivalence of the VSG in series with the LCL filter and the SG is proven. For the conventional SG, since the output active power is immune to the value of the field current, it is feasible to adjust the reactive power with excitation control while keeping the active power constant; that is:

$$P = \frac{3}{2}UI\cos\varphi = C \tag{14}$$

The equivalent phasor diagram of the excitation control for shaded-pole SGs is shown in Figure 5a, from which it can be learnt that, with the variation of the field current, the induced electromotive force $\dot{E}_q$ of the SG changes along the line AA', and the output current value $\dot{I}$ changes along the line BB'. Therefore, the variation of the field current can only adjust the reactive power and power angle δ.

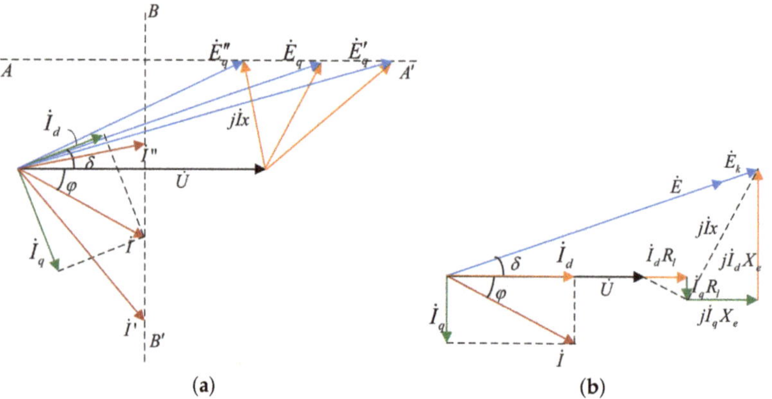

Figure 5. (**a**) Phasor diagram of SG voltage and current; (**b**) phasor diagram of VSG voltage and current.

For the VSG, whose equivalent output circuit is shown as Figure 6a, and the equivalent circuit of LCL filter (single phase) at its output ports is shown as Figure 6b. As demonstrated in [22], the VSG with L- and LCL-type filters corresponds to the shaded-pole and salient-pole SG, respectively. Therefore, the reactive power of a VSG can be controlled independently, like SGs. However, for the VSG with a low voltage feeder, whose R/X ratio usually is non-negligible, coupling between the active and reactive power makes this control scheme unrealizable. A proper method should be adopted to address this problem. In this part the imitation of the excitation control method will be proposed for the power sharing and voltage stable control of VSGs with non-negligible feeder resistance.

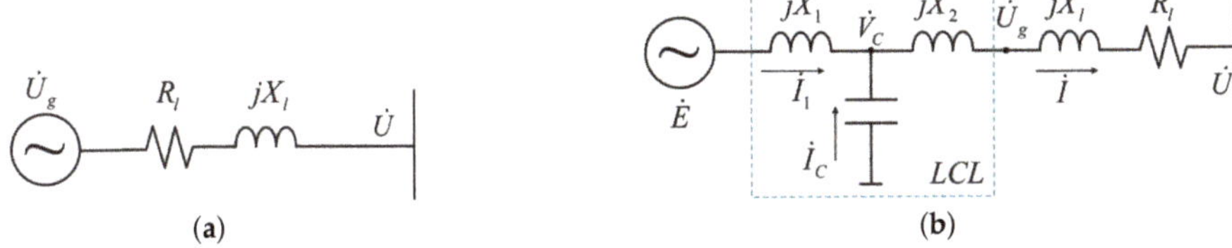

Figure 6. (**a**) Equivalent circuit of a VSG ; (**b**) equivalent circuit of an LCL filter with line impedance.

For the proposed method in this paper, the LCL filter is utilized. Moreover, together with the feeder impedance, it is equivalent to the stator of the SG, where the total impedance of the filter and line impedance corresponds to the armature reactance of the SG. As is shown in Figure 6b, the phaser relation of currents and voltages of the equivalent stator yields:

$$\dot{E} = \dot{V}_C + j\dot{I}_1 X_1 \tag{15}$$

$$\dot{V}_C = j\dot{I}_C X_C \tag{16}$$

$$\dot{V}_C = \dot{U} + \dot{I} \cdot [R_l + j(X_2 + X_l)] \tag{17}$$

$$\dot{I} = \dot{I}_1 + \dot{I}_C \tag{18}$$

It can be derived from the above equations that, $\dot{U}$ and $\dot{E}$ meet:

$$\dot{U} = k\dot{E} - (jX_e + R_l)\dot{I} \tag{19}$$

where $k = \dfrac{X_C}{X_C - X_1}$, $X_e = kX_1 + X_2 + X_l$. Thus, the following can be derived:

$$\dot{E} = \frac{1}{k}[\dot{U} + (jX_e + R_l)\dot{I}] \tag{20}$$

Taking the d-axis alignments to $\dot{U}$, the phasor diagram of the equivalent stator in the VSG is shown in Figure 5b. It would simplify the calculation process, and its vector relations are equivalent to that in Figure 5a. According to Figure 5b, the VSG terminal EMF, output voltage and current meet:

$$U + I_d R_l + I_q X_e = E_k \cos \delta \tag{21}$$

$$I_d X_e - I_q R_l = E_k \sin \delta \tag{22}$$

where $E_k = kE$. Moreover, taking the norm of the equivalent total impedance as $Z = \sqrt{R_l + X_e^2}$. Therefore, the active and reactive power without modulation are expressed as:

$$P = \frac{3}{2}UI\cos\varphi = \frac{3}{2}U \cdot \frac{E_k(X_e \sin\delta + R_l \cos\delta) - UR_l}{Z^2} \tag{23}$$

$$Q = \frac{3}{2}UI\sin\varphi = \frac{3}{2}U \cdot \frac{E_k(R_l \sin\delta - X_e \cos\delta) + UX_e}{Z^2} \tag{24}$$

From Equation (23), it can be learnt that, for the existence of line resistance, the active and reactive powers couple with each other, thus making them unable to adjust the reactive power independently. In consequence, a proper decoupling method should be utilized for the imitation excitation control (IEC) method. An orthogonal linear rotational transformation matrix **T** is used, which can transform active power P_i and reactive power Q_i into 'modified' powers P' and Q' [29]. Moreover, the good thing is that the information related to P, Q and $\dot{E}$ are reserved completely.

$$
\begin{bmatrix} P' \\ Q' \end{bmatrix} = \begin{bmatrix} sin\theta & cos\theta \\ cos\theta & -sin\theta \end{bmatrix} \cdot \begin{bmatrix} P \\ Q \end{bmatrix} = \frac{3}{2} \cdot \frac{U}{Z} \begin{bmatrix} E_k \sin \delta \\ E_k \cos \delta - U \end{bmatrix} \tag{25}
$$

where $cos\theta = X_e/R_l$, $sin\theta = X_e/Z$. From P' and Q', the direct axis component and quadrature axis component of $\dot{E}$ can be acquired by:

$$
\begin{bmatrix} E_{dk} \\ E_{qk} \end{bmatrix} = \begin{bmatrix} \dfrac{Z}{U}P' \\ \dfrac{Z}{U}Q' + U \end{bmatrix} \tag{26}
$$

To eliminate the system errors, the close-loop control system is constituted, which is shown as Figure 2. In many studies, the VSG's filter capacitor voltage, which is acquired by the PLL, is used as a feedback signal. Yet, for the voltage drop and power loss on the filter, it is unsuitable to use the capacitor voltage as the feedback signal to eliminate the EMF error. Fortunately, the information of the VSG output current is undistorted, thus the current-controlled inverter mode is used in this paper. According to the phasor diagram of the voltage and current in Figure 5b and Equation (21), the d-coordinate and q-coordinate components of current I_d, I_q can be calculated by:

$$
\begin{bmatrix} I_d \\ I_q \end{bmatrix} = \frac{1}{Z^2} \begin{bmatrix} R_l & X_e \\ X_e & -R_l \end{bmatrix} \cdot \begin{bmatrix} E_{dk} - U \\ E_{qk} \end{bmatrix} = \mathbf{Y} \cdot \begin{bmatrix} E_{dk} - U \\ E_{qk} \end{bmatrix} \tag{27}
$$

The block diagram of the equivalent power model of the VSG is shown in Figure 7. It can be seen from Equation (27) and Figure 7 that both the control loops of the d-axis and q-axis have paths that cross the equivalent reactance X_e, which leads to coupling between active and reactive currents. Moreover, it may also cause a phase lag in the system transfer function, thereby causing instability in the system. To solve these problems, a feedforward decoupling method is utilized, in which a decoupling matrix D is designed and meets:

$$
\mathbf{D} \cdot \mathbf{Y} = \begin{bmatrix} d_{11} & d_{12} \\ d_{21} & d_{22} \end{bmatrix} \cdot \mathbf{Y} = \begin{bmatrix} y'_1 & 0 \\ 0 & y'_2 \end{bmatrix}
$$

Taking $d_{12} = d_{22} = 1$, the unknown parameters are obtained as $y'_1 = 1/X_e$, $y'_2 = -1/R_l$, and the decoupled incidence matrix $\mathbf{Y}$ is converted to a diagonal matrix. The active and reactive power of the VSG that are under the control of the proposed IEC method are:

$$
P = \frac{3}{2}UI \cos \varphi = \frac{3}{2}U \cdot y'_1(E_{dk} - U) = \frac{3}{2}\frac{U(E_k \cos \delta - U)}{X_e} \tag{28}
$$

$$
Q = \frac{3}{2}UI \sin \varphi = \frac{3}{2}U \cdot y'_2 E_{qk} = -\frac{3}{2}\frac{UE_k \sin \delta}{R_l} \tag{29}
$$

Equation (28) shows that, under the control of the IEC method proposed in this paper, the active and reactive powers of the VSG that takes line resistance into consideration are equivalent to that of the conventional SG in form. However, the of VSG power leading the SG power for 90 electrical degrees due to the effect of the rotational transformation matrix $\mathbf{T}$. Therefore, it is evident that VSGs with the line resistance correspond to SGs. The IEC control diagram with inverter output current feedback is shown in Figure 8, in which the parameters with $*$ represent their reference value.

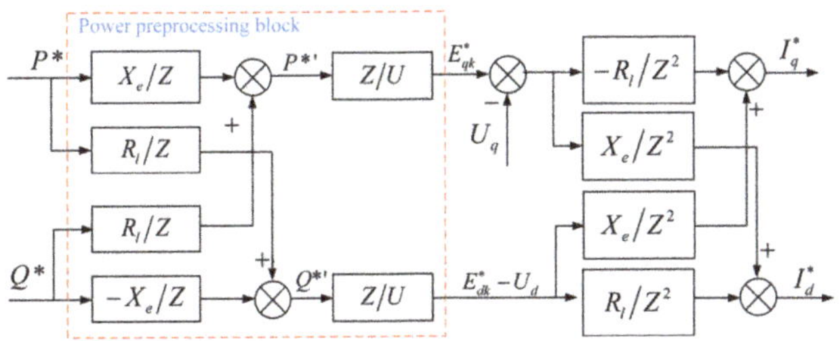

Figure 7. Block diagram of a VSG equivalent power model.

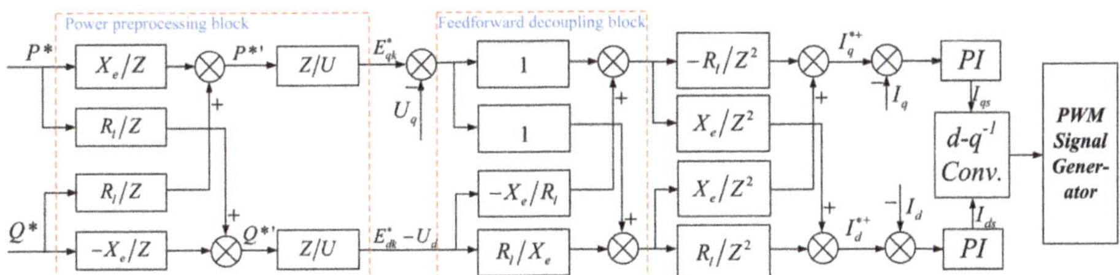

Figure 8. The imitation excitation control (IEC) based control diagram with closed loop control of current.

5. Steady-State Stability Analysis and Transient Energy Support Evaluation

5.1. Small Signal Model for System Steady-State Stability Analysis

When there are small disturbances in the system, power fluctuations may occur in DG units. Whether the system is able to maintain stability under disturbances must be analyzed. Thus the small-signal model of the system is built to investigate the static stability of each DG unit, and analyzes the influence of parameter variations on system stability.

In this part, supposing that the system is symmetric and balanced, so that the output power can be constructed as single phase model. And assume that the power angle δ is small enough, so that it is reasonable to equate the output active and reactive power flows of each DG unit with Equations (23) and (22):

$$\overline{P} = \frac{U}{Z^2}[E_k(X_e\delta + R_l) - UR_l] \tag{30}$$

$$\overline{Q} = \frac{U}{Z^2}[E_k(R_l\delta - X_e) + UX_e] \tag{31}$$

The small signal models of active and reactive power are obtained by linearizing the power flows through a Jacobi matrix at the point of stable equilibrium (δ_s, E_{ks}):

$$\begin{bmatrix} \Delta\overline{P} \\ \Delta\overline{Q} \end{bmatrix} = \begin{bmatrix} \partial\overline{P}/\partial\delta & \partial\overline{P}/\partial E_k \\ \partial\overline{Q}/\partial\delta & \partial\overline{Q}/\partial E_k \end{bmatrix}_{|\delta_s,E_{ks}} \cdot \begin{bmatrix} \Delta\delta \\ \Delta E_k \end{bmatrix}$$

that is:

$$\Delta\overline{P} = \frac{U}{Z}[E_{ks}X_e \cdot \Delta\delta + (\delta_s X_e + R_l) \cdot \Delta E_k] \tag{32}$$

$$\Delta\overline{Q} = \frac{U}{Z}[E_{ks}R_l \cdot \Delta\delta + (\delta_s R_l - X_e) \cdot \Delta E_k] \tag{33}$$

Suppose that E' is the equivalent EMF of a VSG after the LCL lower pass filter in S domain, the transfer function of the LCL filter $K(s)$ is deduced in Appendix A. Derived from Equations (25) and (26):

$$E'(s) = K(s)E = \frac{RP^* - XQ^*}{U} + U \tag{34}$$

where P^* and Q^* are the set values of active and reactive power, respectively. From the swing Equation (5), it can be learnt that the transfer function of the power angle and active power is:

$$\delta(s) = N(s) \cdot (P^* - P) = \frac{K/T_c}{s^2 + KD_p/T_c s}(P^* - P) \tag{35}$$

Therefore, there are:

$$\Delta\delta = N(s)(\Delta P^* - \Delta P) \tag{36}$$

$$\Delta E' = \frac{R\Delta P^* - X\Delta Q^*}{U} \tag{37}$$

Combining Equations (31), (32), (35) and (36), the small signal models of output power flow are described as Equations (37) and (38):

$$\Delta\overline{P} = \frac{A_1(s)}{C(s)}\Delta P^* - \frac{B_1(s)}{C(s)}\Delta Q^* \tag{38}$$

$$\Delta\overline{Q} = \frac{A_2(s)}{C(s)}\Delta P^* - \frac{B_2(s)}{C(s)}\Delta Q^* \tag{39}$$

where

$$A_1(s) = UE_sK(s)N(s)\sin\theta/Z + \delta_s\sin\theta\cos\theta + \cos\theta^2$$
$$A_2(s) = UE_sK(s)N(s)\cos\theta/Z + \delta_s\cos\theta^2 - \sin\theta\cos\theta$$
$$B_1(s) = \delta_s\sin\theta^2 + \sin\theta\cos\theta$$
$$B_2(s) = \delta_s\sin\theta\cos\theta - \sin\theta^2$$

Moreover, C(s) is the characteristic polynomial equation, which is expressed as:

$$C(s) = 1 + \frac{U}{Z}E_sK(s)N(s)\sin\theta$$

According to Routh-Hurwitz Stability Criterion [30], the system is small-signal stable if and only if the characteristic roots of C(s) are distributed on the left half of the complex plane. The characteristic equation $C(s) = 0$ is equivalent to:

$$as^5 + bs^4 + cs^3 + ds^2 + es + f = 0 \tag{40}$$

where $a = T_cL_1C(L_2 + L_l)$, $b = T_cL_{1c}CR_l + KD_pL_1C(L_2 + L_l)$, $c = T_c(L_1 + L_2 + L_l) + KD_pL_1CR_l$, $d = T_cR_l + KD_p(L_1 + L_2 + L_l)$, $e = KD_pR_l + E_sUK\sin\theta L_l$, $f = E_sUK\sin\theta R_l$, $\theta = \arctan(X/R)$ is dependent on the resistance to inductance ratio. The root loci of a small signal stability model for a specific microgrid system will be illustrated in next section.

5.2. Energy Analysis and Stable Condition for DC-Side Capacitor

For the frequency control loop of a VSG, the frequency is adjusted and stabilized by the power-to-frequency closed loop. Moreover, for the inverter with a DC-side capacitor, the energy and active power adjustment in the transient process is realized by this capacitor. Therefore, the analysis of capacitor energy and its capacity requirement is quite necessary. Based on Equation (5), the transfer function of the active power loop can be expressed as:

$$N(s) = \delta(s)/(P_m - P_e) = \frac{K/T_c}{s^2 + KD_p/T_c s} \tag{41}$$

According to the analysis in Section 5.1, the variation of output active power in unit time is linearly related to the unit time power angle variation, so that it can be simply expressed as $\Delta P_e = M\Delta\delta$. Thus the active power closed loop transfer function, which is a second order system, could be expressed as:

$$\Phi(s) = \frac{\Delta P_e}{P_m} = \frac{\omega_n^2}{s^2 + 2\zeta\omega_n s + \omega_n^2} \tag{42}$$

where $\omega_n = \sqrt{\dfrac{KM}{T_c}}$ is the undamped oscillation frequency; $\zeta = \dfrac{D_p}{2}\sqrt{\dfrac{K}{MT_c}}$ is the damping ratio of the system, which decides the response speed of the active power system. The value of ζ can be adjusted by changing D_p and K, while T_c and M are unchangeable, for they relate to the inherent character of the DC-side capacitor and the correlation of δ and active power.

As discussed in Section 2, the inertia support to the VSG is realized by the charging–discharging process of the DC-side capacitor, during which the excess energy should be disposed of by the capacitor. The mechanism of energy support is shown in Figure 9a. When DC-side power changes, the step changes of inverter output power are inhibited by the rapid charging and discharging of the DC-side capacitor. Therefore, the active power response of the inverter is shown as the curve in Figure 9a. In this diagram, the area in green under the horizontal line $y = \Delta P_m$ indicated the energy that is absorbed by capacitor, while the hatched area in yellow above the horizontal line means the energy is released by the capacitor. In the following part of this section, the capacity requirement of the DC-side capacitor that provides energy support to inverter and stabilize the inverter during transient process will be analyzed. Moreover, the relationship between the required capacitor capacity and the system parameters will also be discussed.

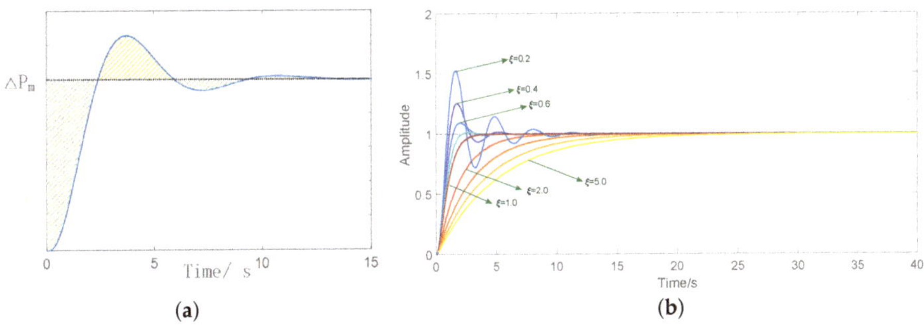

Figure 9. (a) Energy of the DC-side capacitor during step response; (b) Step responses of system output power with different values of ζ.

As it is known that the system would be unstable if the damping ratio $\zeta < 0$, thus such a condition will be eliminated in this paper. The minimum capacity requirement of the DC-side capacitor will be discussed in the following situations:

(1) $0 < \zeta < 1$. Supposing that there is a step change in the reference active power P_m with a variation quantity of ΔP_m, the step response of output active power P_e will be [30]:

$$\Delta P_e(t) = \Delta P_m \left(1 - \frac{e^{-\zeta \omega_n t}}{\sqrt{1 - \zeta^2}} \sin(\omega_d t + \beta) \right)$$

where $\omega_d = \omega_n \sqrt{1 - \zeta^2}$ and $\beta = \arccos \zeta$. The energy that is absorbed or released by the capacitor during this transient process is [23]:

$$\Delta E(t) = \int_0^t (\Delta P_m - \Delta P_e(t)) dt \tag{43}$$

$$= -\frac{\Delta P_m}{\omega_d} \left(e^{-\zeta \omega_n t} \sin(\omega_d t + \gamma) - \sin \gamma \right)$$

where $\gamma = \arccos(2\zeta^2 - 1)$, taking the steady state of the system as when ΔP_e comes into and remains in the range of $(1 \pm 5\%)\Delta P_m$. Therefore, the integral time of ΔE can be taken as the adjusting time $t_s = \frac{3.5}{\zeta \omega_n}$; thus, it has:

$$\Delta E = \frac{\Delta P_m}{\omega_d} \left[\sin \gamma - e^{-3.5} \sin \left(\frac{3.5\sqrt{1 - \zeta^2}}{\zeta} + \gamma \right) \right]$$

(2) $\zeta = 1$. Under this condition, when P_m has a step increase by ΔP_m, the response of ΔP_e is:

$$\Delta P_e(t) = \Delta P_m \left[1 - (1 + \omega_n t)e^{-\omega_n t} \right] \tag{44}$$

Thus the energy that is absorbed or released by the DC-side capacitor under this condition is:

$$\Delta E(t) = \frac{\Delta P_m}{\omega_n} \left(2 - 2e^{-\omega_n t} - \omega_n t e^{-\omega_n t} \right) \tag{45}$$

where $t \to +\infty$, $\Delta E(t) \to \frac{2\Delta P_m}{\omega_n}$.

(3) $\zeta > 1$. The damping ratio $\zeta > 1$, the time response of ΔP_e to input signal step change ΔP_m, is expressed as [30]:

$$\Delta P_e(t) = \Delta P_m \left(1 + \frac{e^{-t/T_1}}{T_2/T_1 - 1} + \frac{e^{-t/T_2}}{T_1/T_2 - 1} \right) \tag{46}$$

where $T_1 = \dfrac{1}{\omega_n(\zeta - \sqrt{\zeta^2 - 1})}$ and $T_2 = \dfrac{1}{\omega_n(\zeta + \sqrt{\zeta^2 - 1})}$.

It can be learnt from Equations (44) and (46) that, when $\zeta \geq 1$, there are only constant terms and exponential decay terms and no periodic oscillation terms in the step response of $\Delta P_e(t)$. Therefore, there are no overshoots in the response process. The energy that is absorbed by the capacitor during this process, when $\zeta > 1$, is:

$$\Delta E(t) = \Delta P_m \left(k_1 \left(e^{-t/T_1} - 1 \right) + k_2 \left(e^{-t/T_2} - 1 \right) \right) \tag{47}$$

where $k_1 = \dfrac{T_1}{T_2/T_1 - 1}$ and $k_2 = \dfrac{T_2}{T_1/T_2 - 1}$; $t \to +\infty$, $\Delta E(t) \to \Delta P_m(-k_1 - k_2)$.

When taking $\omega_n = 2$, the step responses of the system output power with different ζ values are shown in Figure 9b. With the increase of damping ratio ζ (that is the increase of $D_p\sqrt{K}$), the response speed and overshoot of the system dynamic decrease simultaneously. Therefore, the energy that is absorbed by the capacitor is increased. Figure 10 shows the relationship between $\Delta E(\infty)$ and the parameters ζ and ω_n. It can be learnt that the energy that is supported by the capacitor is in approximately direct proportion to the damping ratio ζ, and the slope of the line is in approximately inverse proportion to the ω_n. In order to stabilize the system, the capacitor capacity of each DG unit must meet:

$$E > \frac{2\zeta}{\omega_n}\Delta P_m \tag{48}$$

Taking the rated active power capacity as the biggest variation quantity of a DG, the capacity of it is DC-side capacitor should be no less than $\frac{2\zeta}{\omega_n}\Delta P_m^*$.

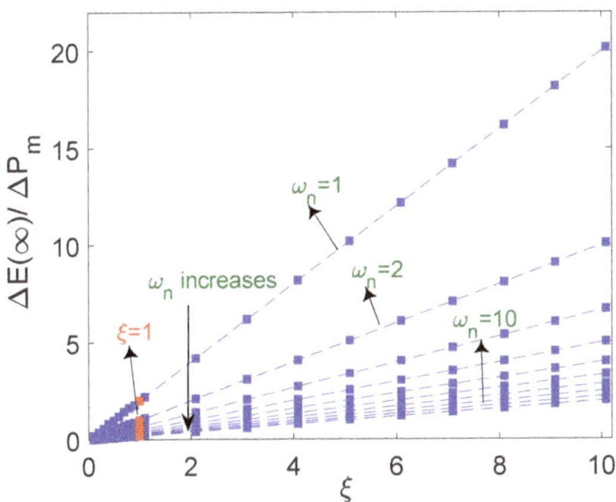

Figure 10. Capacitor energy support values with a changing ζ and ω_n.

6. Case Study and Discussion

6.1. System Configuration and Parameters

To verify the validity of the proposed novel control method for VSGs in this paper, simulations are carried out in MATLAB/Simulink. The simulations are based on the microgrid configuration in Figure 11, which operates in islanded mode. There are two inverters in a microgrid; they are connect in parallel, and are connected to the PCC with their feeders. For each inverter device, a super-capacitor is installed for energy adjustment and inertia support. Besides, there are two load groups in MG as a form of common load; thus, the two VSG units share the load. The detailed system parameters and control parameters of the proposed VSG control method are shown in Table 1.

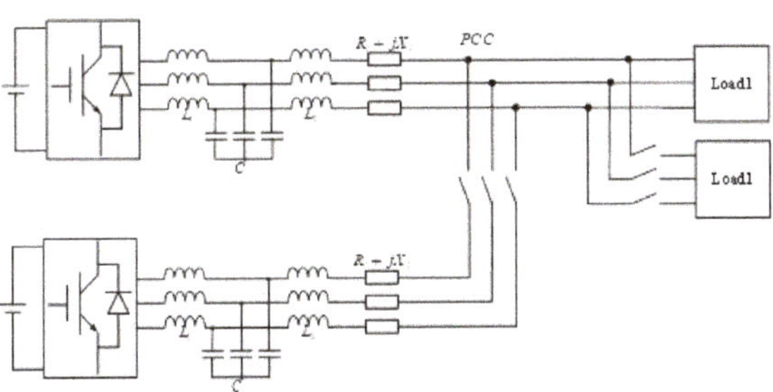

Figure 11. Microgrid configuration under study.

Table 1. Parameters of VSG and distributed generations (DGs).

	Parameters	Values
DC source parameters	DC link voltage	$V_{DG1} = 650$ V $V_{DG2} = 650$ V
	DC source capacities	$S_{DG1} = 120$ kV· A $S_{DG2} = 70$ kV· A
	DC-side capacitors	$C_{DG1} = 0.02$ μF $C_{DG2} = 0.02$ μF
System parameters	Switching frequency	10 kHz
	AC voltage at PCC	380 V/50 Hz (line to line RMS)
Impedance	LCL filter	$L_{f1} = 0.4$ mH $C_f = 4000$ μF $L_{f2} = 0.09$ mH
	VSG feeders	$R_{DG1} = 0.3\ \Omega$ $L_{DG1} = 1.5$ mH $R_{DG2} = 0.4\ \Omega$ $L_{DG2} = 2$ mH
Control parameters	Active power-frequency parameters	$K = 1.5 \times 10^{-4}$ $D_p = 2 \times 10^6$ $T_c = 10.86$

For the traditional droop control method, the influence of filters and resistances on feeders and the couplings between active and reactive power are omitted. Other physical parameters of inverters are the same as those in the proposed VSG method. Besides, the droop parameters of P-f and Q-V are $k_P = 1 \times 10^{-6}$ and $k_Q = 3 \times 10^{-5}$, respectively.

6.2. Time Domain Case Study

First of all, the time domain comparative simulations are conducted for the proposed IEC based VSG control method and the droop control method in the same scenarios: (1) The simulation starts at $t = 0$ s with the load capacity as $P = 120$ kW, $Q = 9$ kVar. The power allocation is: $P_1 = 75$ kW, $Q_1 = 6$ kVar for DG1, and $P_2 = 45$ kW, $Q_2 = 3$ kVar for DG2. (2) At $t = 3.0$ s, there is a load increase to 175 kW, 16.5 kVar; accordingly, the power allocation for each DG is adjusted to: $P_1 = 110$ kW, $Q_1 = 10$ kVar, and $P_2 = 65$ kW, $Q_2 = 6.5$ kVar. (3) At $t = 6.0$ s, there are some loads quit from the microgrid, which makes the load demand decrease to 155 kW and 13.5 kVar. Thus the output of each DG should be adjusted in timely way: $P_1 = 100$ kW, $Q_1 = 8$ kVar and $P_2 = 55$ kW, $Q_2 = 5.5$ kVar.

The simulation results are shown in Figures 12 and 13. The specific simulation results analyses about the two methods are given as follows:

The active and reactive power responses of the two methods are shown in Figures 12a,b and 13a,b. From which it can be learnt that the proposed method works better than the conventional droop control.

Firstly, for the proposed method, the two DGs can all accurately follow the set points of active and reactive power, with the power error under 5%. While the conventional droop control cannot output the specific power. For DG1, the steady state active power errors of droop control during the three periods are 13%, 20% and 11%, respectively; while the corresponding reactive power errors are 17%, 17% and 25%, respectively. Worse data is shown in DG2, with 22%, 32% and 23% active power negative offsets and 50%, 21% and 22% reactive power negative offsets.

Secondly, from Figure 12a,b it can be learnt that, compared with the conventional droop control, the proposed method in this paper has faster response speed and shorter transient process time. Last but not least, the overshoots of output power under the proposed method are smaller than those of the droop control method during the transient process. The biggest overshoot of reactive power of droop control is even 1.3 times its rated value, while the overshoots of active and reactive power of the proposed method are within 10% and 20%, respectively. Moreover, it can be learnt from Figures 12b and 13b that, under the droop control, there is still a large scale of power fluctuation span in the stable state, about 3% of the rated value.

Figure 12c,d illustrates the dynamic performance of frequency under the control of the two kinds of methods. From Figure 12d, it can be learnt that the droop control has a better performance on the maximum frequency deviation, which is about 0.13 Hz, while the maximum frequency deviation of the proposed method is 0.26 Hz. That is because, in order to prevent the system breakdown, the P-f, Q-V droop coefficient must be set sufficiently small. The frequency performance in this paper, however, can also meet the demand of system stability. The steady state deviation can be maintained under 0.1% of the rated value, and the maximal deviation of the transient process is under 0.5% of the rated value. Furthermore, compared with the continuous random fluctuation in the droop control method, the fluctuation of frequency for the proposed method is smaller in degree. Moreover, approximative periodic fluctuation can be modified or even eliminated by adjusting the PI controller parameters or adding a resistance in the LCL filter.

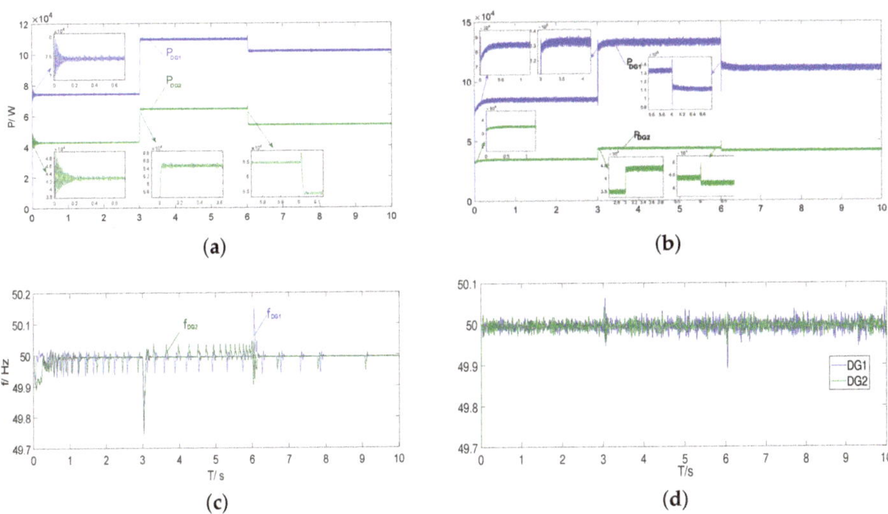

Figure 12. Simulation results of active power and frequency under different control methods. (**a**) Output active power of each DG under the proposed method; (**b**) output active power of each DG under the droop control; (**c**) frequency of each DG under the proposed method ; (**d**) frequency of each DG under the droop control.

The voltage Root Mean Square (RMS) value of PCC and output current RMS of the two DGs in the two methods are shown in Figure 13c,d, respectively. It can be learnt that, although both the methods can realize the voltage stability of the PCC, the voltage and current of the proposed method outperform that of the droop control method, for the proposed method has faster system response velocity and shorter transient time. What is more, when there are operation modes switching in the system, the maximum voltage sag of the proposed method is much smaller than that under the droop control.

In conclusion, the proposed method is valid for the operation of a microgrid with drastic power variation. Moreover, it outperforms the conventional droop control.

6.3. Small Signal Stability Verification

The static stability of the system under the proposed method is verified by the root loci diagram. The small signal model of the system is constructed based on the method in Section 5.1. Taking the parameter values from Table 1, and setting the R/X variation range as 0.01 to 100, the characteristic root loci of C(s) with the variable parameters K and R/X in each feeder are shown in Figures 14–16 respectively.

From the zoomed in diagram of root loci in Figure 15a–d, it can be learnt that, under the control strategy proposed in this paper, the small signal model of VSG output power is statically stable, for its roots always relay in the negative plane. Figure 16 shows the variation tendency of characteristic roots while the R/X changes from 0.01 to 100. Since, with the changing of R/X, there are nonlinear variations of parameters, it is difficult to plot the root locus directly. Therefore, the R/X variation range is set as 0.01–100, and we believe that this range could involve most of the R/X cases of feeders' impedance in real MGs. From the zoomed in views of Figure 16, it can be learnt that, with the changing of R/X, the small signal model of VSG power is also stable, whereas with the increasing of R/X, the system is unstable; that is, the increasing of line resistance does influence the stability of DG power, so, to some extent, it proves reasonable for taking the influence of line resistance into consideration.

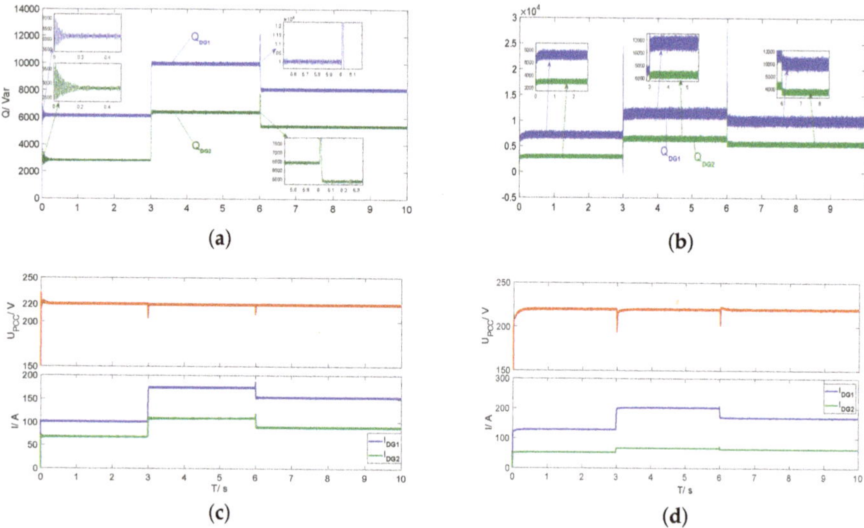

Figure 13. Simulation results of reactive power and voltage under different control methods. (**a**) Output reactive power of each DG under the proposed method; (**b**) output reactive power of each DG under the droop control; (**c**) voltage and current RMS of each DG under the proposed method; (**d**) voltage and current RMS of each DG under the droop control.

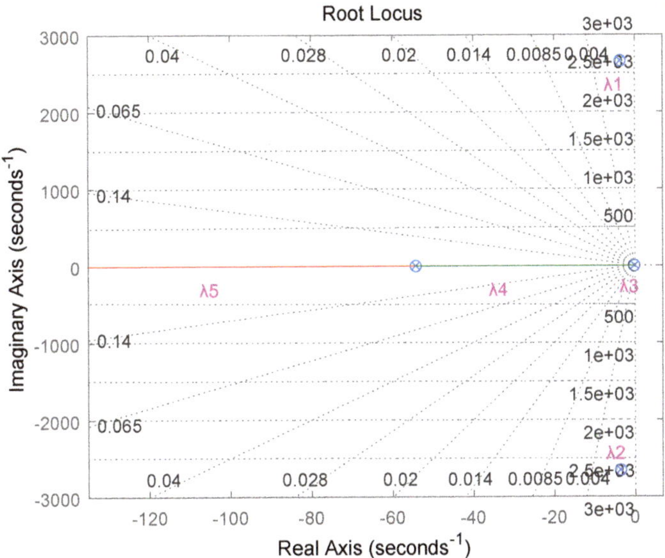

Figure 14. The root loci of the small signal model.

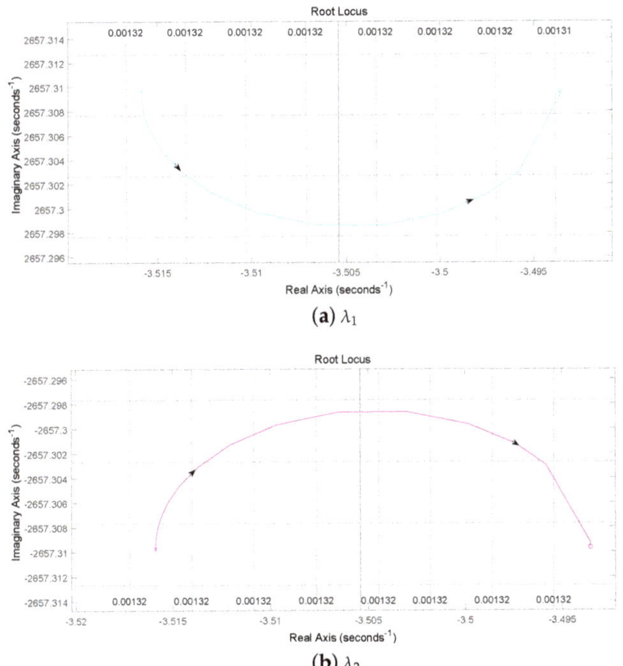

(**a**) λ_1

(**b**) λ_2

Figure 15. *Cont.*

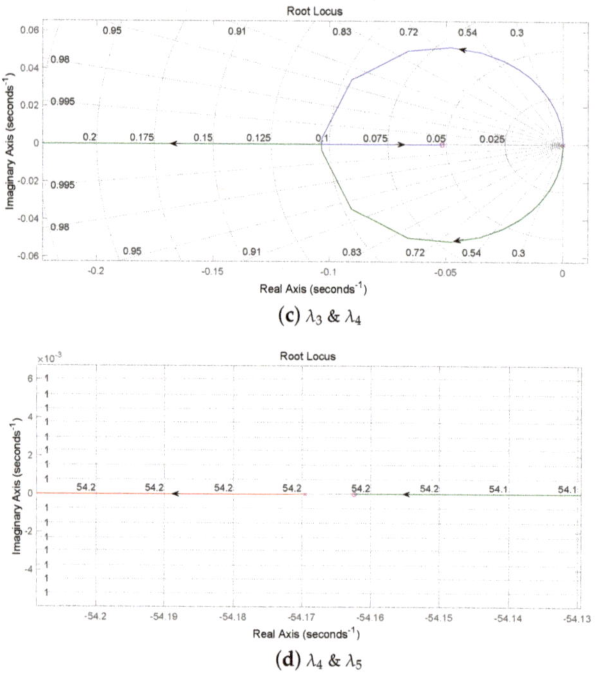

(**c**) λ_3 & λ_4

(**d**) λ_4 & λ_5

Figure 15. The zoomed-in diagrams of root loci.

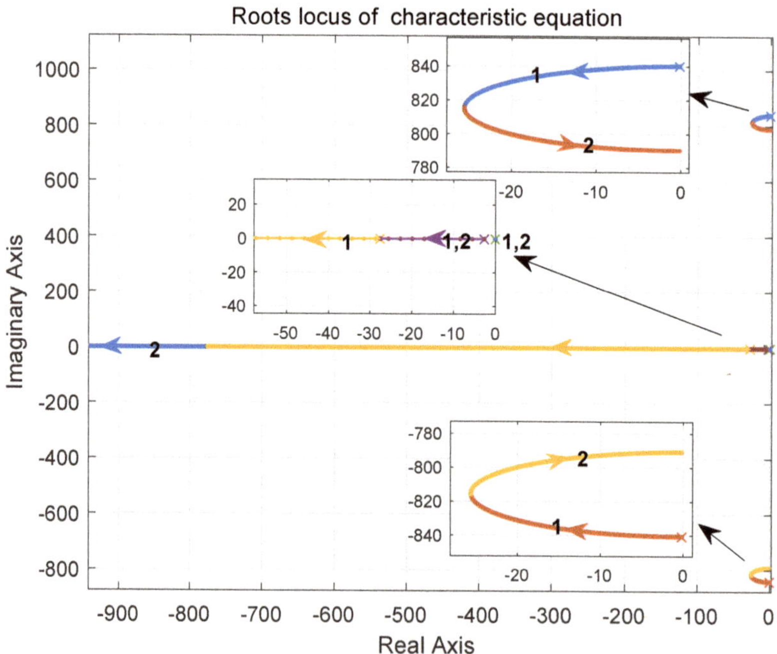

Figure 16. The root loci diagram with open loop coefficient R/X.

7. Conclusions

A novel VSG control method for accurate reactive power tracking/sharing and PCC voltage stabilization has been proposed and validated in this paper. The communication bus is employed to transmit information between the central controller and distributed controllers of each DG. Moreover, the optimized reference power values which are determined by the central controller are sent to each DG. The VSG based IEC method is proposed for DG units to compensate the voltage drop across feeders by tuning the EMF of the inverters. Therefore, the output reactive power can be regulated. A small-signal model is built to evaluate the static stability of the proposed method. It shows that the system under the control of the proposed method is small-signal stable, but an increase in the R/X ratio can influence the stability. The proposed method is verified in MATLAB/Simulink models, and it outperforms the conventional droop control method.

Author Contributions: The main idea and methodology of paper is conceived by B.Z., L.M., and D.Y. The feasibility analysis is proceeded by G.X.; the simulations and their verification are conducted by B.Z., L.M. and Z.M. The article logic and structure is organized by B.Z. The draft is written by L.M., and is revised by B.Z. All of the other authors help rectify the paper.

Funding: This research was funded by the National Natural Science Foundation of China (61703081), the Liaoning Revitalization Talents Program (XLYC1801005), Natural Science Foundation of Liaoning Province (20170520113), and the State Key Laboratory of Alternate Electrical Power System with Renewable Energy Sources (LAPS19005).

Acknowledgments: This work was partially supported by the National Natural Science Foundation of China (61703081), the Liaoning Revitalization Talents Program (XLYC1801005), Natural Science Foundation of Liaoning Province (20170520113), and the State Key Laboratory of Alternate Electrical Power System with Renewable Energy Sources (LAPS19005).

Conflicts of Interest: The authors declare no conflict of interest.

Appendix A

In this section, the dynamic model and transfer function of the LCL filter is constructed. The filter topological graph is shown in Figure 6b. Taking i_1, v_c and i as the state variables, the state function of the LCL filter can be expressed as:

$$\begin{cases} \dfrac{di_1}{dt} = \dfrac{1}{L_1}(E - V_c) \\[2mm] \dfrac{dv_c}{dt} = \dfrac{1}{C}(i_1 - i) \\[2mm] \dfrac{di}{dt} = \dfrac{1}{L_2 + L_l}(V_c - U - iR_l) \end{cases}$$

It can be rewritten as:

$$\dot{x} = \begin{bmatrix} 0 & -\dfrac{1}{L_1} & 0 \\[2mm] \dfrac{1}{C} & 0 & -\dfrac{1}{C} \\[2mm] 0 & \dfrac{1}{L_2 + L_l} & -\dfrac{R_l}{L_2 + L_l} \end{bmatrix} \begin{bmatrix} i_1 \\ v_c \\ i \end{bmatrix} + \begin{bmatrix} \dfrac{1}{L_1} & 0 \\[2mm] 0 & 0 \\[2mm] 0 & -\dfrac{1}{L_2 + L_l} \end{bmatrix} \begin{bmatrix} E \\ U \end{bmatrix} \tag{A1}$$

$$= Ax + Bu$$

where $u = [E \quad U]^T$ is the control input of the system. When v_c, i_{L2} are taken as the system output variables, there is:

$$y = Cx = \begin{bmatrix} 0 & 1 & 0 \\ 0 & 0 & 1 \end{bmatrix} x \tag{A2}$$

Thus the transfer function between output and input variables is:

$$\frac{Y(s)}{U(s)} = C(sI - A)^{-1}B = \frac{1}{\Delta}\begin{bmatrix} \dfrac{(L_2 + L_l)s + R_l}{CL_1(L_2 + L_l)} & -\dfrac{s}{C(L_2 + L_l)} \\ \dfrac{1}{CL_1(L_2 + L_l)} & -\dfrac{CL_l s^2 + 1}{CL_1(L_2 + L_l)} \end{bmatrix} \tag{A3}$$

where $\Delta = s^3 + \dfrac{R_l}{L_2 + L_l}s^2 + \dfrac{L_1 + L_2 + L_l}{CL_1(L_2 + L_l)}s + \dfrac{R_l}{CL_1(L_2 + L_l)}$ Therefore, the frequency-domain expression of voltage E' can be expressed as:

$$E' = V_c(s) - sL_2 I_{L2}(s) = \frac{(L_l s + R_l)E + (CL_1 s^2 + L_1 s + 1)U}{CL_1(L_2 + L_l)s^3 + CL_1 R_l s^2 + (L_1 + L_2 + L_l)s + R_l} \tag{A4}$$

$$= K(s) \cdot E$$

References

1. Bidram, A.; Davoudi, A. Hierarchical structure of microgrids control system. *IEEE Trans. Smart Grid* **2012**, *3*, 1963–1976. [CrossRef]
2. Ma, T.; Yahoui, H.; Vu, H.; Siauve, N.; Morel, H. A Control Strategy of DC Building Microgrid Connected to the Neighborhood and AC Power Network. *Buildings* **2017**, *7*, 42. [CrossRef]
3. Munir, H.; Zou, J.; Xie, C.; Guerrero, J. Cooperation of voltage controlled active power filter with grid-connected DGs in microgrid. *Sustainability* **2019**, *11*, 154. [CrossRef]
4. Karaagac, U.; Mahseredjian, J.; Jensen, S.; Gagnon, R.; Fecteau, M.; Kocar, I. Safe operation of DFIG-based wind parks in series-compensated systems. *IEEE Trans. Power Deliv.* **2017**, *33*, 709–718. [CrossRef]
5. Liu, J.; Miura, Y.; Ise, T. Comparison of dynamic characteristics between virtual synchronous generator and droop control in inverter-based distributed generators. *IEEE Trans. Power Electron.* **2015**, *31*, 3600–3611. [CrossRef]
6. Kerdphol, T.; Rahman, F.; Mitani, Y. Virtual inertia control application to enhance frequency stability of interconnected power systems with high renewable energy penetration. *Energies* **2018**, *11*, 981. [CrossRef]
7. Li, Y.W.; Kao, C. An Accurate Power Control Strategy for Power-Electronics-Interfaced Distributed Generation Units Operating in a Low-Voltage Multibus Microgrid. *IEEE Trans. Power Electron.* **2009**, *24*, 2977–298.
8. Wu, T.; Liu, Z.; Liu, J.; Wang, S.; You, Z. A unified virtual power decoupling method for droop-controlled parallel inverters in microgrids. *IEEE Trans. Power Electron.* **2015**, *31*, 5587–5603. [CrossRef]
9. Rowe, C.N.; Summers, T.J.; Betz, R.E.; Cornforth, D.J.; Moore, T.G. Arctan power–frequency droop for improved microgrid stability. *IEEE Trans. Power Electron.* **2012**, *28*, 3747–3759. [CrossRef]
10. Kerdphol, T.; Rahman, F.; Mitani, Y.; Hongesombut, K.; Küfeoğlu, S. Virtual inertia control-based model predictive control for microgrid frequency stabilization considering high renewable energy integration. *Sustainability* **2017**, *9*, 773. [CrossRef]
11. Zhong, Q.C.; Nguyen, P.L.; Ma, Z.; Sheng, W. Self-synchronized synchronverters: Inverters without a dedicated synchronization unit. *IEEE Trans. Power Electron.* **2013**, *29*, 617–630. [CrossRef]
12. Shintai, T.; Miura, Y.; Ise, T. Oscillation damping of a distributed generator using a virtual synchronous generator. *IEEE Trans. Power Deliv.* **2014**, *29*, 668–676. [CrossRef]
13. Hirase, Y.; Sugimoto, K.; Sakimoto, K.; Ise, T. Analysis of resonance in microgrids and effects of system frequency stabilization using a virtual synchronous generator. *IEEE J. Emerg. Sel. Top. Power Electron.* **2016**, *4*, 1287–1298. [CrossRef]
14. Hirase, Y.; Abe, K.; Sugimoto, K.; Shindo, Y. A grid connected inverter with virtual synchronous generator model of algebraic type. *IEEJ Trans. Power Energy* **2012**, *132*, 371–380. [CrossRef]
15. Chen, Y.; Hesse, R.; Turschner, D.; Beck, H.P. Improving the grid power quality using virtual synchronous machines. In Proceedings of the 2011 International Conference on Power Engineering, Energy and Electrical Drives, Malaga, Spain, 11–13 May 2011; pp. 1–6.

16. Soni, N.; Doolla, S.; Chandorkar, M.C. Improvement of transient response in microgrids using virtual inertia. *IEEE Trans. Power Deliv.* **2013**, *28*, 1830–1838. [CrossRef]

17. Guerrero, J.M.; De Vicuna, L.G.; Matas, J.; Castilla, M.; Miret, J. A wireless controller to enhance dynamic performance of parallel inverters in distributed generation systems. *IEEE Trans. Power Electron.* **2004**, *19*, 1205–1213. [CrossRef]

18. Mahmood, H.; Michaelson, D.; Jiang, J. Accurate reactive power sharing in an islanded microgrid using adaptive virtual impedances. *IEEE Trans. Power Electron.* **2014**, *30*, 1605–1617. [CrossRef]

19. Zhu, Y.; Zhuo, F.; Wang, F.; Liu, B.; Gou, R.; Zhao, Y. A virtual impedance optimization method for reactive power sharing in networked microgrid. *IEEE Trans. Power Electron.* **2015**, *31*, 2890–2904. [CrossRef]

20. Zhang, H.; Kim, S.; Sun, Q.; Zhou, J. Distributed adaptive virtual impedance control for accurate reactive power sharing based on consensus control in microgrids. *IEEE Trans. Smart Grid* **2016**, *8*, 1749–1761. [CrossRef]

21. He, J.; Li, Y.W.; Blaabjerg, F. An enhanced islanding microgrid reactive power, imbalance power, and harmonic power sharing scheme. *IEEE Trans. Power Electron.* **2014**, *30*, 3389–3401. [CrossRef]

22. Xiong, L.; Zhuo, F.; Wang, F.; Liu, X.; Chen, Y.; Zhu, M.; Yi, H. Static synchronous generator model: A new perspective to investigate dynamic characteristics and stability issues of grid-tied PWM inverter. *IEEE Trans. Power Electron.* **2015**, *31*, 6264–6280. [CrossRef]

23. Zhang, X.; Mao, F.; Xu, H.; Liu, F.; Li, M. An optimal coordination control strategy of micro-grid inverter and energy storage based on variable virtual inertia and damping of VSG. *Chin. J. Electrical Eng.* **2017**, *3*, 25–33.

24. Micallef, A.; Apap, M.; Spiteri-Staines, C.; Guerrero, J.M.; Vasquez, J.C. Reactive power sharing and voltage harmonic distortion compensation of droop controlled single phase islanded microgrids. *IEEE Trans. Smart Grid* **2014**, *5*, 1149–1158. [CrossRef]

25. Mohamed, Y.A.R.I.; El-Saadany, E.F. Adaptive decentralized droop controller to preserve power sharing stability of paralleled inverters in distributed generation microgrids. *IEEE Trans. Power Electron.* **2008**, *23*, 2806–2816. [CrossRef]

26. Pogaku, N.; Prodanovic, M.; Green, T.C. Modeling, analysis and testing of autonomous operation of an inverter-based microgrid. *IEEE Trans. Power Electron.* **2007**, *22*, 613–625. [CrossRef]

27. Barklund, E.; Pogaku, N.; Prodanovic, M.; Hernandez-Aramburo, C.; Green, T.C. Energy management in autonomous microgrid using stability-constrained droop control of inverters. *IEEE Trans. Power Electron.* **2008**, *23*, 2346–2352. [CrossRef]

28. Majumder, R.; Chaudhuri, B.; Ghosh, A.; Majumder, R.; Ledwich, G.; Zare, F. Improvement of stability and load sharing in an autonomous microgrid using supplementary droop control loop. *IEEE Trans. Power Syst.* **2009**, *25*, 796–808. [CrossRef]

29. De Brabandere, K.; Bolsens, B.; Van den Keybus, J.; Woyte, A.; Driesen, J.; Belmans, R. A voltage and frequency droop control method for parallel inverters. *IEEE Trans. Power Electron.* **2007**, *22*, 1107–1115. [CrossRef]

30. Hu, S. *Automatic Control Principles*, 5th ed.; Science Press: Beijing, China, 2007.

© 2019 by the authors. Licensee MDPI, Basel, Switzerland. This article is an open access article distributed under the terms and conditions of the Creative Commons Attribution (CC BY) license (http://creativecommons.org/licenses/by/4.0/).

 *sustainability*

Article

Improvement of MPPT Control Performance Using Fuzzy Control and VGPI in the PV System for Micro Grid

Jong-Chan Kim [1], Jun-Ho Huh [2] and Jae-Sub Ko [3,*]

[1] Department of Computer Engineering, Sunchon National University, 255 Jungang-ro, Suncheon-city Jeollanam do 57922, Korea; seaghost@sunchon.ac.kr
[2] Department of Data Informatics, Korea Maritime and Ocean University, Busan 49112, Korea; 72networks@kmou.ac.kr
[3] Department of Electrical Engineering, Sunchon National University, 255 Jungang-ro, Suncheon-city Jeollanam do 57922, Korea
* Correspondence: kokos22@sunchon.ac.kr; Tel.: +82-61-750-3540

Received: 13 August 2019; Accepted: 15 October 2019; Published: 23 October 2019

Abstract: This paper proposes the method for maximum power point tracking (MPPT) of the photovoltaic (PV) system. The conventional PI controller controls the system with fixed gains. Conventional PI controllers with fixed gains cannot satisfy both transient and steady-state. Therefore, to overcome the shortcomings of conventional PI controllers, this paper presents the variable gain proportional integral (VGPI) controllers that control the gain value of PI controllers using fuzzy control. Inputs of fuzzy control used in the VGPI controller are the slope from the voltage-power characteristics of the PV module. This paper designs fuzzy control's membership functions and rule bases using the characteristics that the slope decreases in size, as it approaches the maximum power point and increases as it gets farther. In addition, the gain of the PI controller is adjusted to increase in transient-state and decrease in steady-state in order to improve the error in steady-state and the tracking speed of maximum power point of the PV system. The performance of the VGPI controller has experimented in cases where the solar radiation is constant and the solar radiation varies, to compare with the performance of the P&O method, which is traditionally used most often in MPPT, and the performance of the PI controller, which is used most commonly in the industry field. Finally, the results from the experiment are presented and the results are analyzed.

Keywords: PV system; PI controller; fuzzy control; MPPT; tracking speed; error; Micro Grid

---

1. Introduction

Renewable energy is drawing much attention as an energy source that can replace fossil fuels. Solar energy is the most representative renewable energy and an infinite, eco-friendly energy but is highly dependent on temperature and solar radiation. Temperature affects voltage, and solar radiation affects current [1–3]. Temperature and solar radiation have a direct impact on solar power output and, in particular, cause a change in MPP (maximum power point). In order to track the MPP of the PV system efficiently, an appropriate DC-DC converter and a tracking algorithm (method) must be integrated and configured, and the following conditions must be satisfied [4]:

- Fast tracking response (transient response).
- No vibration around the MPP (steady-state response).
- Response performance against insolation and temperature change.
- Simple structure and low cost.

Typical methods for MPPT (maximum power point tracking) are the Constant Voltage method using a fixed ratio of the open voltage, P&O (Perturb & Observe) method using power and voltage perturbation, and IncCond (Incremental Conductance) method using slope and conductance that can be obtained in the current (I_{pv})-voltage (V_{pv}) and power (P)-voltage (V_{pv}) curves [5–13]. In addition, These methods use reference voltage [14–21], reference current [22,23], or duty ratio [24] for maximum power point tracking. Among them, the P&O method has the advantages of simple structure and low calculation, whereas the IncCond method has the advantage of tracking the MPP faster and more accurately than the P&O. These methods are most generally used for the MPPT of the photovoltaic (PV) system due to the aforesaid advantages. Since the P&O and IncCond methods track the MPP while varying the voltage or current by a fixed size, however, vibration may occur near the MPP, and performance may deteriorate; thus rapidly changing the solar radiation conditions. Although used to solve these problems, artificial neural networks also have problems such as long learning time and high computational complexity [25,26].

Therefore, this paper proposes a method of tracking the MPP of the PV system using the PI controller, which is widely used to control the industrial field [27–33]. The PI controller is a controller that uses proportional gain and integral gain. The proportional gain and the integral gain of the PI controller are closely related to the rise time, settling time, and steady-state error, and there is clear relationship between the gain and the control amount. In addition, it has a simple structure and a small amount of calculation, which enables a quick response. Since the PI controller is generally controlled with a fixed gain value, however, it is difficult to satisfy both transient and steady states. The response performance of the steady state is degraded when the gain value is increased for the fast response of the transient state, whereas, the performance of the transient state is degraded if the gain value is reduced to improve the steady-state response performance. Therefore, it is necessary to control the gain so as to satisfy both transient and steady states by automatically adjusting the gain value of the PI controller according to the operation state. In this paper, fuzzy control is used to adjust the gain of the PI controller. Methods for adjusting the gain value of PI controller using fuzzy control were presented through several studies [34–38]. However, existing studies depend on designer knowledge for rule base and membership function designs and do not suggest how the gain value of a PI controller changes with its operational state. Thus, the paper proposes simple and clear fuzzy control membership function and rule base design according to the characteristics of the PV system and shows the gains of PI controller, which are changing in the transient and steady-state of the system.

Fuzzy control does not require accurate modeling and has the advantage of controlling nonlinear systems [39]. The VGPI (variable gain proportional integral) controller proposed in this paper uses the voltage and current of photovoltaic power generation, as inputs to control the gain value of the PI controller with a fuzzy controller, and the PI controller tracks the MPP of the PV system.

The VGPI controller compares the error at tracking speed and steady state with the most commonly used P&O controller and PI controller with a fixed gain value. It also shows the characteristics of gain values of the PI controller controlled by fuzzy control in the VGPI controller.

The rest of this paper is organized as follows: Section 2 discusses the DC-DC converter and the conventional MPPT method; Section 3 presents the MPPT control by the VGPI controller; Section 4 shows the comparison and analysis results of the MPPT control characteristics with the method proposed in the paper and the existing method; Finally, Section 5 presents the conclusion.

2. Conventional MPPT Method

2.1. MPPT Control by DC-DC Converter

Figure 1 shows the MPPT control of photovoltaic power generation using the DC-DC converter [40–43]. The PV module supplies voltage and current to load R through the DC-DC converter. Figure 2 shows the relationship curves of the current (I_{pv}) - voltage (V_{pv}) of the PV module.

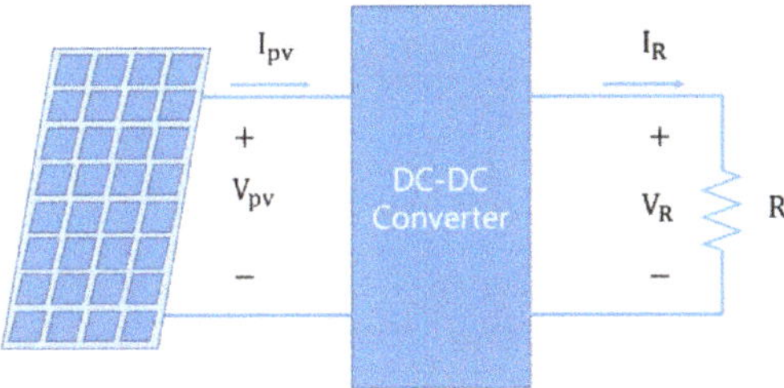

Figure 1. Maximum power point tracking (MPPT) control of the photovoltaic (PV) system by DC-DC converter.

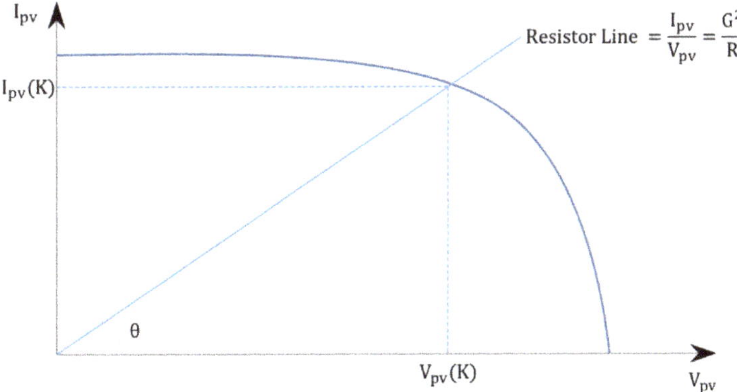

Figure 2. I-V curve of the PV module.

If the gain of the converter is G, the relationship between input and output in Figure 2 can be obtained as follows:

$$V_R = R \times I_R \tag{1}$$

$$G = \frac{V_R}{V_{pv}} \tag{2}$$

$$G = \frac{I_{pv}}{I_R} \tag{3}$$

$$\frac{V_{pv}}{I_{pv}} = \frac{R}{G^2} \tag{4}$$

$$V_{pv} = \frac{R}{G^2} \times I_{pv} \tag{5}$$

In Figure 2, the inclination angle (θ) of the resistor line can be calculated as follows:

$$\theta = \mathrm{atan}\left(\frac{G^2}{R}\right) \tag{6}$$

In this paper, a buck converter is used as DC-DC converter. Figure 3 shows the structure of the buck converter.

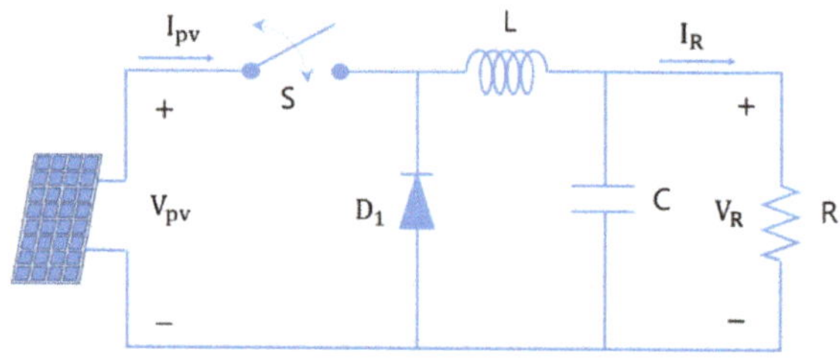

Figure 3. Structure of the buck converter.

The diode (D_1) is reverse-biased during on-time and forward-biased during off-time in the buck converter. The relationship between the input and output of the Buck converter is expressed by the duty ratio (D) [44–46].

$$\frac{V_R}{V_{pv}} = D \tag{7}$$

$$\frac{I_R}{I_{pv}} = \frac{1}{D} \tag{8}$$

Since the buck converter has the same gain (G) and duty ratio, it can be expressed as:

$$D = G \tag{9}$$

$$\theta = \operatorname{atan}\left(\frac{G^2}{R}\right) = \operatorname{atan}\left(\frac{D^2}{R}\right) \tag{10}$$

Since the duty ratio of the buck converter is $0 \le D \le 1$, the range of the inclination angle can be determined as follows, and Table 1 shows the maximum and minimum inclination angles of the buck converter:

$$\theta|_{D=0} = \operatorname{atan}\left(\frac{0^2}{R}\right) = 0 \tag{11}$$

$$\theta|_{D=1} = \operatorname{atan}\left(\frac{1^2}{R}\right) = \operatorname{atan}\left(\frac{1}{R}\right) \tag{12}$$

Table 1. Maximum and minimum inclination angles.

Minimum Inclination Angle	Maximum Inclination Angle
$\theta|_{D=0} = 0$	$\theta|_{D=1} = \operatorname{atan}\left(\frac{1}{R}\right)$

Figure 4 shows the tracking region and non-tracking region of the MPP according to the inclination angle. In order to track the MPP efficiently, the resistance value must be selected such that the inclination angle is lower than that at the MPP. When the duty ratio of the buck converter changes from 0 to 1, the buck converter changes from the maximum voltage (V_{oc}), the open voltage, to the load voltage (V_R) [4].

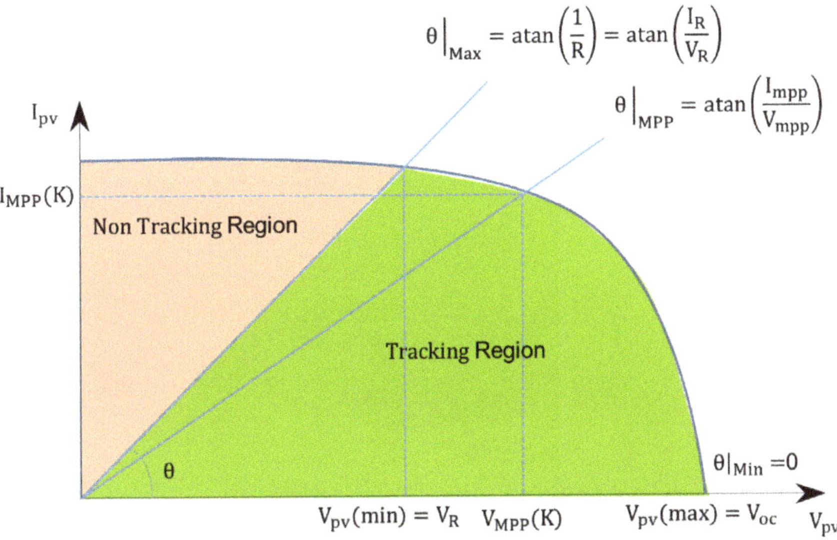

Figure 4. MPPT range according to the inclination angle.

2.2. P&O Method

The P&O MPPT method tracks the MPP by varying the voltage of the solar cell and observing the power and increasing or decreasing the PV voltage in the direction wherein the current power is greater than the previous power.

Table 2 shows the operating state of the P&O method according to the voltage and power states. In cases 1 and 3, when voltage change (ΔV) is increasing (positive) or decreasing (negative), power change (ΔP) is increasing (positive), and control is continued in the same direction. In cases 2 and 4, since power change (ΔP) is negative, it tracks the maximum power by performing control in the direction opposite to the change in voltage. Figure 5 shows the flow chart of Table 2 [5,6,10,15].

Table 2. Operating state of the P&O method according to the voltage and power states.

Case	Perturbation $[\Delta V_{pv}=V_{pv}(k)-V_{pv}(k-1)]$	Change in Power $[\Delta P=P(k)-P(k-1)]$	Next Perturbation $[\Delta V_{ref}(k)]$
1	Positive	Positive	Positive
2	Positive	Negative	Negative
3	Negative	Positive	Negative
4	Negative	Negative	Positive

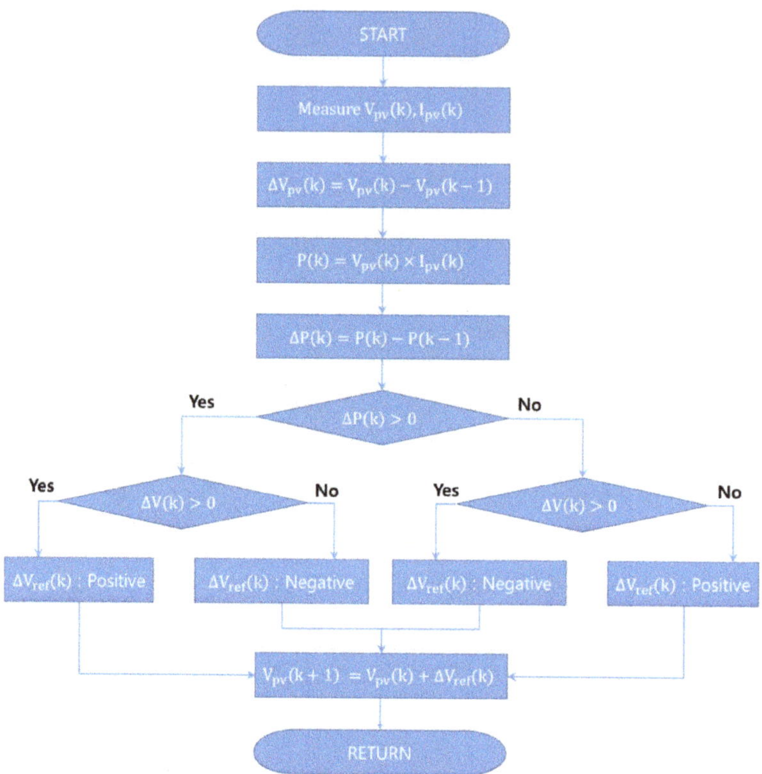

Figure 5. Flowchart of the Perturb & Observe (P&O) MPPT method.

2.3. IncCond Method

The IncCond (Incremental Conductance) method is a method of using the slope of the power-voltage curve of the solar cell, and the slope can be expressed by Equation (20). It is one of the most widely used methods in the field along with the P&O method because of its stable characteristics and simple implementation method. Figure 6 shows the control principle of IC MPPT. As shown in the characteristic curve of Figure 6, IC MPPT finds the MPP by using the fact that the slope of the characteristic curve is 0 (zero) in MPP. The slope of the output curve of the solar cell can be expressed as dP/dV_{pv}. In Figure 6, the MPP is point B, and the slope is zero. Based on the MPP, we can see that the left side has a positive slope, and the right side has a negative slope. The conditions at each point are shown in Equations (14)–(16). In Equations (14)–(16), I_{pv}/V_{pv} is the conductance of the inverse of the resistance, and dI_{pv}/dV_{pv} is the incremental conductance of the change in conductance. Therefore, the method of using the slope of power and voltage is called the incremental conductance (IncCond) method.

$$\frac{dP}{dV_{pv}} = \frac{d(V_{pv}I_{pv})}{dV_{pv}} = \frac{dV_{pv} \times I_{pv}}{dV_{pv}} + \frac{V_{pv} \times dI_{pv}}{dV_{pv}} = I_{pv} + V_{Pv}\frac{dI_{pv}}{dV_{pv}} \tag{13}$$

$$A: \frac{dP}{dV_{pv}} = I_{Pv} + V_{pv}\frac{dI_{pv}}{dV_{pv}} < 0 \rightarrow \frac{dI_{pv}}{dV_{pv}} < -\frac{I_{pv}}{V_{pv}} \tag{14}$$

$$B: \frac{dP}{dV_{pv}} = I_{Pv} + V_{pv}\frac{dI_{pv}}{dV_{pv}} = 0 \rightarrow \frac{dI_{pv}}{dV_{pv}} = -\frac{I_{pv}}{V_{pv}} \tag{15}$$

$$C: \frac{dP}{dV_{pv}} = I_{Pv} + V_{Pv}\frac{dI_{pv}}{dV_{pv}} > 0 \rightarrow \frac{dI_{pv}}{dV_{pv}} > -\frac{I_{pv}}{V_{pv}} \tag{16}$$

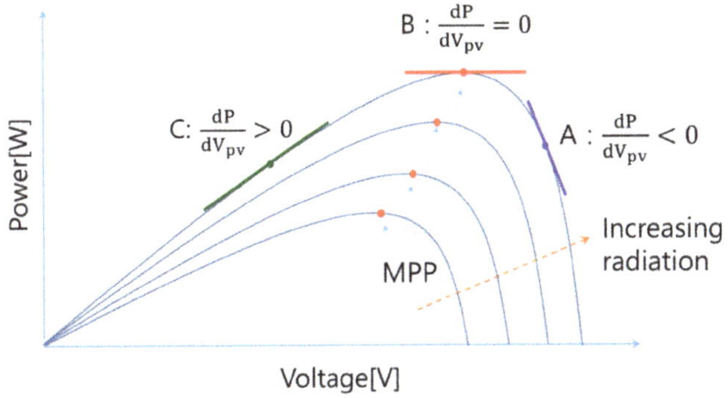

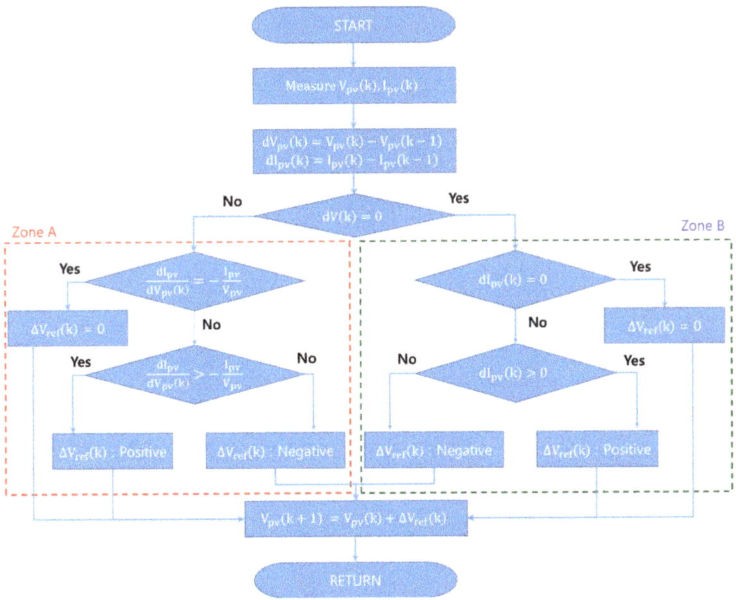

Figure 6. Control principle of the IncCond MPPT method.

In Figure 6, the MPP of the photovoltaic power generation moves to the left as the solar radiation increases [47,48]. Therefore, if the solar radiation is increased, the voltage is increased; if the solar radiation is decreased, however, the voltage is decreased, and the MPP change due to the solar radiation change can be tracked more quickly.

The flow chart of the IncCond method is shown in Figure 7 using the slope condition of the P-V curve and MPP variation according to the changing solar radiation in Figure 6 [12,14].

Figure 7. Flow chart of the IncCond method.

In Figure 7, Zone A shows the part that tracks the MPP along the slope in the P-V curve, and Zone B shows the part that tracks the MPP for the change in solar radiation. Since solar radiation greatly affects the current of the photovoltaic power generation, the change of solar radiation is caused by the change of current. Therefore, the amount of solar radiation can be judged to have changed when only the change of current occurs without a change in voltage.

3. Proposed MPPT Method

In this paper, a PI controller is used to track the MPP of the PV system. The PI controller is a method that uses proportional control and integral control. The PI controller uses the gain for proportional control and the gain for integral control. Table 3 shows the effect of these gain values on the system. If large values of proportional gain and integral gain are selected to reduce the rise time and steady-state error, the overshoot may increase greatly, and the settling time may be longer. In addition, the error may increase in steady state. Generally, in the case of PI control, since two gain values are fixed, it is very important to select the gains value corresponding to the control state [27,49].

Table 3. Influence of the proportional integral (PI) controller gain value on the system.

Parameter	Rise Time	Overshoot	Settling Time	Steady-State Error
K_p Increase	Decrease	Increase	Small Change	Decrease
K_i Increase	Decrease	Increase	Increase	Decrease Significantly

This paper proposes a method of adjusting the gain value of the PI controller using fuzzy control to improve this characteristic of the PI controller. Fuzzy control does not require accurate system modeling and has the advantage of handling nonlinear systems. Fuzzy control is controlled by using the rules of the "IF THEN" structure in simple language. The fuzzy controller inputs the error and the error change value to perform control through fuzzification and inferential engine defuzzification [28]. The most common reasoning method used in the fuzzy controller is Mamdani's MIN-MAX method. The "IF THEN" rule for multiple inputs has "AND" and "OR" operations, and it can be expressed as follows [4]:

$$\text{IF } x \text{ is } A_1 \text{ AND } x \text{ is } A_2 \ldots \text{ AND } x \text{ is } A_n \text{ THEN } y \text{ is } B_s$$

$$\text{IF } x \text{ is } A_s \text{ THEN } y \text{ is } B_s$$

$$A_s = A_1 \cap A_2 \cap A_3 \cap \ldots A_n \tag{17}$$

$$\mu A_s(x) = \text{Min}[\mu A_1(x), \ \mu A_2(x), \cdots, \mu A_n(x)]$$

$$\text{IF } x \text{ is } A_1 \text{ OR } x \text{ is } A_2 \ldots \text{ OR } x \text{ is } A_n \text{ THEN } y \text{ is } B_s$$

$$\text{IF } x \text{ is } A_s \text{ THEN } y \text{ is } B_s$$

$$A_s = A_1 \cup A_2 \cup A_3 \cup \ldots A_n \tag{18}$$

$$\mu A_s(x) = \text{max}[\mu A_1(x), \ \mu A_2(x), \cdots, \mu A_n(x)]$$

where $\mu A_n(x)$ represents the membership strength of the fuzzy membership function for input A_n. Various methods of adjusting the gain value of the PI controller using fuzzy control have been proposed. These methods are based on user knowledge in designing Fuzzy Control's membership functions and rule base and do not represent the background to design. This approach has the problem of redesigning membership functions and rule bases for other users to use.

Therefore, in this paper, using the operating characteristics for the gain value of the PI controller, a simpler and easier-to-understand controller is designed. Member functions and rule bases designed in the paper are based on the following:

1. Error and changing error which is the input of fuzzy control, are as shown in expressions (19) and (20).
2. The error and the changing error decrease in size as solar power is closer to the MPP.
3. If the input value is large, the tracking speed should be fast because it is far from the MPP. This increases the gain value of the PI controller.

4. If the input value is small, it is close to the MPP and the error in steady state must be reduced.

This reduces the gain value of the PI controller.

The input of fuzzy control, error and error variation, are divided into seven sections: Negative big (NB), Negative medium (NM), Negative small (NS), zero(ZE), Positive big (PB), Positive medium (PM) and positive small (PS). The output of the fuzzy control is designed to perform three actions: increase (P: positive), hold (ZE: zero) and decrease (N: negative).

Tables 4 and 5 show the rule base for proportional gain (K_p) and integral gain (K_i) designed in the paper, and Figures 8–10 show membership function for the input and output of fuzzy control.

Table 4. Rule base to adjust gain K_p.

ce \ e	NB	NM	NS	ZE	PS	PM	PB
NB	P	P	ZE	ZE	ZE	P	P
NM	P	ZE	ZE	N	ZE	ZE	P
NS	P	P	N	N	N	P	P
ZE	P	ZE	N	N	N	ZE	P
PS	P	P	N	N	N	P	P
PM	P	ZE	ZE	N	ZE	ZE	P
PB	P	P	ZE	ZE	ZE	P	P

Table 5. Rule base to adjust gain K_i.

ce \ e	NB	NM	NS	ZE	PS	PM	PB
NB	P	P	ZE	ZE	ZE	P	P
NM	P	P	ZE	N	ZE	P	P
NS	P	P	N	N	N	P	P
ZE	P	ZE	N	N	N	ZE	P
PS	P	P	N	N	N	P	P
PM	P	P	ZE	N	ZE	P	P
PB	P	P	ZE	ZE	ZE	P	P

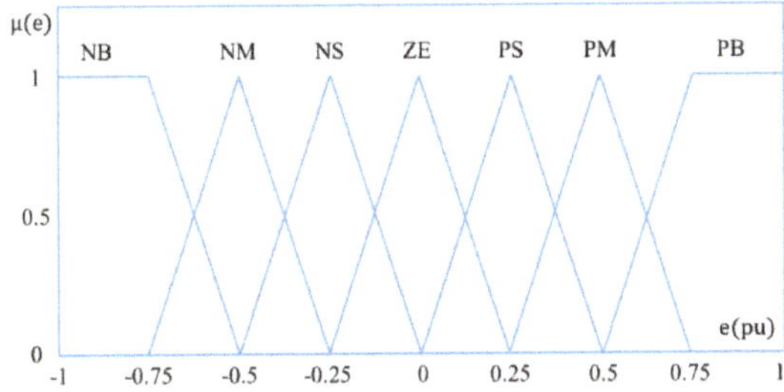

Figure 8. Member function for error.

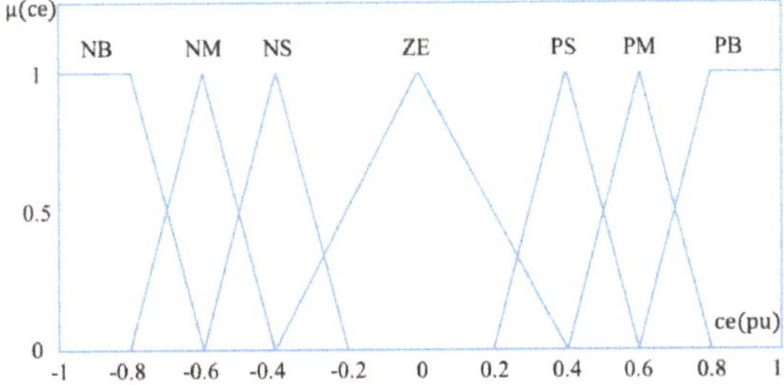

Figure 9. Member function for changing error (ce).

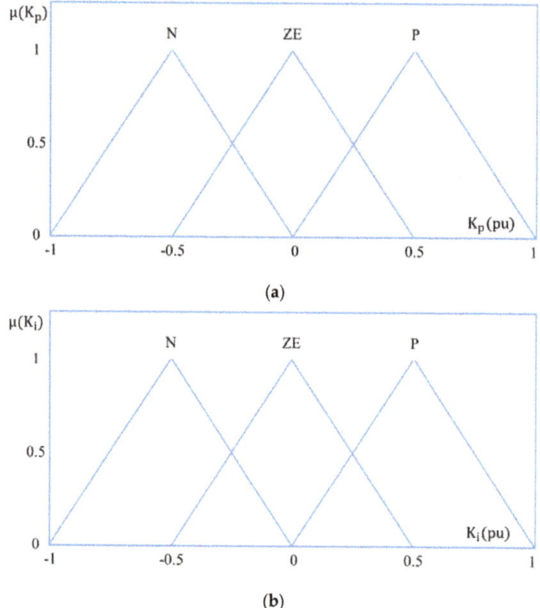

Figure 10. Member function for output. (**a**) Proportional gain (K_p); (**b**) Integral gain (K_i).

$$E(k) = \frac{P(k) - P(k-1)}{V_{pv}(k) - V_{pv}(k-1)} \tag{19}$$

$$CE(k) = E(k) - E(k-1) \tag{20}$$

Equations (21)–(24) show the gain of the PI controller adjusted by the fuzzy controller. Outputs ΔK_p and ΔK_i of the fuzzy controller are calculated using the center of gravity (COG) [27,50]:

$$K_p(k) = K_p(k-1) + \Delta K_p \tag{21}$$

$$K_i(k) = K_i(k-1) + \Delta K_i \tag{22}$$

$$\Delta K_p = \frac{\sum_{j=1}^{n} \mu(K_p)_j \cdot (K_p)_j}{\sum_{j=1}^{n} \mu(K_p)_j} \tag{23}$$

$$\Delta K_i = \frac{\sum_{j=1}^{n} \mu(K_i)_j \cdot (K_i)_j}{\sum_{j=1}^{n} \mu(K_i)_j} \tag{24}$$

Figure 11 shows an example of the input error (0.7) and changing error (0.3) of fuzzy control. When the error and error change values are calculated, the membership strength is calculated from the membership functions shown in Figures 8 and 9. In the membership function for error 0.7, the membership strength is 0.8 for PM and 0.2 for PB. For changing error 0.3, the strength of membership is calculated for the membership function, which is 0.5 for ZE and 0.25 for PS. Four output values are calculated through the AND operation of Equation (17) and the rule base of K_p in Table 4, and 0.413 can be obtained by calculating the final values through Equation (23).

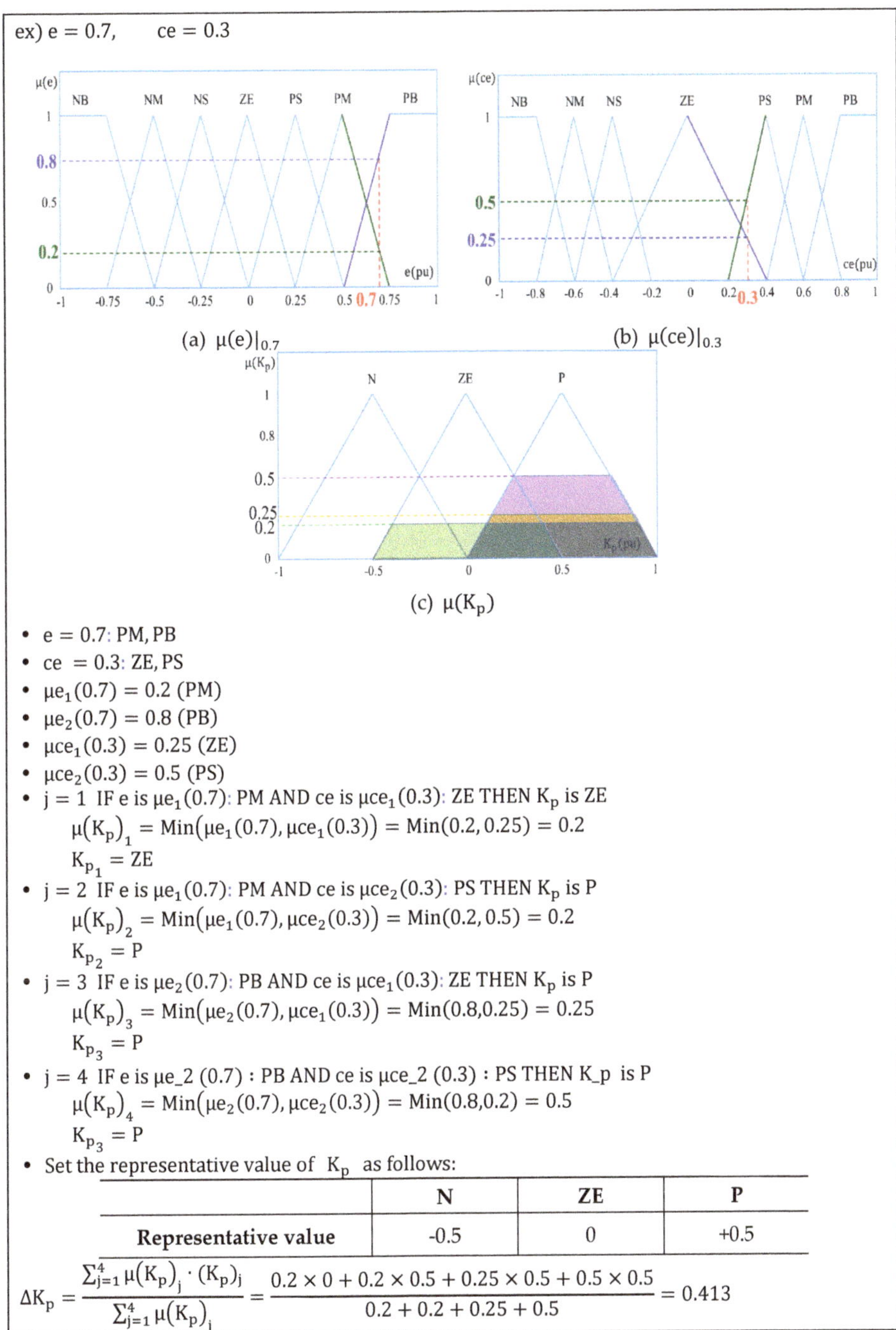

- $e = 0.7$: PM, PB
- $ce = 0.3$: ZE, PS
- $\mu e_1(0.7) = 0.2$ (PM)
- $\mu e_2(0.7) = 0.8$ (PB)
- $\mu ce_1(0.3) = 0.25$ (ZE)
- $\mu ce_2(0.3) = 0.5$ (PS)
- $j = 1$ IF e is $\mu e_1(0.7)$: PM AND ce is $\mu ce_1(0.3)$: ZE THEN K_p is ZE

    $\mu(K_p)_1 = Min(\mu e_1(0.7), \mu ce_1(0.3)) = Min(0.2, 0.25) = 0.2$

    $K_{p_1} = ZE$
- $j = 2$ IF e is $\mu e_1(0.7)$: PM AND ce is $\mu ce_2(0.3)$: PS THEN K_p is P

    $\mu(K_p)_2 = Min(\mu e_1(0.7), \mu ce_2(0.3)) = Min(0.2, 0.5) = 0.2$

    $K_{p_2} = P$
- $j = 3$ IF e is $\mu e_2(0.7)$: PB AND ce is $\mu ce_1(0.3)$: ZE THEN K_p is P

    $\mu(K_p)_3 = Min(\mu e_2(0.7), \mu ce_1(0.3)) = Min(0.8, 0.25) = 0.25$

    $K_{p_3} = P$
- $j = 4$ IF e is $\mu e\_2(0.7)$: PB AND ce is $\mu ce\_2(0.3)$: PS THEN K_p is P

    $\mu(K_p)_4 = Min(\mu e_2(0.7), \mu ce_2(0.3)) = Min(0.8, 0.2) = 0.5$

    $K_{p_3} = P$
- Set the representative value of K_p as follows:

	N	**ZE**	**P**
Representative value	-0.5	0	+0.5

$$\Delta K_p = \frac{\sum_{j=1}^{4} \mu(K_p)_j \cdot (K_p)_j}{\sum_{j=1}^{4} \mu(K_p)_j} = \frac{0.2 \times 0 + 0.2 \times 0.5 + 0.25 \times 0.5 + 0.5 \times 0.5}{0.2 + 0.2 + 0.25 + 0.5} = 0.413$$

Figure 11. K_p calculation for error and changing error.

Figure 12 shows the structure of the VGPI controller for MPPT control of the PV system. The inputs are the voltage (V_{pv}) and current (I_{pv}) of the PV system, and the output is the change value of the PI controller gain value ($\Delta K_p, \Delta K_i$) through the fuzzy controller. The PI controller outputs the PWM (Pulse Width Modulation) signal for MPPT control using the adjusted gains by fuzzy control, and this signal controls the DC-DC converter.

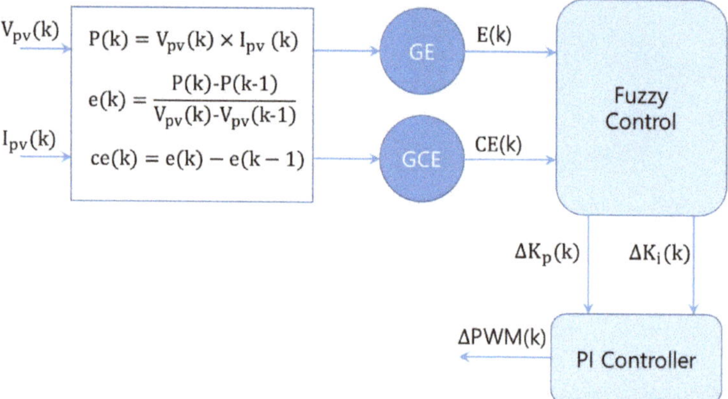

Figure 12. Variable gain proportional integral (VGPI) controller for the MPPT control of the PV system.

4. Experiment Result

The control performance of MPPT is verified by the speed at which maximum power is tracked and the magnitude of the error in steady state using voltage, current and power output from the PV module. In order to verify MPPT's performance in the paper, the experimental device was constructed with a PV module, a Buck converter, a DC-DC step down converter and a battery.

Experiments in solar power use solar simulators or use artificial light sources to construct a constant experimental environment. In the paper, a constant experimental environment was constructed using artificial lighting. Artificial light sources in the experimental environment can be used to maintain or change the test conditions. In addition, the same environmental conditions can be configured for different methods, so that the performance of the proposed method and the conventional method can be compared. The proposed method and conventional method compare the speed at which the maximum power point is tracked and the error at steady state. Since the environment is constructed using the same artificial light source, comparisons of output power, voltage and current can be a valid method for verifying peak power point tracking performance [51–54].

Figure 13 shows the circuit diagram and control system for the MPPT control performance test of solar power generation. In this paper, MPPT control is controlled by the buck converter, and voltage and current are measured using the INA219 voltage current sensor. Switching of the buck converter was performed using P-channel MOSFET (F9530N) for high-side switching of the buck converter. P-channel MOSFET has a switching state of "on" when a "low" signal is inputted to the gate, so the NPN transistor (2N3904) and pull-up resistor (1 kΩ) are used to control the buck converter.

DC-DC step down converter (KIS-3R33S) was used to maintain constant voltage for changes in the voltage of solar power, and the cell phone battery was charged.

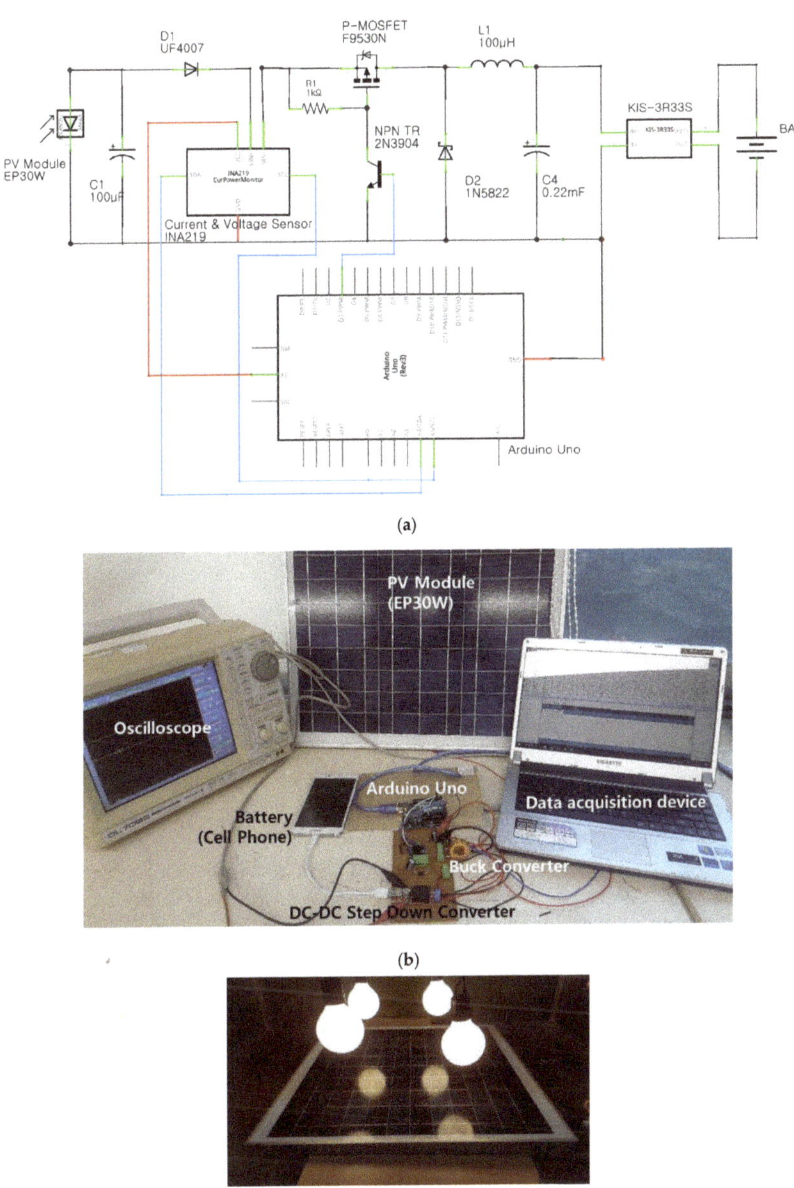

(a)

(b)

(c)

Figure 13. Experimental setup for the MPPT control performance test of the PV system. (**a**) Circuit diagram for experiments; (**b**) System for experiments; (**c**) Artificial light.

Figure 14 shows the change of output voltage according to PWM signal of Buck Converter. When MPPT control is performed using a buck converter, the voltage gradually decreases from the open-circuit voltage to the load voltage according to the PWM signal. In Figure 14, CH1 represents the PWM signal output from the controller, and CH2 denotes the voltage change of the PV module. The switching frequency of the controller used is 3.9 [kHz].

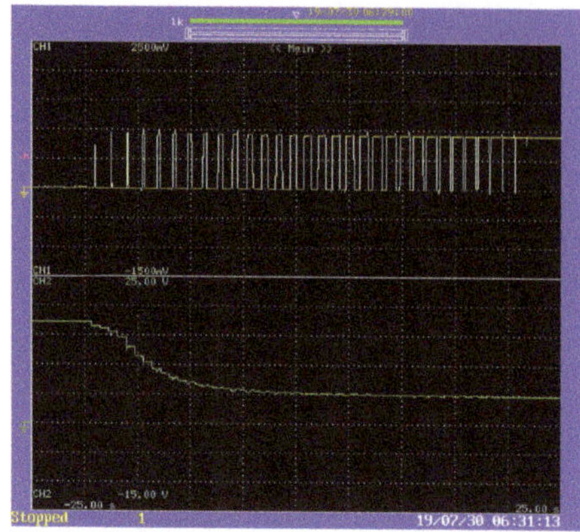

Figure 14. PWM signal and voltage PV module (V_{pv}).

Figures 15–17 show the response characteristics of the VGPI, PI, and P&O methods under constant solar irradiation conditions.

The PI controller used for comparison with VGPI uses 0.035 for proportional gain (K_p) and 0.005 for integral gain (K_i). Arduino's PWM ranges from 0 to 255, with 0 representing 0% and 255 representing 100% duty ratio. The P&O controller adjusts the PWM to a fixed size 3 to regulate the voltage at a constant rate.

In Figure 15, Figure 15a shows the voltage and current, Figure 15b presents the output power, Figure 15c illustrates the gain of the PI controller controlled by fuzzy control, Figure 15d shows the control value (C_p) and PWM signal for switching control of the DC-DC converter and Figure 15e is output voltage controlled by step down converter. The gain of the PI controller in (C) is increased for fast tracking in transient state, and the gain value is decreased for improving accuracy and stability in steady state. The control value (C_p) for tracking the MPP increases as the gain of the PI controller is adjusted according to the operating state and decreases in steady state. As a result, the variation of the PWM signal for switching of the DC-DC converter is reduced, and the power ripple is reduced; thus enabling more accurate MPPT. The output voltage in Figure 15e remains constant even as the voltage of solar power changes.

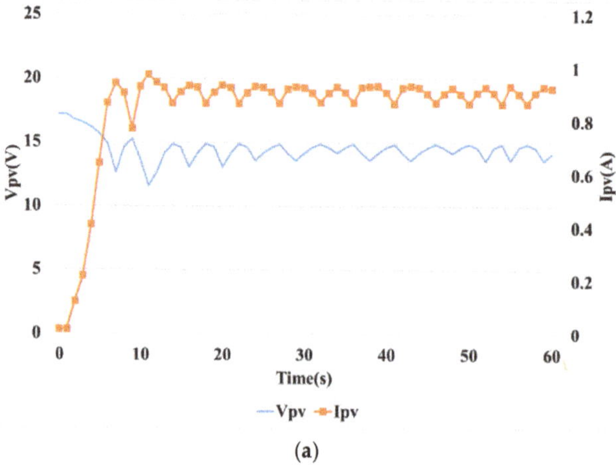

(a)

Figure 15. *Cont.*

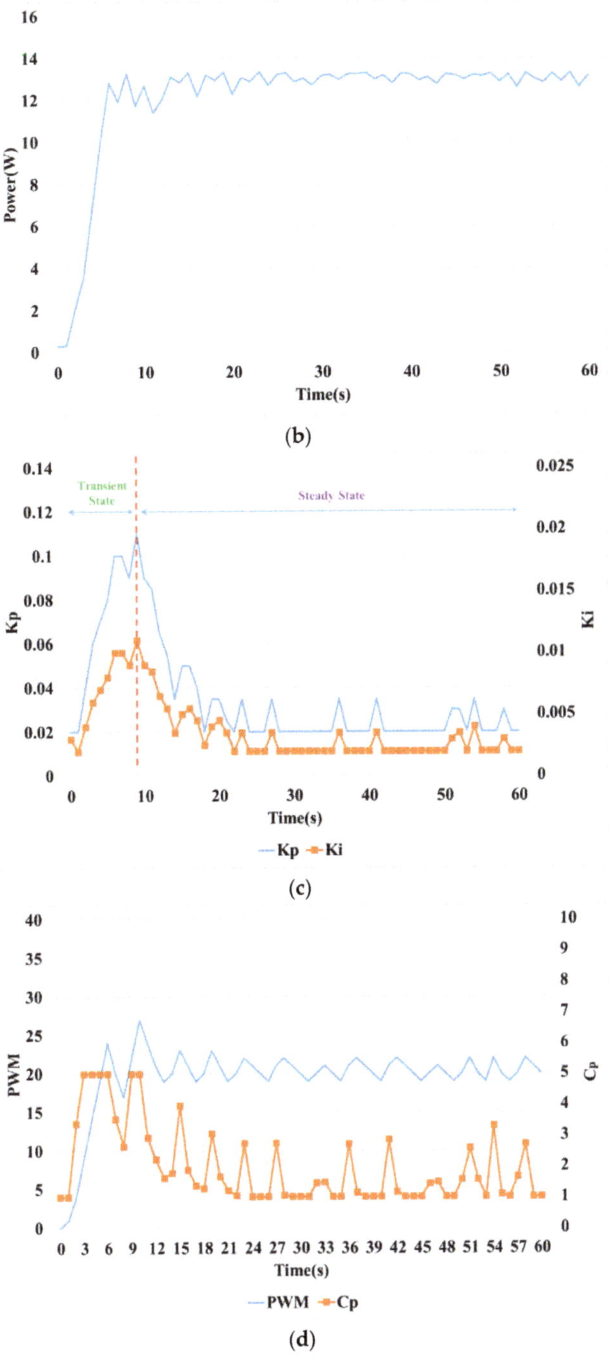

(**b**)

(**c**)

(**d**)

Figure 15. *Cont.*

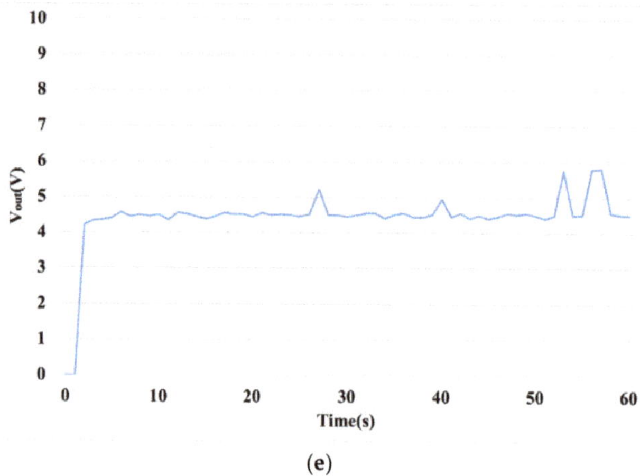

(**e**)

Figure 15. Response characteristics of the VGPI MPPT method. (**a**) Voltage (V_{pv}) and Current (I_{pv}) of the PV module; (**b**) Output Power of the PV module; (**c**) Proportional gain (K_p) and Integral gain (K_i) of the PI controller; (**d**) PWM signal and control value (C_p) (**e**) Output Voltage.

Figure 16 shows the response performance of the MPPT control of photovoltaic power generation using the PI controller. In particular, Figure 16c shows the fixed gain of the PI controller. Although the PWM signal and the control value (C_p) of Figure 16d are controlled according to operating state by PI control, the ripple of the output power increases because it is larger than the value of Figure 15d.

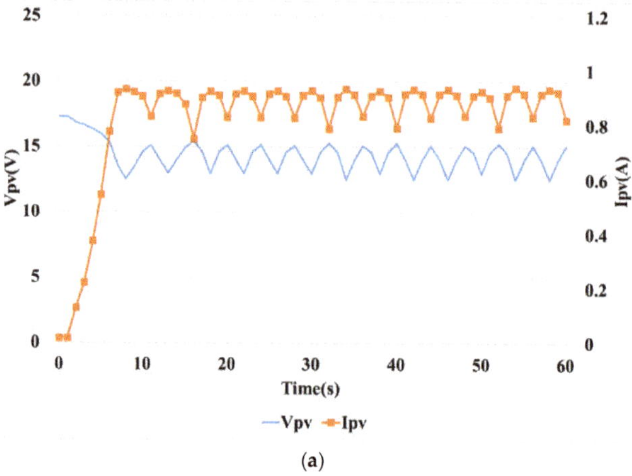

(**a**)

Figure 16. *Cont.*

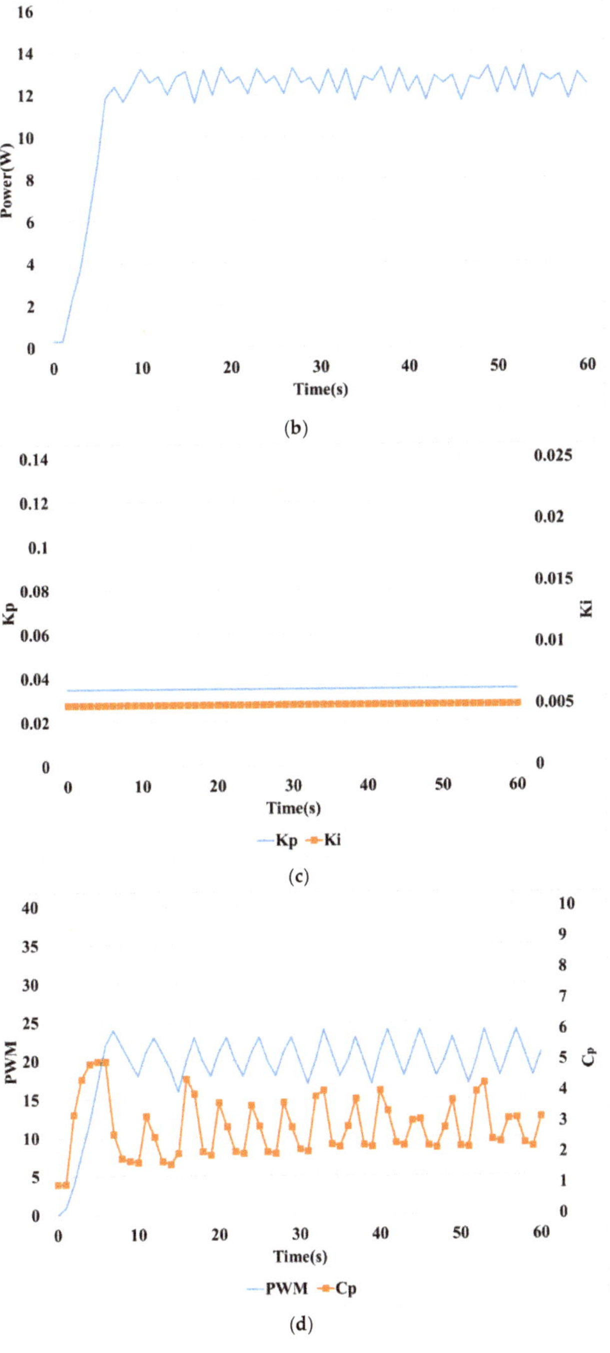

(b)

(c)

(d)

Figure 16. *Cont.*

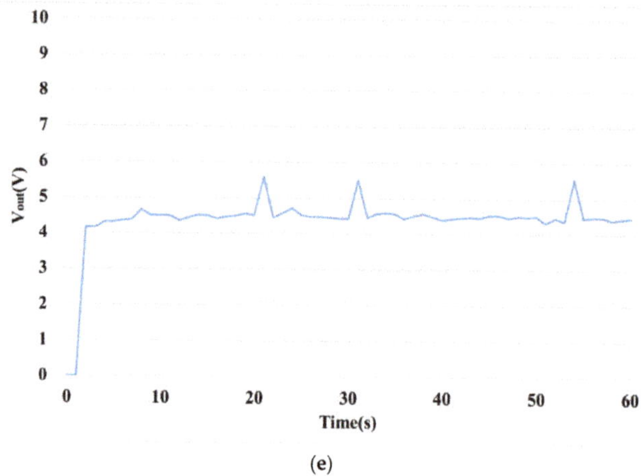

(e)

Figure 16. Response characteristics of the PI MPPT method. (**a**) Voltage (V_{pv}) and Current (I_{pv}) of the PV module; (**b**) Output Power of the PV module; (**c**) Proportional gain (K_p) and Integral gain (K_i) of the PI controller; (**d**) PWM signal and control value (C_p); (**e**) Output Voltage.

Figure 17 shows the response characteristics of the most commonly used P&O method for MPPT control. In particular, Figure 17a shows the voltage and current, Figure 17b presents the output power, and Figure 17c shows the PWM signal and control value (C_p). Since the P&O method uses the fixed control value (C_p) in both transient state and steady state, voltage in Figure 17a, power in Figure 17b, and PWM signal in Figure 17c have a constant ripple magnitude.

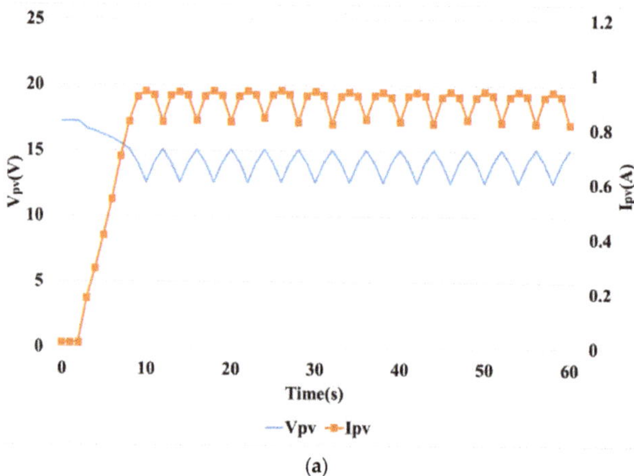

(a)

Figure 17. *Cont.*

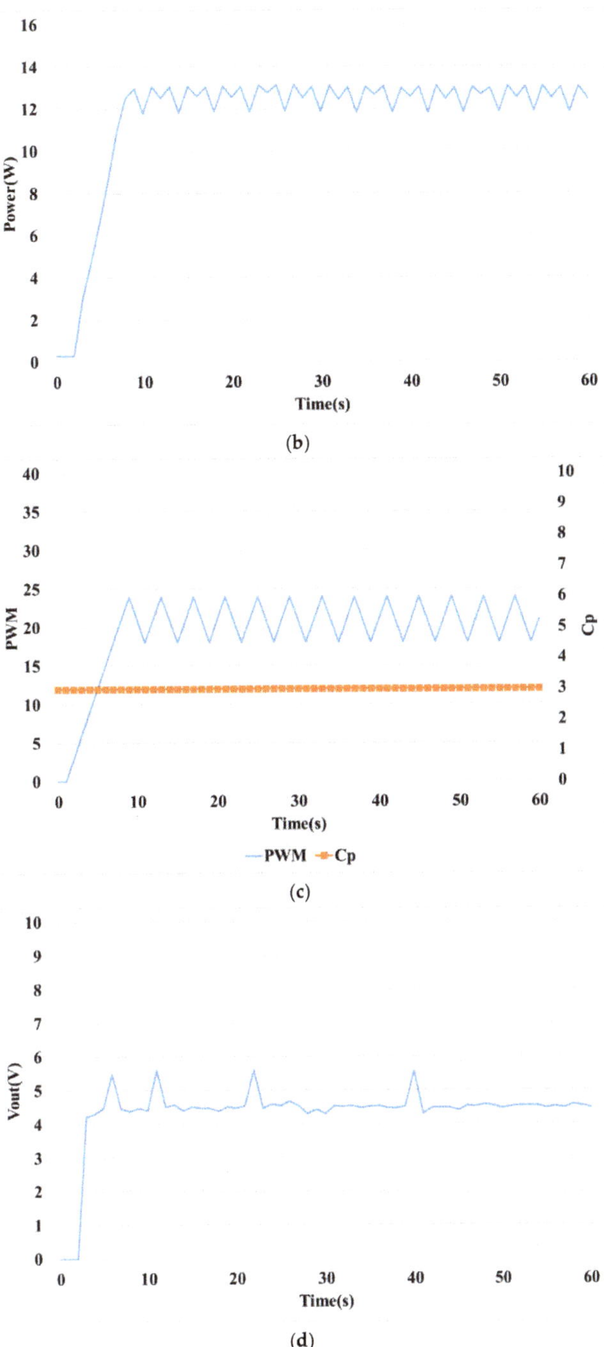

Figure 17. Response characteristics of the P&O MPPT method. (**a**) Voltage (V_{pv}) and Current (I_{pv}) of the PV module; (**b**) Output Power of the PV module; (**c**) PWM signal and control value (C_p); (**d**) Output Voltage.

Figure 18 and Table 6 show a comparison of the power response characteristics in transient states of the VGPI, PI, and P&O methods in Figures 15–17. The time to trace the maximum power point in transient state was measured as the time to reach the average power (12.5 [W]) of steady state, the results are shown in Table 6. As the results in Table 6 show, the VGPI controller has the most tracking speed with a high gain value in transient.

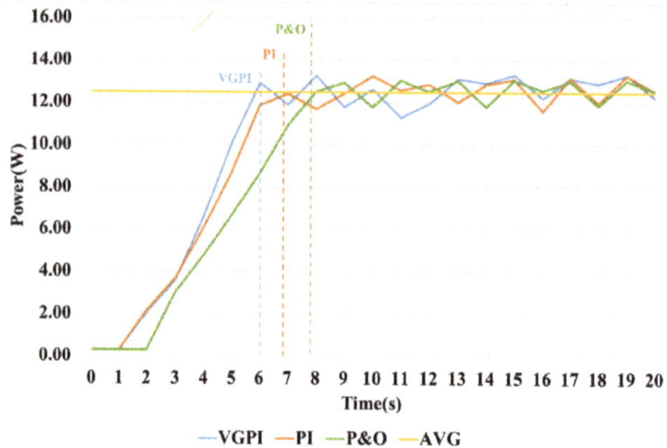

Figure 18. Comparison of response characteristics in transient state.

Table 6. Comparison of rising time in Figure 18.

	VGPI	**PI**	**P&O**
Rising Time(sec)	5.94 (73.7%)	7.13 (88.4%)	8.06 (100%)

Figure 19 and Table 7 represent the magnified picture and characteristics of the steady-state portion of Figures 15–17. The VGPI controller had the lowest error because it had lower gain values in a steady state, and the ripple was about 50% lower than the P&O method. Since the PI controller used a high gain value for fast tracking speed in transient conditions, steady-state error was rather higher than the P&O method.

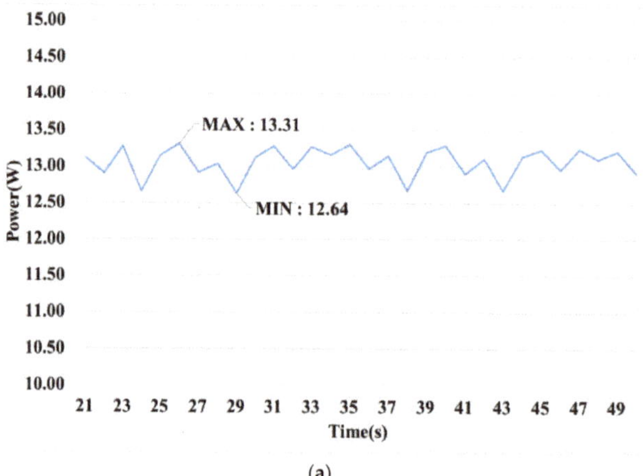

(a)

Figure 19. *Cont.*

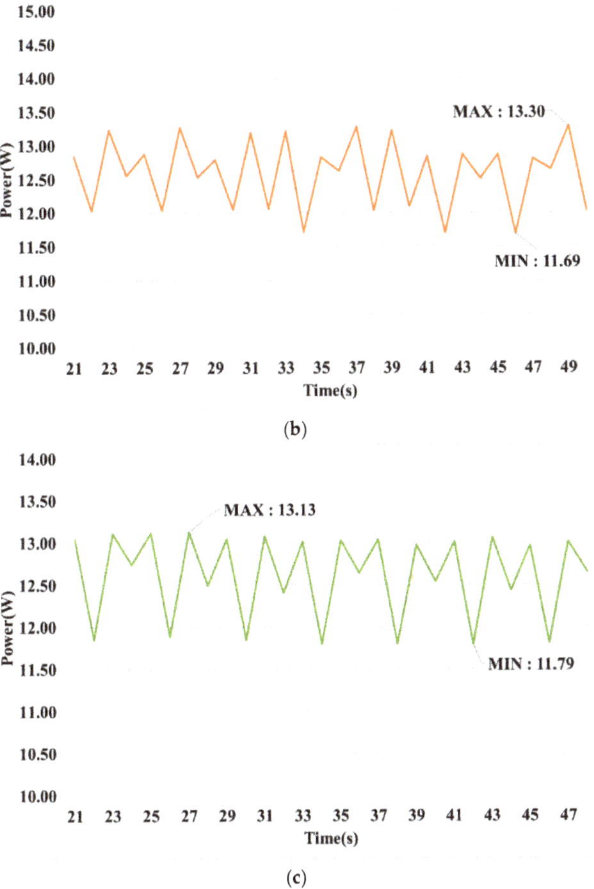

Figure 19. Comparison of response characteristics in steady state. (**a**) VGPI MPPT method; (**b**) PI MPPT method; (**c**) P&O MPPT method.

Table 7. Comparison of peak to peak in Figure 5.

	VGPI	**PI**	**P&O**
Min(W)	12.64	11.69	11.79
Max(W)	13.31	13.30	13.13
Peak to peak(W)	0.67(50.3%)	1.60(119.8%)	1.13(100%)

Figure 20 and Table 8 show MPPT control characteristics for changing conditions of solar radiation. Figure 20a represents the output of the VGPI, and Figure 20b represents the change in the gain value of the VGPI controller. The gain value of the VGPI controller represents a characteristic that is increasing in a transient state and is decreasing in steady state. Table 8 shows comparison of steady-state error for conditions with different solar irradiance. VGPI controllers show low steady-state error across all sections. Figure 20d represents the output voltage of the VGPI, PI, and P&O methods, with constant voltage output for the changing conditions of the solar radiation.

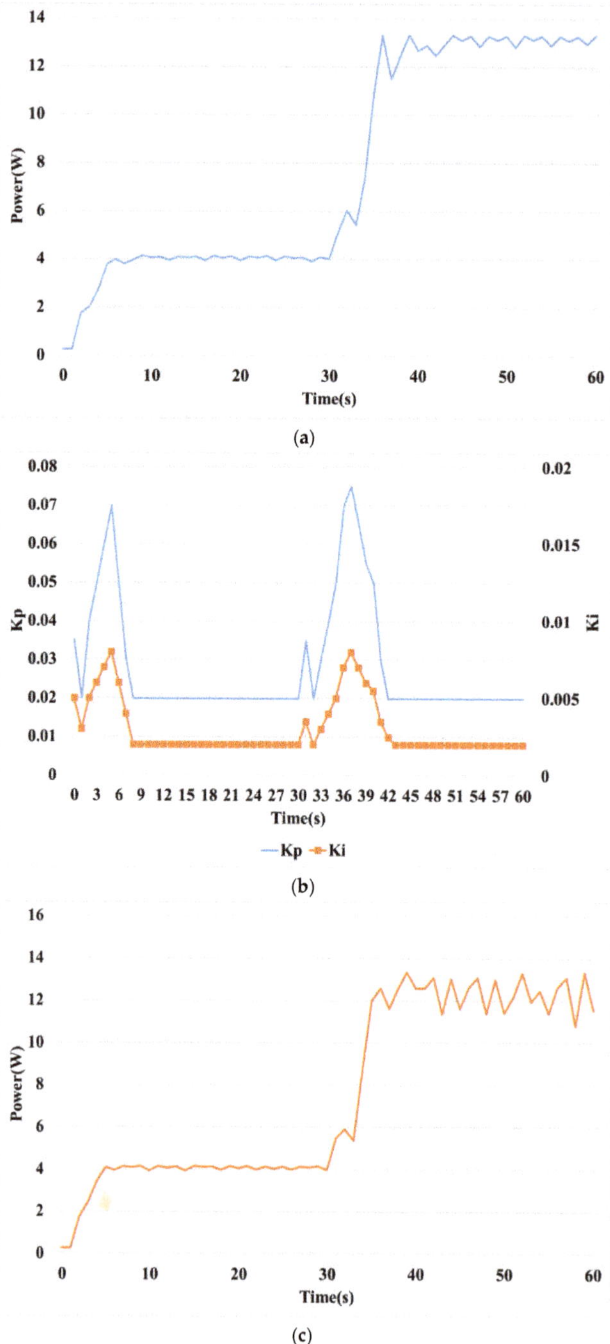

(a)

(b)

(c)

Figure 20. *Cont.*

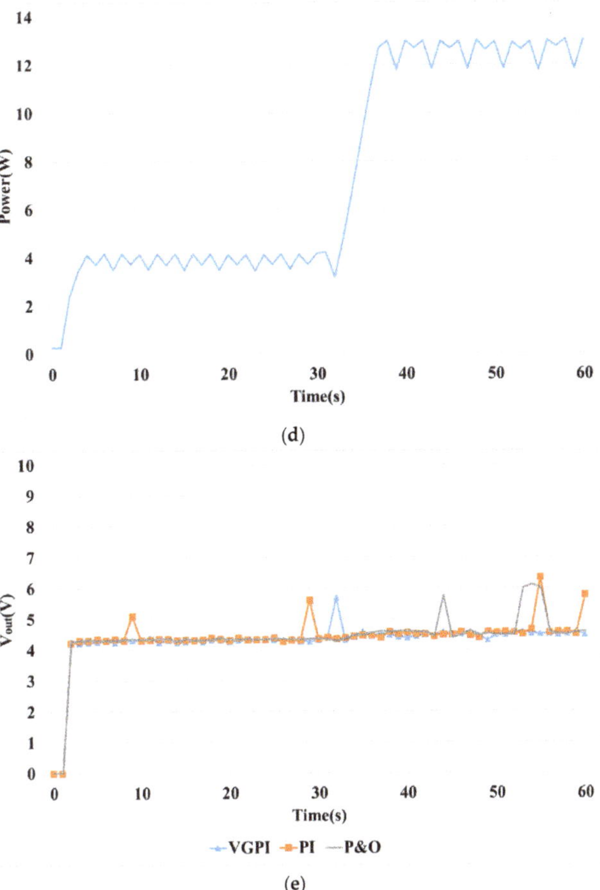

Figure 20. Comparison of MPPT control response characteristics for solar radiation variation. (**a**) Power of VGPI MPPT method; (**b**) Gain of VGPI method; (**c**) Power of PI MPPT method; (**d**) P&O MPPT method; (**e**) Output Voltage.

Table 8. Comparison of steady-state error to solar radiation variation.

		VGPI	PI	P&O
11–30[sec]	MIN [W]	3.95	3.96	3.46
	MAX [W]	4.15	4.16	4.17
	Peak to Peak [W]	0.20 (29.6%)	0.20 (29.6%)	0.71 (100%)
41–60[sec]	MIN [W]	12.47	10.82	11.78
	MAX [W]	13.31	13.35	13.07
	Peak to Peak [W]	0.84 (65.0%)	2.53 (195.9%)	1.29 (100%)

5. Conclusions

This paper proposes a method for tracking the MPP of solar power generation. P&O and IncCond methods, which are commonly used MPPT methods, have limitations in performance improvement because they track a MPP by varying a voltage or a current with a constant magnitude.

In this paper, the VGPI controller is proposed to solve this problem. The characteristics of the VGPI controller presented in the paper are as follows.

1. Based on the PI controller, which is most commonly used in industrial sites.
2. The gain value of the PI controller adjusts according to the operating state using fuzzy control.
3. Design a fuzzy control membership function and rule base in accordance with the basic operation of the PV module.
4. The membership function of the fuzzy control output consists of three types: increase (P: Positive), hold (ZE: zero) and decrease (N: negative).

The gain value of the PI controller is increased for a quick response in transient conditions and reduced to reduce steady-state error in normal conditions. The VGPI proposed in this paper compares the tracking time at transient-state and the error in steady state with conventional MPPT methods for two scenarios (constant and changing conditions of solar radiation).

Under constant or varying conditions, the proportional gain (K_p) and integral gain (K_i) of the VGPI controller increased to the maximum value in the transient state and decreased to the minimum value in the steady state by the fuzzy control, and were continuously adjusted.

As a result, the VGPI controller proposed in this paper is about 14% better than the conventional PI and P&O in tracking speed, and the error in steady-state shown respectively 36.5% and 40% lower than PI and P&O. Even in conditions with varying solar radiation, the VGPI controller had excellent MPPT performance than other controllers because of continuous gain adjustment. VGPI controllers are able to adjust continuous gain values according to changing environments, and both transient and steady-state response performance was improved.

This method is expected to be applicable to various variable systems, as well as MPPT for solar power. Fuzzy control used in this paper requires continuous calculation according to changing environmental conditions. In addition, the calculation of fuzzy control depends on the membership function and rule base, and the calculation amount can be greatly increased according to the environmental change conditions. This phenomenon degrades the MPPT performance.

Author Contributions: Conceptualization, J.-C.K. and J.-S.K.; Data curation, J.-C.K. and J.-S.K.; Formal analysis, J.-C.K. and J.-H.H.; Funding acquisition, J.-H.H.; Investigation, J.-H.H.; Methodology, J.-C.K., J.-H.H. and J.-S.K.; Project administration, J.-H.H.; Software, J.-H.H. and J.-S.K.; Visualization, J.-S.K.; Writing—original draft, J.-C.K., J.-H.H. and J.-S.K.; Writing—review & editing, J.-H.H. and J.-S.K.

Funding: This research was supported by Energy Cloud R&D Program through the National Research Foundation of Korea (NRF) funded by the Ministry of Science, ICT (NRF-2019M3F2A1073385).

Conflicts of Interest: The authors declare no conflict of interest.

References

1. Dincer, F. The analysis on photovoltaic electricity generation status, potential and policies of the leading countries in solar energy. *Renew. Sustain. Energy Rev.* **2011**, *15*, 713–720. [CrossRef]
2. Eltawil, M.A.; Zhao, Z. Grid-connected photovoltaic power systems: Technical and potential problems—A review. *Renew. Sustain. Energy Rev.* **2010**, *14*, 112–129. [CrossRef]
3. Vergura, S. A Complete and Simplified Datasheet-Based Model of PV Cells in Variable Environmental Conditions for Circuit Simulation. *Energies* **2016**, *9*, 326. [CrossRef]
4. Coelho, R.F.; Martins, D.C. An Optimized Maximum Power Point Tracking Method Based on PV Surface Temperature Measurement. Available online: https://www.intechopen.com/books/sustainable-energy-recent-studies/an-optimized-maximum-power-point-tracking-method-based-on-pv-surface-temperature-measurement/ (accessed on 7 August 2019).
5. Hohm, D.P.; Ropp, M.E. Comparative Study of Maximum Power Point Tracking Algorithms Using an Experimental, Programmable, Maximum Power Point Test Bed. In Proceedings of the IEEE Photovoltaic Specialists Conference, Anchorage, AK, USA, 15–22 September 2000; pp. 1699–1702.
6. Tan, C.W.; Green, T.C.; Hernandez-Aramburo, C.A. Analysis of Perturb and Observe Maximum Power Point Tracker Algorithm for Photovoltaic Applications. In Proceedings of the IEEE 2nd International Power and Energy Conference, Johor Bahru, Malaysia, 1–3 December 2008; pp. 237–242.

7. Boico, F.; Lahman, B. Study of Different Implementation Approaches for a Maximum power Point Tracker. In Proceedings of the IEEE Computers in Power Electronics, Troy, NY, USA, 16–19 July 2006; pp. 15–21.

8. Liu, B.; Duan, S.; Liu, F.; Xu, P. Analysis and Improvement of a Maximum Power Point Tracking Algorithm Based on Incremental Conductance Method for Photovoltaic Array. In Proceedings of the IEEE International Conference on Power Applications, Bangkok, Thailand, 27–30 November 2007; pp. 637–641.

9. Yuvarajan, S.; Shoeb, J. A Fast and Accurate Maximum Power Point Tracker for PV Systems. In Proceedings of the IEEE Applied Power Electronics Conference and Exposition, Austin, TX, USA, 24–28 February 2008; pp. 167–172.

10. Femia, N.; Petrone, G.; Spagnuolo, G.; Vitelli, M. Optimizing Sampling Rate of P&O MPPT Technique. In Proceedings of the IEEE Power Electronics Specialist Conference, Aachen, Germany, 20–25 June 2004; Volume 3, pp. 1945–1949.

11. Pandey, A.; Dasgupta, N.; Mukerjee, A. Design Issues in Implementing MPPT for Improved Tracking and Dynamic Performance. In Proceedings of the IEEE Conference on Industrial Electronics, Paris, France, 6–10 November 2006; pp. 4387–4391.

12. Nevzat, O. Recent developments in maximum power point tracking technologies for photovoltaic systems. *Int. J. Photoenergy* **2010**. [CrossRef]

13. De Brito, M.A.G.; Junior, L.G.; Sampaio, L.P.; Melo, G.A.; Canesin, C.A. Main maximum power point tracking strategies intended for photovoltaics. In Proceedings of the XI Power Electronics Brazilian Conference, Praiamar, Brazil, 11–15 September 2011; pp. 524–530.

14. Kjaer, S.B. Evaluation of the "hill climbing" and the incremental conductance maximum power point trackers for photovoltaic power systems. *IEEE Trans. Energy Convers.* **2012**, *27*, 922–929. [CrossRef]

15. Elgendy, M.A.; Zahawi, B.; Atkinson, D.J. Assessment of perturb and observe MPPT algorithm implementation techniques for PV pumping applications. *IEEE Trans. Sustain. Energy* **2012**, *3*, 21–33. [CrossRef]

16. Abdelsalam, A.; Ahmed, S.; Massoud, A.; Enjeti, P. High performance adaptive perturb and observe MPPT technique for photovoltaic-based microgrids. *IEEE Trans. Power Electron.* **2011**, *26*, 1010–1021. [CrossRef]

17. Piegari, L.; Rizzo, R. Adaptive perturb and observe algorithm for photovoltaic maximum power point tracking. *IET Renew. Power Gen.* **2010**, *4*, 317–328. [CrossRef]

18. Wasynczuk, O. Dynamic behavior of a class of photovoltaic power systems. *IEEE Trans. Power Appl. Syst.* **1983**, *9*, 3031–3037. [CrossRef]

19. Sera, D.; Teodorescu, R.; Hantschel, J.; Knoll, M. Optimized maximum power point tracker for fast-changing environmental conditions. *IEEE Trans. Ind. Electron.* **2008**, *55*, 2629–2637. [CrossRef]

20. Pandey, A.; Dasgupta, N.; Mukerjee, A.K. High-performance algorithms for drift avoidance and fast tracking in solar MPPT system. *IEEE Trans. Energy Convers.* **2008**, *23*, 681–689. [CrossRef]

21. Fortunato, M.; Giustiniani, A.; Petrone, G.; Spagnuolo, G.; Vitelli, M. Maximum power point tracking in a one-cycle-controlled single-stage photovoltaic inverter. *IEEE Trans. Ind. Electron.* **2008**, *55*, 2684–2693. [CrossRef]

22. Patel, H.; Agarwal, V. MPPT scheme for a PV-fed single-phase single-stage grid-connected inverter operating in CCM with only one current sensor. *IEEE Trans. Energy Convers.* **2009**, *24*, 256–263. [CrossRef]

23. Ropp, M.E.; Gonzalez, S. Development of a MATLAB/Simulink model of a single-phase grid-connected photovoltaic system. *IEEE Trans. Energy Convers.* **2009**, *24*, 195–202. [CrossRef]

24. Koutroulis, E.; Kalaitzakis, K.; Voulgaris, N.C. Development of a microcontroller-based, photovoltaic maximum power point tracking control system. *IEEE Trans. Power Electron.* **2001**, *16*, 46–54. [CrossRef]

25. Veerachary, M.; Senjyu, T.; Uezato, K. Neural-network-based maximum-power-point tracking of coupled-inductor interleavedboost-converter-supplied PV system using fuzzy controller. *Ind. Electron. IEEE Trans.* **2003**, *50*, 749–758. [CrossRef]

26. Mirbagheri, S.Z.; Aldeen, M.; Saha, S. A Comparative Study of MPPT Algorithms for Sandalone PV Systems under RCIC. In Proceedings of the 2015 IEEE PES Asia-Pacific Power and Energy Engineering Conference (APPEEC), Brisbane, Australia, 15–18 November 2015; pp. 1–5.

27. Ko, J.S.; Huh, J.H.; Kim, J.C. Improvement of Temperature Control Performance of Thermoelectric Dehumidifier Used Industry 4.0 by the SF-PI Controller. *Processes* **2019**, *7*, 98. [CrossRef]

28. Ko, J.S.; Huh, J.H.; Kim, J.C. Improvement of Energy Efficiency and Control Performance of Cooling System Fan Applied to Industry 4.0 Data Center. *Electronics* **2019**, *8*, 582. [CrossRef]

29. Ang, K.H.; Chong, G.; Li, Y. PID control system analysis, design, and technology. *IEEE Trans. Control Syst. Technol.* **2005**, *13*, 559–576.

30. Li, Y.; Ang, K.H.; Chong, G.C.Y. Patents, software, and hardware for PID control: An overview and analysis of the current art. *IEEE Control Syst.* **2006**, *26*, 42–54.

31. Qin, Y.; Sun, L.; Hua, Q.; Liu, P. A Fuzzy Adaptive PID Controller Design for Fuel Cell Power Plant. *Sustainability* **2018**, *10*, 2438. [CrossRef]

32. Sun, L.; Li, D.; Lee, K.Y. Optimal disturbance rejection for PI controller with constraints on relative delay margin. *ISA Trans.* **2016**, *63*, 103–111. [CrossRef] [PubMed]

33. Wang, J.-S.; Yang, G.H. Data-Driven Approach to Accommodating Multiple Simultaneous Sensor Faults in Variable-Gain PID Systems. *IEEE Trans. Ind. Electron.* **2019**, *66*, 3117–3126. [CrossRef]

34. Pal, A.K.; Mudi, R.K. Self-Tuning Fuzzy PI Controller and its Application to HVAC System. *Int. J. Comput. Cogn. (IJCC)* **2008**, *1*, 25–30.

35. Kassem, A.M. Fuzzy-logic Based Self-tuning PI Controller for High-Performance Vector Controlled Induction Motor Fed by PV-Generator. *Wseas Trans. Syst.* **2013**, *12*, 22–31.

36. Wahyunggoro, O.; Saad, N. Development of Fuzzy-logic-based Self Tuning PI Controller for Servomotor. In *Advanced Strategies for Robot Manipulators*; INTECH: Shanghai, China, 2010; pp. 311–328.

37. Anantwar, H.; Lakshmikantha, B.R.; Sundar, S. Fuzzy self tuning PI controller based inverter control for voltage regulation in off-grid hybrid power system. In Proceedings of the International Conference on Power Engineering, Computing and CONtrol (PECCON), Chennai, India, 2–4 March 2017; Volume 117, pp. 409–416.

38. Mudi, R.K.; Pal, N.R. A self-tuning fuzzy PI controller. *Fuzzy Sets Syst.* **2000**, *115*, 327–338. [CrossRef]

39. Ibarra, L.; Webb, C. Advantages of Fuzzy Control While Dealing with Complex/Unknown Model Dynamics: A Quadcopter Example. Available online: https://www.intechopen.com/books/new-applications-of-artificial-intelligence/advantages-of-fuzzy-control-while-dealing-with-complex-unknown-model-dynamics-a-quadcopter-example/ (accessed on 7 August 2019).

40. Walker, G.R.; Sernia, P.C. Cascaded dc-dc converter connection of photovoltaic modules. *IEEE Trans. Power Electron.* **2004**, *19*, 1130–1139. [CrossRef]

41. Ghasemi, A.; Eilaghi, S.F.; Adib, E. A new non-isolated high step up SEPIC converter for photovoltaic applications. In Proceedings of the 3rd Power Electronics and Drive Systems Technology Conference, Tehran, Iran, 15–16 February 2012; pp. 51–56.

42. Coelho, R.F.; Concer, F.M.; Martins, D.C. A Simplified Analysis of DC-DC Converters Applied as Maximum Power Point Tracker in Photovoltaic Systems. In Proceedings of the IEEE International Symposium on Power Electronics for Distributed Generation Systems, Hefei, China, 16–18 June 2010; pp. 1–14.

43. Taghvaee, M.; Radzi, M.A.M.; Moosavain, S.; Hizam, H.; Marhaban, M.H. A current and future study on non-isolated DC-DC converters for photovoltaic applications. *Renew. Sustain. Energy Rev.* **2013**, *17*, 216–227. [CrossRef]

44. Ramki, T.; Tripathy, L.N. Comparison of Different DC-DC Converter for MPPT Application of Photovoltaic System. In Proceedings of the International Conference on Electrical, Electronics, Signals, Communication and Optimization, Visakhapatnam, India, 24–25 January 2015; pp. 1–6.

45. Baharudin, N.H.; Mansur, T.M.N.T.; Hamid, F.A.; Misrun, M.I. Topologies of DC-DC Converter in Solar PV Applications. *Indones. J. Electr. Eng. Comput. Sci.* **2017**, *8*, 368–374. [CrossRef]

46. Umashankar, S.; Srikanth, P.; Vijay Kumar, D.; Kothari, D.P. Comparative Study of Maximum Power Point Tracking Algorithms with DC-DC Converters for Solar PV System. *Int. J. Electr. Comput. Eng.* **2011**, *3*, 11–20.

47. Mahanta, J.; Sharma, B.; Sarmah, N. A Review of Maximum Power Point Tracking Algorithm for Solar Photovoltaic Applications. *J. Electr. Electron. Eng.* **2018**, *13*, 1–13.

48. Vuksic, M.; Kovacevic, T.; Mise, J. Solar Climber: A Problem Solving Approach in Power Electronics and Control Systems Teaching. Available online: https://www.researchgate.net/publication/263470253_Solar_Climber_A_Problem_Solving_Approach_in_Power_Electronics_and_Control_Systems_Teaching/ (accessed on 23 July 2019).

49. Rao, K.S.; Mishra, R. Comparative study of P, PI and PID controller for speed control of VSI-fed induction motor. *Int. J. Eng. Dev. Res.* **2014**, *2*, 2740–2744.

50. Saletic, D.Z.; Velasevic, D.M.; Mastorakis, N.E. Analysis of Basic Defuzzification Techniques. Available online: https://www.researchgate.net/publication/264874571_Analysis_of_Basic_Defuzzification_Techniques/ (accessed on 7 August 2019).

51. Haque, A. Maximum Power Point Tracking (MPPT) Scheme for Solar Photovoltaic System. *Energy Technol. Policy* **2014**, *1*, 115–122. [CrossRef]
52. Na, W.; Chen, P.; Kim, J.H. An Improvement of a Fuzzy Logic-Controlled Maximum Power Point Tracking Algorithm for Photovoltaic Applications. *Appl. Sci.* **2017**, *7*, 326. [CrossRef]
53. Li, C.; Chen, Y.; Zhou, D.; Liu, J.; Zeng, J. A High-Performance Adaptive Incremental Conductance MPPT Algorithm for Photovoltaic Systems. *Energies* **2016**, *9*, 288. [CrossRef]
54. Piegari, L.; Rizzo, R.; Spina, I.; Tricoil, P. Optimized Adatpvie Perturb and Observe Maximum Power Point Tracking Control for Photovoltaic Generation. *Energies* **2015**, *8*, 3418–3436. [CrossRef]

© 2019 by the authors. Licensee MDPI, Basel, Switzerland. This article is an open access article distributed under the terms and conditions of the Creative Commons Attribution (CC BY) license (http://creativecommons.org/licenses/by/4.0/).

 *sustainability*

Article

A Novel Command-Filtered Adaptive Backstepping Control Strategy with Prescribed Performance for Photovoltaic Grid-Connected Systems

Weiming Zhang , Tinglong Pan *, Dinghui Wu and Dezhi Xu

School of Internet of Things Engineering, Jiangnan University, Wuxi 214122, China;
wmzhang21@163.com (W.Z.); wdh123@jiangnan.edu.cn (D.W.); lutxdz@126.com (D.X.)
* Correspondence: tlpan@jiangnan.edu.cn

Received: 8 July 2020; Accepted: 7 August 2020; Published: 10 September 2020

Abstract: With the aim of solving the power fluctuation and bus voltage instability problems caused by external environment variations in the photovoltaic grid-connected system, a prescribed performance-based adaptive backstepping controller is proposed for the system to regulate the bus voltage and the inverter current. First, the mathematical model of the grid-connected inverter is established, in which the uncertain system parameters are estimated via a designed projection-based adaptive law. Then, the command-filtered backstepping sliding mode control method is applied to the system for power regulation. In order to achieve favorable tracking performance, the prescribed performance technique is introduced in the voltage regulation strategy by constraining the compensated voltage tracking error within a certain range from a novel point of view. Finally, the simulation is carried out considering the variations of environmental situations, and the obtained results demonstrate the sound performance of the prescribed performance-based control strategy with respect to the photovoltaic grid-connected system.

Keywords: photovoltaic grid-connected system; power fluctuation; DC bus voltage stabilization; prescribed performance; command-filtered adaptive backstepping control

1. Introduction

With the depletion of fossil fuels and the call for environmental protection, more and more green power generation modes are being integrated into the power grid to supply power, among which the photovoltaic (PV) power generation is widely used with large-scale installed capacity for its high income, simple operation, and maintenance, as well as reasonable resource allocation [1–3]. However, the increasing share of these intermittent renewable electricity generations bring about great challenges to grid-balancing and power supply security [4–6]. Besides, considering the uncertain variations of the external environment containing the solar irradiation and temperature which results in the power fluctuation problem, it is essential to draft some measures to ensure the stability of the system [6].

There exist many investigations on the stabilization of the PV grid-connected system by researchers around the world [7–11]. From the point of view of the model of the PV grid-connected system, the Reference [7] studies the influence of nonlinearities like dead time and the phase-locked loop with regard to power quality and stability via the generalized Nyquist stability criterion, and verifies the theory with a simulation and experiment. In [8], a feedback linearization technique was introduced for the partial linearization of the three-phase PV grid-connected system, which promotes the design process of the controller, and the zero dynamic stability of the system was acquired under the designed controller, which helps transport the grid power. In terms of the control method with respect to the grid-connected system, a H_∞ control method was brought to the grid-connected inverter in [9] to achieve the stable operation of the system when the grid impedance possibly varies, and the control performance proved better than conventional PID control. In [10], an adaptive sliding mode control method was used for the inverter of a single-phase PV grid-connected system for bus voltage stabilization, and the fuzzy neural network was employed for the estimation of system uncertainties. In [11], a command-filtered backstepping control (CBC) method was proposed to solve the voltage tracking and power fluctuation problems in the grid-connected system, and possessed better performance than PID. However, the concrete numerical control objectives are absent in [11] and the bus voltage is regulated indirectly by adjusting active power, which results in the lack of voltage tracking precision, and the simulation situations are set from the theoretical level without considering the actual conditions. Thus, the existed problems motivate us to continue research based on the backstepping control method to seek better performance.

The backstepping control method was first proposed by Farrell et al. in [12] to solve the tracking control problem in nonlinear systems, and the command filter was introduced to avoid the differential expansion problem. The corresponding controller design process was provided with certain procedures based on the Lyapunov stability criterion, which is much simpler than other advanced control approaches and attracts much attention. The theory of backstepping control has been studied widely, including several aspects like finite time, input saturation, and distributed control [13–16]. In terms of its applications, many advanced control strategies have been integrated with backstepping control to obtain better control performance and achieve a higher control objective in various fields, such as with the robot [17], ship [18], wind turbine [19], unmanned aerial vehicle [20,21]. On the other hand, the prescribed performance control (PPC) technique was introduced by Bechlioulis et al. in [22], which focuses on the tracking control accuracy and boundedness. The output tracking error is constrained within the prescribed range upon when the design process begins, and then converts to an equivalent error, which turns the restriction problem into an unrestricted one. On account of its boundedness with respect to the tracking error, the prescribed performance control gained extensive investigations [23–27]. According to the above investigation, the backstepping control integrated with prescribed performance control was selected for the PV grid-connected system for the voltage stabilization and power fluctuation reduction.

In general, the command-filtered adaptive backstepping sliding mode control approach was designed accompanied by the prescribed performance technique for the stable operation of the PV grid-connected system. The control objectives contain some aspects, namely, how the active power fluctuation shall be reduced, and the DC bus voltage shall be maintained stable around the reference with adequate tracking precision. Besides, the three-phase voltage of the inverter shall be kept stable and the harmonics reduced. The main contributions and innovations of this paper are listed in the following.

- The prescribed performance technique is introduced in the command-filtered backstepping sliding mode controller to constrain the voltage tracking error within the preset range, namely, enable the DC bus voltage which possesses the prescribed performance, which constitutes to the stable operation of the PV grid-connected system.
- The integral sliding mode control method is considered in the controller design process to improve the robustness of the system, and the stability of the prescribed performance-based control strategy is proved via the Lyapunov stability criterion.
- The environmental variations of PV arrays, including solar irradiation and temperature, are reasonably considered in the system to test the stability of the control system.

The rest of the paper is organized as follows: Section 2 provides the preliminaries, including the mathematical modeling of the PV grid-connected system and basic knowledge of prescribed performance control. Based on the model, the controller is designed with the prescribed performance technique and the stability proof is offered. In Section 3, the simulation is carried out on the PV grid-connected system with the battery energy storage system (BESS), and the results are discussed to test the effectiveness of the proposed controller. Ultimately, some conclusions are drawn in Section 4.

2. Materials and Methods

2.1. PV Arrays Modeling

The circuit topology of the PV cell is shown in Figure 1. The parameters in Figure 1 are explained as follows: I_s is the photocurrent, I_D the diode current, R_{sh} is the shunt resistance, and I_R is the corresponding current, R_s is the series resistance, I_{pv} is the output current of the cell, and U_{pv} is the output voltage. First, the diode current I_D can be expressed as [11,28]:

$$I_D = I_0 \left(\exp\left(\frac{q\left(U_{pv} + R_s I_{pv}\right)}{AkT} \right) - 1 \right), \tag{1}$$

where I_0 is the reverse saturation current, A is the dimensionless coefficient, q is the elementary charge ($q = 1.6 \times 10^{-19}C$), k is the Boltzmann constant, and T is the temperature of the cell, which is counted in the Kelvin scale.

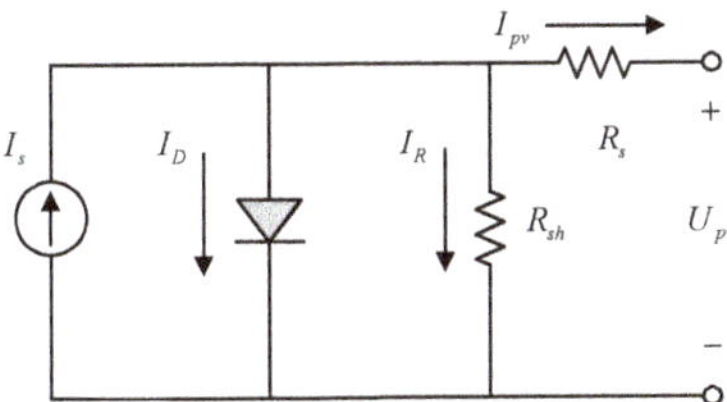

Figure 1. Circuit topology of a PV cell.

Referred to as Kirchhoff's Current Law, the output current I_{pv} generated by the PV cell in Figure 1 is represented as

$$I_{pv} = I_s - I_0 \left(\exp\left(\frac{q\left(U_{pv} + R_s I_{pv}\right)}{AkTN_s} \right) - 1 \right) - \frac{U_{pv} + R_s I_{pv}}{R_{sh}}, \tag{2}$$

and its photocurrent I_s can be represented as

$$I_s = \left(I_{sc} + k_i \left(T - T_n \right) \right) \frac{R_0}{R_n}, \tag{3}$$

where I_{sc} means the short-circuit current, k_i denotes the thermal coefficient of the current, T_n expresses the reference temperature of the PV cell, R_0 represents the solar irradiation, and R_n describes the reference solar irradiation. Meanwhile, the saturation current I_0 affected by the temperature is presented as

$$I_0 = I_{RS} \left(\frac{T}{T_n} \right)^3 \exp\left[\frac{qE_g}{Ak} \left(\frac{1}{T_n} - \frac{1}{T} \right) \right], \tag{4}$$

where E_g represents the band-gap power of the semiconductor, and I_{RS} denotes the reverse saturation current when operating under the reference environment.

By and large, PV modules consist of multiple PV cells which are linked together in series and parallel, and multiple PV modules constitute the PV array in series and parallel as well, so that the power of the grid can be supplied. Figure 2 denotes the structure of a PV system, including the structure of a PV array and the circuit topology of a PV module, where N_s and N_p are the number of PV cells in the series in and a parallel way, respectively [11,29]. Then, the output current I_{pv} can be rewritten as below:

$$I_{pv} = N_p I_s - N_p I_0 \left(\exp\left(\frac{q\left(U_{pv} + R_s I_{pv}\right)}{AkTN_s} \right) - 1 \right) - N_p \frac{U_{pv} + R_s I_{pv}}{N_s R_{sh}}. \tag{5}$$

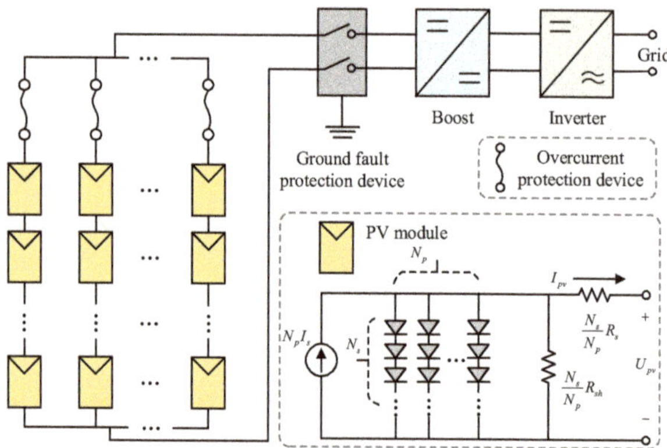

Figure 2. Configuration of the PV system.

The basic parameters of each PV module are provided in Table 1. Moreover, 64 parallel strings are arranged in a PV array with five series-connected modules per string, which illustrates how the total power of each PV array is $5 \cdot 64 \cdot 0.315072$ kW ≈ 100.8 kW.

Table 1. The basic parameters of a PV module.

Definition	Parameter	Value
Maximum output power	P_p	315.072 W
Open-circuit voltage	V_{oc}	64.6 V
Short-circuit current	I_{sc}	6.14 A
Voltage at maximum power point	V_{mp}	54.7 V
Current at maximum power point	I_{mp}	5.76 A

According to [11], the maximum power point varies in the light of the external environment, and the maximum power point tracking (MPPT) approach was employed to control the power converter switch so that the maximum output power of the PV array can be obtained. Here, the incremental conductance algorithm is utilized as one of the MPPT control approaches. Besides, the output power P, the current I, and the voltage U of the PV array satisfies the basic equation, which is described as

$$P = UI. \tag{6}$$

Notice that the slope at the maximum power point would be zero under the steady solar irradiance and temperature, namely,

$$\frac{dP}{dU} = I + U\frac{dI}{dU} = 0. \tag{7}$$

It can be further expressed as

$$\frac{dI}{dU} = -\frac{I}{U}. \tag{8}$$

According to (8), the PV array would operate at the maximum power point if the derivative of the current with respect to voltage gets equal to the negative conductance, which only relies on the voltage and current data of PV arrays. Besides, considering the irregularly variable environment, the power fluctuation of the grid-connected system tends to be inevitable. Hence, the BESS is integrated to the DC bus to compensate for the acquired power of the grid.

2.2. Mathematical Modeling of PV Grid-Connected Systems

The PV grid-connected system consists of four PV arrays, filter capacitors, R-L filters, inverters, and a three-phase grid. Then, the mathematical model of the grid-connected inverter is presented as below [11]:

$$\frac{di_d}{dt} = -\frac{R}{L}i_d + wi_q - \frac{E_d}{L} + \frac{u_{dc}}{L}k_d$$
$$\frac{di_q}{dt} = -\frac{R}{L}i_q - wi_d - \frac{E_d}{L} + \frac{u_{dc}}{L}k_q,$$

(9)

where E_d, E_q, i_d, and i_q denote the grid voltages and currents in the d-q axis, respectively; and R, L, and C represent the resistance, inductance, and capacitance of the system, respectively. k_d and k_q denote the switching function in the d-q axis.

On the other hand, it is acquired referring to the Kirchhoff's Voltage and Current Laws that

$$C\frac{du_{dc}}{dt} = i_0 - i_{dc},$$

(10)

where u_{dc} denotes the DC bus voltage, i_0 represents the output current of the boost circuit, and i_{dc} expresses the input current of the inverter. Besides, regardless of the power loss of inverters, the following equation is provided to achieve power balance as

$$u_{dc}i_0 = \frac{3}{2}\left(E_d i_d + E_q i_q\right),$$

(11)

where E_q is considered as 0 when the system is operating at a steady state. Then, substituting (11) to (10), it can be derived further as follows:

$$\frac{du_{dc}}{dt} = \frac{3E_d i_d}{2Cu_{dc}} - \frac{i_{dc}}{C}.$$

(12)

To summarize, the complete mathematical model of the grid-connected inverter is described as follows [11]:

$$\frac{du_{dc}}{dt} = \frac{3E_d i_d}{2Cu_{dc}} - \frac{i_{dc}}{C}$$
$$\frac{di_d}{dt} = -\frac{R}{L}i_d + wi_q - \frac{E_d}{L} + \frac{u_d}{L}$$
$$\frac{di_q}{dt} = -\frac{R}{L}i_q - wi_d - \frac{E_q}{L} + \frac{u_q}{L}.$$

(13)

Since it is tough to measure the value of the system parameter accurately, the parameters could be considered uncertain during the controller design process, which is described as

$$\eta_1 = \frac{1}{C}, \ \eta_2 = \frac{R}{L}, \ \eta_3 = \frac{1}{L}.$$

(14)

Taking (14) into account, the mathematical model (13) can be rewritten as [11]

$$\frac{du_{dc}}{dt} = \eta_1\left(\frac{3E_d i_d}{2u_{dc}} - i_{dc}\right)$$
$$\frac{di_d}{dt} = -\eta_2 i_d + wi_q - \eta_3 E_d + \eta_3 u_d$$
$$\frac{di_q}{dt} = -\eta_2 i_q - wi_d - \eta_3 E_q + \eta_3 u_q.$$

(15)

2.3. Basic Principle of Prescribed Performance Control

In order to obtain the desired performance with high accuracy, the PPC technique is adopted in this paper by pre-setting a range for the tracking error to achieve the prescribed performance.

First, a smooth function $p(t)$: $R^+ \rightarrow R^+$ is introduced to constrain the tracking error, which should be positive and monotonically decreasing and satisfy $\lim_{t\to\infty} p(t) = p_\infty > 0$ [30]. Here, we choose the following function which satisfies the aforementioned conditions, namely,

$$p(t) = (p_0 - p_\infty)\, e^{-lt} + p_\infty, \tag{16}$$

where p_0, p_∞, and l are all positive constants. Then, the defined tracking error $\bar{z}_1$ can be constrained within the following prescribed range as

$$-p(t) < \bar{z}_1(t) < p(t). \tag{17}$$

Considering the smooth function (16) and the error constraint (17), the amplitude and convergence rate of tracking error $\bar{z}_1$ will be determined by the smooth function $p(t)$ if the initial value of the defined tracking error $\bar{z}_1$ satisfies $0 \leq |\bar{z}_1(0)| \leq p_0$.

Furthermore, when it comes to the controller design process, it is necessary to convert the constrained problem into an equivalent unconstrained one. Hence, the constrained tracking error is represented by the combination of smooth function $p(t)$ and transformation function $\zeta(t)$, that is,

$$\bar{z}_1(t) = p(t)\,\zeta(\varepsilon(t)), \tag{18}$$

where $\varepsilon(t)$ is the transformation error and $\zeta(\varepsilon(t))$ should be equipped with the smooth, monotonically increasing, and reversible characteristics [30]. Notice that $\zeta(\varepsilon(t))$ is rewritten as $\zeta(\varepsilon)$ for simplicity in the following contents. Apart from the above characteristics, $\zeta(\varepsilon)$ ought to satisfy another two conditions, that is,

$$-1 < \zeta(\varepsilon) < 1 \tag{19}$$

$$\begin{cases} \lim_{\varepsilon \to -\infty} \zeta(\varepsilon) = -1 \\ \lim_{\varepsilon \to \infty} \zeta(\varepsilon) = 1 \end{cases} \tag{20}$$

Based on the above description, $\zeta(\varepsilon)$ is chosen as

$$\zeta(\varepsilon) = \frac{e^\varepsilon - e^{-\varepsilon}}{e^\varepsilon + e^{-\varepsilon}}. \tag{21}$$

Thus, the transformed error $\varepsilon(t)$ is described in the following form as

$$\varepsilon(t) = \zeta^{-1}\left(\frac{\bar{z}_1(t)}{p(t)}\right) = \frac{1}{2}\ln\frac{\bar{z}_1(t) + p(t)}{p(t) - \bar{z}_1(t)}. \tag{22}$$

2.4. Proposed Controller Design for Grid-Connected Inverter and Stability Proof

In this section, the proposed prescribed performance-based command-filtered backstepping adaptive sliding mode controller is established to gain the stable DC-link voltage and expected power demanded by the PV grid-connected system. More specifically, the control law u_d and u_q are both established to track the reference current, so that the control objective can be achieved. The block of the whole control system is depicted in Figure 3 and the concrete design process is described in the following.

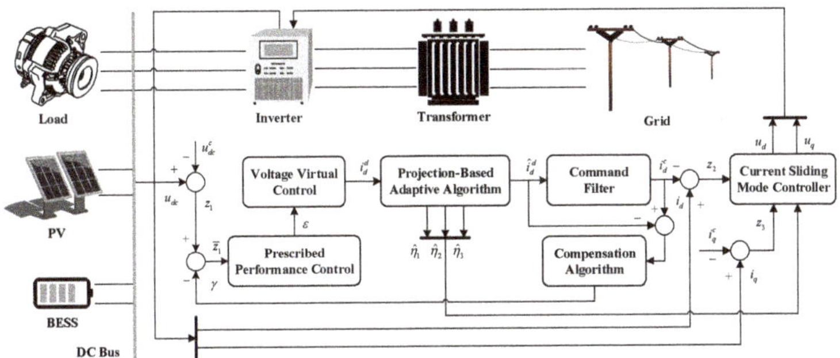

Figure 3. Control block of the proposed controller in the PV grid-connected system.

A series of tracking errors are defined first as follows.

$$z_1 = u_{dc} - u_{dc}^c \tag{23}$$

$$z_2 = i_d - i_d^c \tag{24}$$

$$z_3 = i_q - i_q^c. \tag{25}$$

Considering that the numerous differential processes are inevitable in the controller design process due to the existence of the derivatives of virtual control signals, the command filter is employed in order to avoid the potential differential expansion, whose structure diagram is drawn in Figure 4 and dynamics are given as follows [11,22,31].

$$\begin{bmatrix} \dot{q}_1 \\ \dot{q}_2 \end{bmatrix} = \begin{bmatrix} q_2 \\ 2\zeta\omega_n \left[S_R \left(\dfrac{\omega_n^2}{2\zeta\omega_n} \left(S_A(u) - q_1 \right) \right) - q_2 \right], \end{bmatrix} \tag{26}$$

where

$$\begin{bmatrix} q_1 \\ q_2 \end{bmatrix} = \begin{bmatrix} x^c \\ \dot{x}^c \end{bmatrix}, u = x^d, \tag{27}$$

and other parameters are defined as follows: ω_n and ζ are the bandwidth and damping of the command filter; and $S_A(.)$ and $S_R(.)$ are the amplitude and magnitude constraint, respectively.

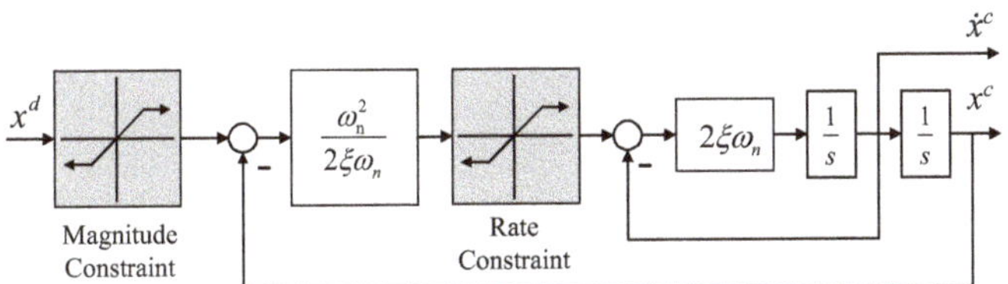

Figure 4. Control structure of the constrained command filter.

Since the filter error is inevitable during the filtering process, a corresponding compensated tracking error is defined as

$$\bar{z}_1 = z_1 - \gamma_1, \tag{28}$$

where γ_1 denotes the compensation signal with respect to the filter error, and has the following form:

$$\dot{\gamma}_1 = -\beta\gamma_1 + \frac{3E_d\eta_1}{2u_{dc}} \left(i_d^c - i_d^d \right), \tag{29}$$

where $\beta > 0$ is the compensation coefficient. Combining (15), (28) and (29), we can achieve the further representation of $\dot{z}_1$ as follows.

$$\dot{z}_1 = \frac{3E_d i_d \eta_1}{2u_{dc}} - \eta_1 i_{dc} - \dot{u}_{dc}^c + \beta\gamma_1 - \frac{3E_d \eta_1}{2u_{dc}} i_d^c + \frac{3E_d \eta_1}{2u_{dc}} i_d^d. \tag{30}$$

Meanwhile, according to (22), the derivative of the transformed error ε is obtained as

$$\dot{\varepsilon} = \frac{\partial \zeta^{-1}}{\partial(z_1/p)} \frac{1}{p} \left(\dot{z}_1 - \frac{\dot{p}z_1}{p} \right) = r\left(\dot{z}_1 - v \right), \tag{31}$$

where $r = \frac{p}{p^2 - z_1^2}$ and $v = \frac{\dot{p}z_1}{p}$ is defined for simplicity. Referring to the characteristics of the smooth function $p(t)$ and transformation function $\zeta(t)$, it can be deduced that $r > 0$.

Then, the first Lyapunov function is given as

$$V_1 = \frac{1}{2}\varepsilon^2. \tag{32}$$

Taking (30) and (31) into account, the further relationship can be obtained as follows.

$$\dot{V}_1 = \varepsilon r \left(\frac{3E_d i_d \eta_1}{2u_{dc}} - \eta_1 i_{dc} - \dot{u}_{dc}^c + \beta\gamma_1 - \frac{3E_d \eta_1}{2u_{dc}} i_d^c + \frac{3E_d \eta_1}{2u_{dc}} i_d^d - v \right), \tag{33}$$

and the virtual control law is selected as follows:

$$i_d^d = \frac{2u_{dc}}{3E_d \eta_1} \left(\eta_1 i_{dc} + \dot{u}_{dc}^c - \beta\gamma_1 + v - k_1\varepsilon \right), \tag{34}$$

where k_1 is a positive constant. Substituting the virtual controller (34) into (33), it is derived as

$$\dot{V}_1 = -k_1 r\varepsilon^2 + \frac{3E_d \eta_1}{2u_{dc}} \varepsilon r z_2. \tag{35}$$

Notice that the parameter η_1 is unknown, and an estimation value $\hat{\eta}_1$ is used here, whose adaptive algorithm would be given thereinafter, and the virtual controller and compensation signal are rewritten as follows, respectively.

$$\hat{i}_d^d = \frac{2u_{dc}}{3E_d \hat{\eta}_1} \left(\hat{\eta}_1 i_{dc} + \dot{u}_{dc}^c - \beta\gamma_1 + v - k_1\varepsilon \right); \tag{36}$$

$$\dot{\gamma}_1 = -k_1\gamma_1 + \frac{3E_d \hat{\eta}_1}{2u_{dc}} \left(i_d^c - \hat{i}_d^d \right). \tag{37}$$

Considering the robustness of the sliding mode control (SMC) method with respect to the disturbance and unknown parameters, SMC is employed in this paper for current control. Moreover, the integral SMC can eliminate the arrival phase by adjusting the reasonable initial state of the integrator so that the robustness can be improved. Thus, the integral sliding surfaces are defined based on the current tracking errors of d-q axis, as follows [11].

$$S_2 = z_2 + \tau_2 \int_0^t z_2 dt \tag{38}$$

$$S_3 = z_3 + \tau_3 \int_0^t z_3 dt, \tag{39}$$

where τ_2 and τ_3 are both positive coefficients of the integral sliding surface.

Then, another Lyapunov function composed of the transformation error, sliding surfaces, and adaptive parameters is defined as

$$V_2 = V_1 + \frac{1}{2}\left(S_2{}^2 + S_3{}^2 + \frac{\tilde{\eta}_1^2}{\lambda_1} + \frac{\tilde{\eta}_2^2}{\lambda_2} + \frac{\tilde{\eta}_3^2}{\lambda_3}\right),\tag{40}$$

where λ_1, λ_2, and λ_3 are the adaptive gains; $\tilde{\eta}_1 = \eta_1 - \hat{\eta}_1$, $\tilde{\eta}_2 = \eta_2 - \hat{\eta}_2$ and $\tilde{\eta}_3 = \eta_3 - \hat{\eta}_3$ are the estimation errors of unknown parameters. In light of (35), the derivation of V_2 is obtained as follows:

$$\dot{V}_2 = -k_1 r\varepsilon^2 + \frac{3E_d\eta_1}{2u_{dc}}\varepsilon r z_2 + S_2\dot{S}_2 + S_3\dot{S}_3 - \frac{\tilde{\eta}_1}{\lambda_1}\dot{\hat{\eta}}_1 - \frac{\tilde{\eta}_2}{\lambda_2}\dot{\hat{\eta}}_2 - \frac{\tilde{\eta}_3}{\lambda_3}\dot{\hat{\eta}}_3,\tag{41}$$

where $\dot{S}_2$ is derived based on (15), (24) and (38) as

$$\begin{aligned}\dot{S}_2 &= \dot{z}_2 + \tau_2 z_2 \\ &= -\eta_2 i_d + \omega i_q + \eta_3\left(u_d - E_d\right) - i_d^{ic} + \tau_2 z_2 \\ &= -\hat{\eta}_2 i_d - \tilde{\eta}_2 i_d + \omega i_q + \hat{\eta}_3\left(u_d - E_d\right) + \tilde{\eta}_3\left(u_d - E_d\right) - i_d^{ic} + \tau_2 z_2,\end{aligned}\tag{42}$$

and $\dot{S}_3$ is derived based on (15), (25) and (39) as

$$\begin{aligned}\dot{S}_3 &= \dot{z}_3 + \tau_3 z_3 \\ &= -\eta_2 i_q - \omega i_d + \eta_3\left(u_q - E_q\right) - i_q^{ic} + \tau_3 z_3 \\ &= -\hat{\eta}_2 i_q - \tilde{\eta}_2 i_q - \omega i_d + \hat{\eta}_3\left(u_q - E_q\right) + \tilde{\eta}_3\left(u_q - E_q\right) - i_q^{ic} + \tau_3 z_3.\end{aligned}\tag{43}$$

Through substitution and simplification, the results can be figured out as follows:

$$\begin{aligned}\dot{V}_2 = -k_1 r\varepsilon^2 &+ S_2\left(\frac{3E_d\hat{\eta}_1\varepsilon r z_2}{2u_{dc}S_2} - \hat{\eta}_2 i_d + \omega i_q + \hat{\eta}_3\left(u_d - E_d\right) - i_d^{ic} + \tau_2 z_2\right) \\ &+ S_3\left(-\hat{\eta}_2 i_q - \omega i_d + \hat{\eta}_3\left(u_q - E_q\right) - i_q^{ic} + \tau_3 z_3\right) \\ &+ \tilde{\eta}_1\left(\frac{3E_d\varepsilon r z_2}{2u_{dc}} - \frac{\dot{\hat{\eta}}_1}{\lambda_1}\right) + \tilde{\eta}_2\left(-\frac{\dot{\hat{\eta}}_2}{\lambda_2} - S_2 i_d - S_3 i_q\right) \\ &+ \tilde{\eta}_3\left(-\frac{\dot{\hat{\eta}}_3}{\lambda_3} + S_2\left(u_d - E_d\right) + S_3\left(u_q - E_q\right)\right).\end{aligned}\tag{44}$$

In order to achieve the asymptotical stability of the system, the projection-based adaptive law is established to eliminate the influence of η_1, η_2, and η_3, which is represented as

$$\begin{aligned}\dot{\hat{\eta}}_1 &= \lambda_1 \text{Proj}\left(\hat{\eta}_1, \frac{3E_d\varepsilon r z_2}{2u_{dc}}\right) \\ \dot{\hat{\eta}}_2 &= \lambda_2 \text{Proj}\left(\hat{\eta}_2, -S_2 i_d - S_3 i_q\right) \\ \dot{\hat{\eta}}_3 &= \lambda_3 \text{Proj}\left(\hat{\eta}_3, S_2\left(u_d - E_d\right) + S_3\left(u_q - E_q\right)\right),\end{aligned}\tag{45}$$

where $\text{Proj}\left(\cdot, \cdot\right)$ represents the projection operator [32]. The projection-based adaptive algorithm is employed here to ensure the boundedness of the estimated parameters, and the following relationship can be obtained as

$$\tilde{\eta}_1\left[\frac{3E_d\varepsilon r z_2}{2u_{dc}} - \text{Proj}\left(\hat{\eta}_1, \frac{3E_d\varepsilon r z_2}{2u_{dc}}\right)\right] \leq 0$$

$$\tilde{\eta}_2\left[-S_2 i_d - S_3 i_q - \text{Proj}\left(\hat{\eta}_2, -S_2 i_d - S_3 i_q\right)\right] \leq 0$$

$$\tilde{\eta}_3\left[S_2\left(u_d - E_d\right) + S_3\left(u_q - E_q\right) - \text{Proj}\left(\hat{\eta}_3, S_2\left(u_d - E_d\right) + S_3\left(u_q - E_q\right)\right)\right] \leq 0.$$

Hence, we can get the following inequality related to $\dot{V}_2$ as

$$\dot{V}_2 \leq -k_1 r\varepsilon^2 + S_2 \left(\frac{3E_d \hat{\eta}_1 \varepsilon r z_2}{2u_{dc} S_2} - \hat{\eta}_2 i_d + \omega i_q + \hat{\eta}_3 (u_d - E_d) - i_d^{ic} + \tau_2 z_2 \right)$$
$$+ S_3 \left(-\hat{\eta}_2 i_q - \omega i_d + \hat{\eta}_3 (u_q - E_q) - i_q^{ic} + \tau_3 z_3 \right). \tag{46}$$

With the aim of guaranteeing the asymptotical stability of the system, $\dot{V}_2$ ought to remain non-positive, that is, $\dot{V}_2 \leq 0$. As a result, the following saturation function is used here for stability.

$$-k_2 \mathrm{Sat}(S_2) = \frac{3E_d \hat{\eta}_1 \varepsilon r z_2}{2u_{dc} S_2} - \hat{\eta}_2 i_d + \omega i_q + \hat{\eta}_3 (u_d - E_d) - i_d^{ic} + \tau_2 z_2$$
$$-k_3 \mathrm{Sat}(S_3) = -\hat{\eta}_2 i_q - \omega i_d + \hat{\eta}_3 (u_q - E_q) - i_q^{ic} + \tau_3 z_3, \tag{47}$$

where k_2 and k_3 are both positive constants, and $\mathrm{Sat}(\cdot)$ denotes the saturation function, which is given as follows.

$$\mathrm{Sat}(x) = \begin{cases} 1, & x > \varphi \\ x/\varphi, & |x| \leq \varphi \\ -1, & x < -\varphi \end{cases} \tag{48}$$

where $\varphi \in (0, 0.5]$ is the sliding layer. In conclusion, the control law u_d and u_q can be obtained in the following.

$$u_d = -\frac{1}{\hat{\eta}_3} \left(\frac{3E_d \hat{\eta}_1 \varepsilon r z_2}{2u_{dc} S_2} - \hat{\eta}_2 i_d + \omega i_q - \hat{\eta}_3 E_d - i_d^{ic} + \tau_2 z_2 + k_2 \mathrm{Sat}(S_2) \right)$$
$$u_q = -\frac{1}{\hat{\eta}_3} \left(-\hat{\eta}_2 i_q - \omega i_d - \hat{\eta}_3 E_q - i_q^{ic} + \tau_3 z_3 + k_3 \mathrm{Sat}(S_3) \right). \tag{49}$$

Under the proposed control law (49), the inequality (46) can be further described as

$$\dot{V}_2 \leq -k_1 r\varepsilon^2 - k_2 S_2 \mathrm{Sat}(S_2) - k_3 S_3 \mathrm{Sat}(S_3) \leq 0. \tag{50}$$

Obviously, $\dot{V}_2 = 0$ holds only if $[\varepsilon, S_2, S_3] = [0, 0, 0]$. Considering the variation of the environmental situation, the output power generated by PV fluctuates constantly, which leads to the fluctuation of the DC bus voltage. That is to say, the voltage tracking error and sliding mode surfaces would not always be zero, which indicates that $\dot{V}_2$ would not always be zero. Therefore, the system is proved asymptotically stable based on the Lyapunov stability criterion and LaSalle's invariance principle.

3. Results and Discussion

In this section, the performance of the designed controller is simulated in the MATLAB/Simulink environment and the simulation model is shown in Figure 5. Considering that the environment varies constantly, including solar irradiance and temperature, which would have an influence on the stable operation of the PV grid-connected system and lead to the power fluctuation, the distributed PV arrays under variable conditions are integrated for power generation. it is worth noting that four photovoltaic arrays, which are not far away from each other share the same set of environmental parameters including the solar irradiation and temperature, which are both shown in Figure 6a, and the external characteristics of each PV array, including output voltage, current, and power are shown in Figure 6b–d, respectively. Under the environment, the proposed controller is applied to the system for stable operation. The corresponding parameters of the PV grid-connected system and controller are both shown in Table 2.

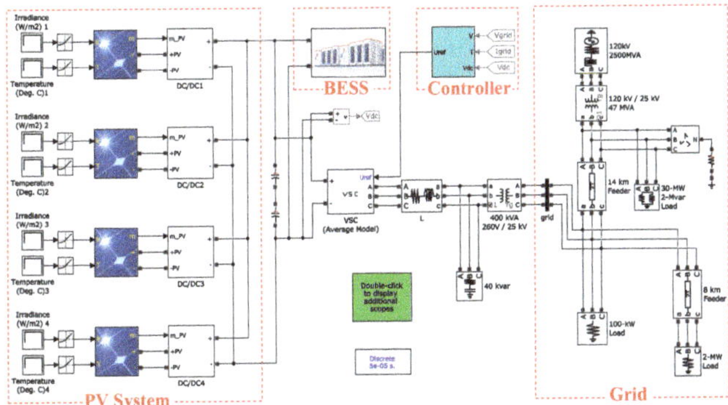

Figure 5. Simulation model of the PV grid-connected system with BESS.

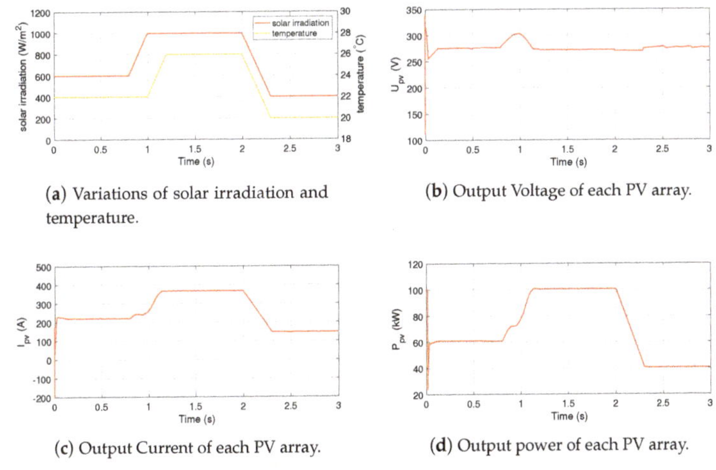

(**a**) Variations of solar irradiation and temperature.

(**b**) Output Voltage of each PV array.

(**c**) Output Current of each PV array.

(**d**) Output power of each PV array.

Figure 6. External characteristics of each PV array.

Table 2. The parameters of the control system.

Classification	Parameter	Value	Parameter	Value	Parameter	Value
	R	1 mΩ	R_1, R_2, R_3, R_4	5 mΩ	P_g	400 kW
PV grid-connected system	L	45 μH	L_1, L_2, L_3, L_4	5 mH	f	60 Hz
	C	50 mF	C_1, C_2, C_3, C_4	0.1 mF	u_{dc}	500 V
	p_0	250	p_∞	5	l	1.5
	k_1	8000	k_2	60,000	k_3	80,000
Proposed controller	λ_1	0.1	λ_2	0.1	λ_3	0.1
	τ_2	3	τ_3	3	β	0.9
	ϕ	0.5	ω_n	300	ς	0.1

As to the parameter adjustment of the proposed control strategy, the prescribed performance parameters rank first owing to their strong constraint on the tracking error. The initial value of the performance function p_0 exhibits great importance at the beginning of the system operation. Considering the inevitable response time in initial stage of the controller operation, p_0 shall be set relatively large, yet a too-large p_0 would lead to large initial overshoot. Besides, l regulates the convergence rate of the tracking error to some extent, and p_∞ shows its effectiveness on the tracking error at the stable state. Hence, an appropriate performance function would be beneficial to the control performance. In terms of other parameters, they shall be adjusted according to the tracking error variation and some certain performance variation; for instance, the adaptive coefficient shall be adjusted in view of the estimation error of the adaptive parameters. Based on the adjusted parameters, the simulation results are given and the concrete analysis is provided as follows.

The concrete objectives of the control performance are listed as follows: The tracking error shall be constrained within ±2% and the power fluctuation shall be limited no more than ±5% at the stable state. The power demanded by grid was 400 kW, yet the maximum output power of all the PV arrays is less than 400 kW at the beginning, which brings about 30% power fluctuation. Therefore, in order to maintain power balance, the BESS needs to be taken into account and introduced in 0.5 s. The power transmission performance of each module in the power grid is shown in Figure 7. As is seen, there exists about 30% power fluctuation when the BESS is absent from the grid-connected system before 0.5 s, and the introduction of BESS compensates for the power difference since 0.5 s, which contributes to the relatively stable and smooth power delivered to the grid with no more than 5% fluctuation. Besides, the output power under the proposed controller is free of overshoot and provided with a quick response.

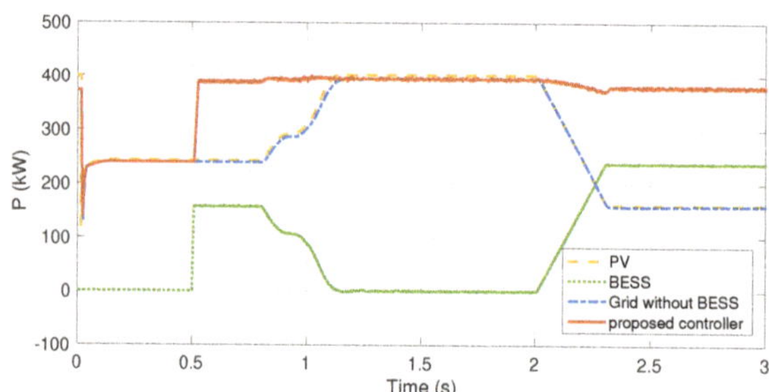

Figure 7. Power transmission performance of each module in power grid.

Moreover, the DC bus voltage tracking performance is given in Figure 8. It can be found that the bus voltage under the proposed controller arrives at 510 V and converges to the reference within nearly 0.01 s at the beginning. Then, the voltage is maintained at 500.5 V under the control of the proposed strategy, while the one under CBC fluctuates at 502 V. Besides, the performance difference can be more clearly observed when BESS is introduced, namely, the fluctuation caused by BESS under the proposed controller appears little, while the one of CBC reaches 4 V, which indicates the bus voltage under the proposed controller has little fluctuation, higher convergence speed, and more precise tracking performance than the one under CBC. In order to show the effect more intuitively, the voltage tracking error is depicted in Figure 9. It indicates that the introduction of BESS brings about little overshoot under the proposed controller, yet causes large overshoot under CBC. Furthermore, the voltage error with the proposed controller holds within the performance range owing to the prescribed performance technique, which further validates the advantages of the proposed controller.

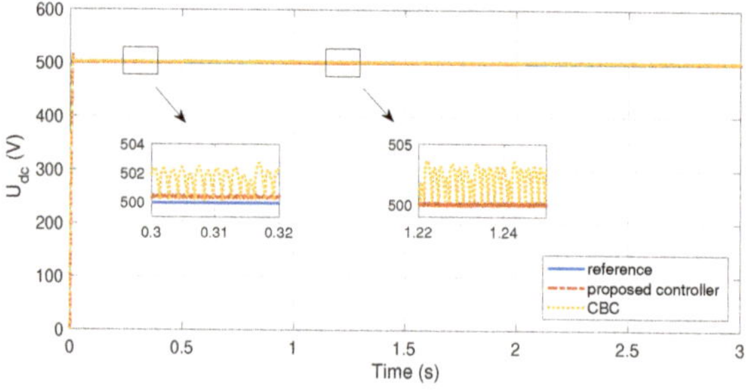

Figure 8. Bus voltage tracking performance between the proposed controller and conventional CBC.

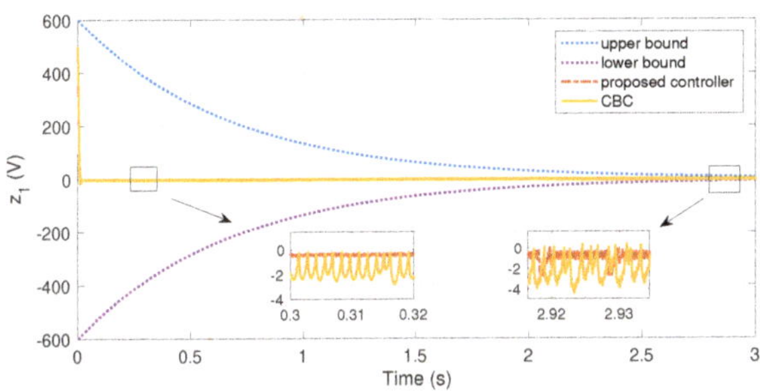

Figure 9. Bus voltage tracking error between the proposed controller and CBC.

Figure 10 presents the estimation performance of the adaptive parameters. It can be found that the adaptive parameters own slow-varying curves, which correspond with their characteristics. Besides, Figure 11 shows the performance of a three-phase voltage and current at the grid side with the partial enlarged drawing. It is vividly seen that the output voltage and current of the inverter own smooth variation curves with little distortion, which indicates how the designed controller can still inject a great sinusoidal current into the grid. In addition, Figure 11c also presents the total harmonic distortion (THD) of the grid current under the proposed controller with the value of 1.88%. In contrast, the grid current is more harmonic when CBC is utilized, as shown in Figure 11d. In this case, the THD value is 2.38%. Consequently, it can proved that the proposed controller is effective for the PV grid-connected system and possesses better performance with little overshoot, a quicker response, higher accuracy, and being less harmonic than CBC.

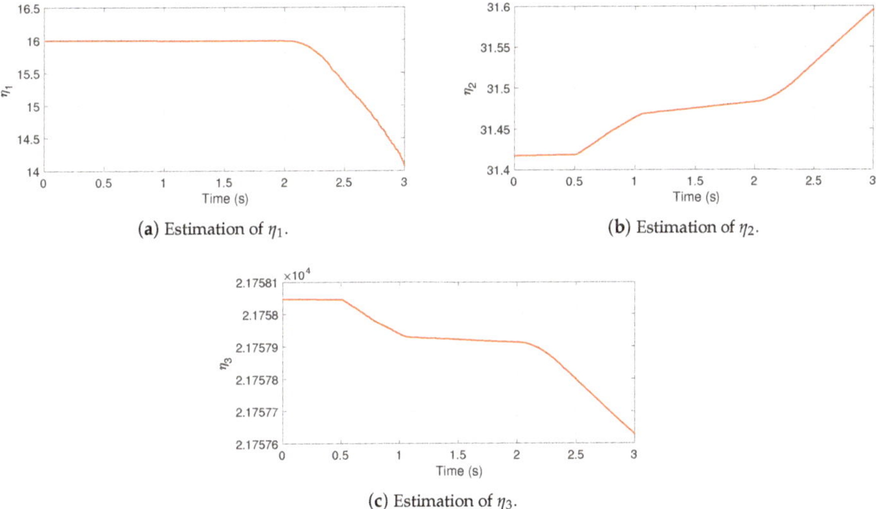

(a) Estimation of η_1.

(b) Estimation of η_2.

(c) Estimation of η_3.

Figure 10. Estimation curves of the adaptive parameters.

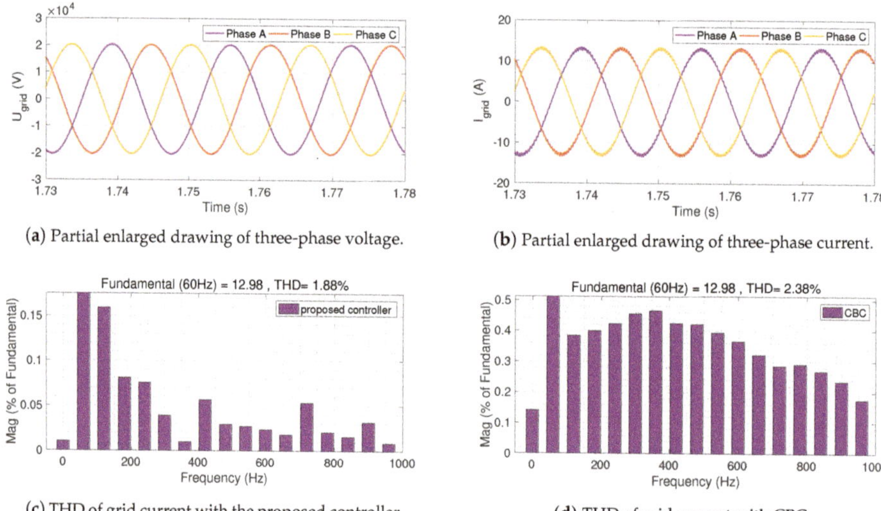

(**a**) Partial enlarged drawing of three-phase voltage.

(**b**) Partial enlarged drawing of three-phase current.

(**c**) THD of grid current with the proposed controller.

(**d**) THD of grid current with CBC.

Figure 11. Three-phase voltage and current performance at grid side.

4. Conclusions

In this paper, a novel command-filtered adaptive backstepping control strategy with a prescribed performance technique was proposed to control the active power of the PV grid-connected system. The concrete conclusions could be further drawn according to the results as follows.

1. The proposed control strategy exhibits great performance for power and voltage control of the PV grid-connected system with unknown system parameters and constantly variable environmental conditions.
2. The DC bus voltage tracking error is constrained within the preset range owing to the introduction of prescribed performance technique, which guarantees high tracking precision and constitutes to the stable operation of the system.
3. The introduced projection-based adaptive law provides accurate estimation for system parameters, which is more in line with the actual system.
4. Comparing with CBC, the prescribed performance-based control strategy owns more stable power delivery and voltage with little fluctuation, which shows better static and dynamic performance of the proposed controller.

In the future, the semi-physical experiment platform of a small-scale PV grid-connected system shall be established to further validate the effectiveness and practicability of the proposed controller.

Author Contributions: Conceptualization, W.Z. and T.P.; methodology, W.Z. and D.W.; software, W.Z. and T.P.; validation, W.Z. and D.W.; investigation, T.P. and D.X.; writing—original draft preparation, W.Z.; writing—review and editing, T.P. and D.W.; funding acquisition, T.P. and D.X. All authors have read and agreed to the published version of the manuscript.

Funding: This work was partially supported by the National Natural Science Foundation of China (61973140, 61672266), National first-class discipline program of Food Science and Technology (JUFSTR20180205).

Conflicts of Interest: The authors declare no conflict of interest.

Abbreviations

The following abbreviations and nomenclatures are used in this manuscript:

A	Dimensionless Coefficient
BESS	Battery Energy Storage System
C	Capacitance of System (F)
CBC	Command-Filtered Backstepping Control
E_d, E_q	Grid Voltages in d-q Axis (V)
E_g	Band-Gap Power of Semiconductor (W)
i_0	Output Current of Boost Circuit (A)
I_0	Reverse Saturation Current (A)
i_d, i_q	Grid Currents in d-q Axis (A)
I_D	Diode Current (A)
i_{dc}	Input Current of Inverter (A)
I_{pv}	Output Current of PV Cell (A)
I_R	Corresponding current (A)
I_{RS}	Reverse Saturation Current (A)
I_s	Photocurrent (A)
I_{sc}	Short-Circuit Current (A)
k	Boltzmann Constant
k_d, k_q	Switching Function in d-q Axis
k_i	Thermal Coefficient of Current
L	Inductance of System (H)
MPPT	Maximum Power Point Tracking
N_p	Number of PV Cells in Parallel Way
N_s	Number of PV Cells in Series Way
PPC	Prescribed Performance Control
PV	Photovoltaic
q	Elementary Charge (C)
R	Resistance of System (Ω)
R_0	Solar Irradiation (W/m^2)
R_n	Reference Solar Irradiation (W/m^2)
R_s	Series Resistance (Ω)
R_{sh}	Shunt Resistance (Ω)
SMC	Sliding Mode Control
T	Temperature of PV Cell ($^\circ$C)
T_n	Reference Temperature of PV Cell ($^\circ$C)
THD	Total Harmonic Distortion
u_{dc}	DC Bus Voltage (V)
U_{pv}	Output Voltage of PV Cell (V)

References

1. Kouro, S.; Leon, J.I.; Vinnikov, D.; Franquelo, L.G. Grid-connected photovoltaic systems: An overview of recent research and emerging PV converter technology. *IEEE Ind. Electron. Mag.* **2015**, *9*, 47–61. [CrossRef]
2. Eltawil, M.A.; Zhao, Z. Grid-connected photovoltaic power systems: Technical and potential problems—A review. *Renew. Sustain. Energy Rev.* **2010**, *14*, 112–129. [CrossRef]
3. Yang, B.; Li, W.; Zhao, Y.; He, X. Design and analysis of a grid-connected photovoltaic power system. *IEEE Trans. Power Electron.* **2010**, *25*, 992–1000. [CrossRef]
4. Li, S.; Wei, Z.; Ma, Y. Fuzzy load-shedding strategy considering photovoltaic output fluctuation characteristics and static voltage stability. *Energies* **2018**, *11*, 779. [CrossRef]
5. Liserre, M.; Teodorescu, R.; Blaabjerg, F. Stability of photovoltaic and wind turbine grid-connected inverters for a large set of grid impedance values. *IEEE Trans. Power Electron.* **2006**, *21*, 263–272. [CrossRef]
6. Li, Y.; Ishikawa, M. An efficient reactive power control method for power network systems with solar photovoltaic generators using sparse optimization. *Energies* **2017**, *10*, 696. [CrossRef]

7.   Zhang, Q.; Zhou, L.; Mao, M.; Xie, B.; Zheng, C. Power quality and stability analysis of large-scale grid-connected photovoltaic system considering non-linear effects. *IET Power Electron.* **2018**, *11*, 1739–1747. [CrossRef]

8.   Mahmud, M.A.; Pota, H.; Hossain, M. Dynamic stability of three-phase grid-connected photovoltaic system using zero dynamic design approach. *IEEE J. Photovoltaics* **2012**, *2*, 564–571. [CrossRef]

9.   Yıldıran, N.; Tacer, E. A new approach to H-infinity control for grid-connected inverters in photovoltaic generation systems. *Electr. Power Components Syst.* **2019**, *47*, 1413–1422. [CrossRef]

10.   Zhu, Y.; Fei, J. Adaptive global fast terminal sliding mode control of grid-connected photovoltaic system using fuzzy neural network approach. *IEEE Access* **2017**, *5*, 9476–9484. [CrossRef]

11.   Xu, D.; Wang, G.; Yan, W.; Yan, X. A novel adaptive command-filtered backstepping sliding mode control for PV grid-connected system with energy storage. *Sol. Energy* **2019**, *178*, 222–230. [CrossRef]

12.   Farrell, J.A.; Polycarpou, M.; Sharma, M.; Dong, W. Command filtered backstepping. *IEEE Trans. Autom. Control* **2009**, *54*, 1391–1395. [CrossRef]

13.   Yu, J.; Shi, P.; Zhao, L. Finite-time command filtered backstepping control for a class of nonlinear systems. *Automatica* **2018**, *92*, 173–180. [CrossRef]

14.   Shen, Q.; Shi, P. Distributed command filtered backstepping consensus tracking control of nonlinear multiple-agent systems in strict-feedback form. *Automatica* **2015**, *53*, 120–124. [CrossRef]

15.   Cui, G.; Xu, S.; Lewis, F.L.; Zhang, B.; Ma, Q. Distributed consensus tracking for non-linear multi-agent systems with input saturation: A command filtered backstepping approach. *IET Control Theory Appl.* **2016**, *10*, 509–516. [CrossRef]

16.   Yu, J.; Zhao, L.; Yu, H.; Lin, C.; Dong, W. Fuzzy finite-time command filtered control of nonlinear systems with input saturation. *IEEE Trans. Cybern.* **2017**, *48*, 2378–2387.

17.   Pan, Y.; Wang, H.; Li, X.; Yu, H. Adaptive command-filtered backstepping control of robot arms with compliant actuators. *IEEE Trans. Control Syst. Technol.* **2017**, *26*, 1149–1156. [CrossRef]

18.   Jin, Z.; Zhang, W.; Liu, S.; Gu, M. Command-filtered backstepping integral sliding mode control with prescribed performance for ship roll stabilization. *Appl. Sci.* **2019**, *9*, 4288. [CrossRef]

19.   Ren, H.; Deng, G.; Hou, B.; Wang, S.; Zhou, G. Finite-time command filtered backstepping algorithm-based pitch angle tracking control for wind turbine hydraulic pitch systems. *IEEE Access* **2019**, *7*, 135514–135524. [CrossRef]

20.   Choi, I.H.; Bang, H.C. Adaptive command filtered backstepping tracking controller design for quadrotor unmanned aerial vehicle. *Proc. Inst. Mech. Eng. Part J. Aerosp. Eng.* **2012**, *226*, 483–497. [CrossRef]

21.   Zhao, S.; Dong, W.; Farrell, J.A. Quaternion-based trajectory tracking control of VTOL-UAVs using command filtered backstepping. In Proceedings of the 2013 American Control Conference, Washington, DC, USA, 17–19 June 2013; IEEE: Piscataway, NJ, USA, 2013; pp. 1018–1023.

22.   Bechlioulis, C.P.; Rovithakis, G.A. Robust adaptive control of feedback linearizable MIMO nonlinear systems with prescribed performance. *IEEE Trans. Autom. Control* **2008**, *53*, 2090–2099. [CrossRef]

23.   Na, J.; Chen, Q.; Ren, X.; Guo, Y. Adaptive prescribed performance motion control of servo mechanisms with friction compensation. *IEEE Trans. Ind. Electron.* **2013**, *61*, 486–494. [CrossRef]

24.   Huang, Y.; Na, J.; Wu, X.; Liu, X.; Guo, Y. Adaptive control of nonlinear uncertain active suspension systems with prescribed performance. *ISA Trans.* **2015**, *54*, 145–155. [CrossRef]

25.   Kostarigka, A.K.; Doulgeri, Z.; Rovithakis, G.A. Prescribed performance tracking for flexible joint robots with unknown dynamics and variable elasticity. *Automatica* **2013**, *49*, 1137–1147. [CrossRef]

26.   Zhou, Q.; Li, H.; Wang, L.; Lu, R. Prescribed performance observer-based adaptive fuzzy control for nonstrict-feedback stochastic nonlinear systems. *IEEE Trans. Syst. Man Cybern. Syst.* **2017**, *48*, 1747–1758. [CrossRef]

27.   Qiu, J.; Sun, K.; Wang, T.; Gao, H. Observer-based fuzzy adaptive event-triggered control for pure-feedback nonlinear systems with prescribed performance. *IEEE Trans. Fuzzy Syst.* **2019**, *27*, 2152–2162. [CrossRef]

28.   Belhachat, F.; Larbes, C. Modeling, analysis and comparison of solar photovoltaic array configurations under partial shading conditions. *Sol. Energy* **2015**, *120*, 399–418. [CrossRef]

29.   Dhoke, A.; Sharma, R.; Saha, T.K. An approach for fault detection and location in solar PV systems. *Sol. Energy* **2019**, *194*, 197–208. [CrossRef]

30. Liu, D.; Yang, G.H. Data-driven adaptive sliding mode control of nonlinear discrete-time systems with prescribed performance. *IEEE Trans. Syst. Man Cybern. Syst.s* **2019**, *49*, 2598–2604. [CrossRef]
31. Jiang, B.; Xu, D.; Shi, P.; Lim, C.C. Adaptive neural observer-based backstepping fault tolerant control for near space vehicle under control effector damage. *IET Control Theory Appl.* **2014**, *8*, 658–666. [CrossRef]
32. Chen, H.; Wang, H. Numerical simulation for conservative fractional diffusion equations by an expanded mixed formulation. *J. Comput. Appl. Math.* **2016**, *296*, 480–498. [CrossRef]

© 2020 by the authors. Licensee MDPI, Basel, Switzerland. This article is an open access article distributed under the terms and conditions of the Creative Commons Attribution (CC BY) license (http://creativecommons.org/licenses/by/4.0/).

Article

Electricity Theft Detection Using Supervised Learning Techniques on Smart Meter Data

Zahoor Ali Khan [1,*], **Muhammad Adil** [2], **Nadeem Javaid** [2], **Malik Najmus Saqib** [3], **Muhammad Shafiq** [4] and **Jin-Ghoo Choi** [4,*]

1. Computer Information Science, Higher Colleges of Technology, Fujairah 4114, UAE
2. Department of Computer Science, COMSATS University Islamabad, Islamabad 44000, Pakistan; adilxyz1@gmail.com (M.A.); nadeemjavaidqau@gmail.com (N.J.)
3. Department of Cybersecurity, College of Computer Science and Engineering, University of Jeddah, Jeddah 21959, Saudi Arabia; mnajam@uj.edu.sa
4. Department of Information and Communication Engineering, Yeungnam University, Gyeongsan, Gyeongbuk 38541, Korea; shafiq@ynu.ac.kr
* Correspondence: zkhan1@hct.ac.ae (Z.A.K.); jchoi@yu.ac.kr (J.-G.C.)

Received: 10 August 2020; Accepted: 23 September 2020; Published: 28 September 2020

Abstract: Due to the increase in the number of electricity thieves, the electric utilities are facing problems in providing electricity to their consumers in an efficient way. An accurate Electricity Theft Detection (ETD) is quite challenging due to the inaccurate classification on the imbalance electricity consumption data, the overfitting issues and the High False Positive Rate (FPR) of the existing techniques. Therefore, intensified research is needed to accurately detect the electricity thieves and to recover a huge revenue loss for utility companies. To address the above limitations, this paper presents a new model, which is based on the supervised machine learning techniques and real electricity consumption data. Initially, the electricity data are pre-processed using interpolation, three sigma rule and normalization methods. Since the distribution of labels in the electricity consumption data is imbalanced, an Adasyn algorithm is utilized to address this class imbalance problem. It is used to achieve two objectives. Firstly, it intelligently increases the minority class samples in the data. Secondly, it prevents the model from being biased towards the majority class samples. Afterwards, the balanced data are fed into a Visual Geometry Group (VGG-16) module to detect abnormal patterns in electricity consumption. Finally, a Firefly Algorithm based Extreme Gradient Boosting (FA-XGBoost) technique is exploited for classification. The simulations are conducted to show the performance of our proposed model. Moreover, the state-of-the-art methods are also implemented for comparative analysis, i.e., Support Vector Machine (SVM), Convolution Neural Network (CNN), and Logistic Regression (LR). For validation, precision, recall, F1-score, Matthews Correlation Coefficient (MCC), Receiving Operating Characteristics Area Under Curve (ROC-AUC), and Precision Recall Area Under Curve (PR-AUC) metrics are used. Firstly, the simulation results show that the proposed Adasyn method has improved the performance of FA-XGboost classifier, which has achieved F1-score, precision, and recall of 93.7%, 92.6%, and 97%, respectively. Secondly, the VGG-16 module achieved a higher generalized performance by securing accuracy of 87.2% and 83.5% on training and testing data, respectively. Thirdly, the proposed FA-XGBoost has correctly identified actual electricity thieves, i.e., recall of 97%. Moreover, our model is superior to the other state-of-the-art models in terms of handling the large time series data and accurate classification. These models can be efficiently applied by the utility companies using the real electricity consumption data to identify the electricity thieves and overcome the major revenue losses in power sector.

Keywords: data pre-processing; electricity theft; imbalance data; parameter tuning; smart grid

---

1. Introduction

1.1. Background and Motivation

The smart grid system is defined as the conventional electricity network with the addition of digital communication technologies, i.e., sensors and smart meters. Recent studies in [1–4] show that the smart grid can help in efficient management of electrical power. The transactive energy framework [5] and short term load scheduling [6] are introduced to ensure optimal use of installed resources in the smart grid system. The hierarchical energy management system is presented in [7] to reduce the peak hours and trade more electricity at lower prices. The information-gap decision theory based solution is utilized to reduce the intermittent nature of renewable energies [8]. In a smart grid, the smart meter exchanges the information between electricity users and the grid. It records a huge amount of data, including the electrical energy consumption of consumers. Exploiting these data, the artificial intelligence techniques can track the energy consumption patterns of consumers and accurately identify the electricity thieves.

The electricity thieves bring major revenue losses to the electric utility. Electricity losses in transmission and distribution can be generally categorized into Technical Losses (TL) and Non-Technical Losses (NTL). TL occurs due to power dissipation in overhead power lines, transformers, and other substation equipment that are used to transfer electricity. NTL primarily consists of electricity theft. Electricity theft is defined as the energy consumed without authorization of power utility [9]. It includes bypassing the electricity meter, energy corruption of unregistered connections, tampering the meter reading, and direct hooking [10]. It is accountable for major revenue losses and decreases power quality [11]. A recent survey estimates that every year the power utility companies lose more than \$20 billion worldwide [12]. The NTL affects both the developed and developing countries. For instance, in Pakistan, the electricity transmission and distribution losses of 17.5% were recorded for the years 2017–2018 [13]. India also loses about \$4.5 billion each year due to electricity theft. A recent survey estimates that 20% of the total electricity is lost in India due to the illegal electricity consumption [14]. This problem also affects the rich nations. In the United States, the losses due to illegal electricity consumption are about \$6 billion annually, while, in the UK, the power losses exceed up to £175 million every year [15]. Moreover, electricity theft behaviors can also affect the operations and reliability of the power system. It decreases the power quality by overloading of transformers and voltage imbalances.

1.2. Literature Review

The researchers have recently implemented various approaches to detect the electricity theft. These approaches can be divided into three categories: state based solutions, game theory, and machine learning. The state based solutions use the additional hardware equipment like wireless sensors, distribution transformers, and smart meters to detect the electricity theft [16]. This method has a high cost of implementation due to the need of additional hardware equipment. In a game theory based method, there is assumed to be a game between the power utility and the electricity thieves. The outcome of a game can be derived from the difference between the electricity consumption behavior of electricity thieves and benign users [17]. However, it needs to define a utility function for all the players in a game, which is quite challenging. The machine learning techniques are widely used for ETD. They can be further categorized into unsupervised techniques (clustering) and supervised techniques (classification) that are later applied to unlabelled datasets in order to classify fraudulent and normal consumers. The existing methods used for ETD are presented in Table 1, which contains their contributions and limitations.

Table 1. Existing methods used for ETD.

Methods	Contributions	Limitations
Hardware based [16]	Its focus is on designing specific hardware devices in order to detect electricity theft	High cost of hardware installation
Game theory [17]	There is a game between the electricity thieves and the utility. The outcome of a game can be derived from the difference between the electricity consumption behavior of electricity thieves and benign users	This method needs to define a utility function for all the players in a game, which is quite challenging.
Machine learning [18–25]	It uses the smart meter data to effectively detect anomalous consumption behavior of dishonest consumers	Performance is poor on highly imbalanced data

1.2.1. Positioning of Our Work in the Literature

Our approach proposes a solution based on supervised learning. Therefore, we will study the details about the recent advances made in supervised learning techniques. The Support Vector Machine (SVM) and Logistic Regression (LR) are mostly used for ETD [18]. These techniques perform better when the dataset is small. However, these techniques are not effective when the dataset is large and extremely imbalanced. Hasan et al. [19] proposed a hybrid model consisting of CNN and Long Short Term Memory (LSTM). The CNN is utilized for feature extraction while LSTM used the refined features to classify the data into honest consumers and electricity thieves. To solve the problem of an imbalanced dataset, the Synthetic Minority Over Sampling Technique (SMOTE) is utilized. It [18] has achieved good results. i.e., precision 90% and recall 87%. However, the overfitting problem is not considered, which is caused by the addition of duplicate information through SMOTE. In [16], the authors proposed a hybrid model based on Multi-Layer Perceptron (MLP) and LSTM for ETD. This model detects the NTL by combining the auxiliary data through MLP and electricity consumption data through LSTM. However, the unbalanced data problem is not solved before classification. Moreover, the FPR of this model is high due to training on less data. It has achieved 54.5% PR-AUC, when 80% data are used for training. In [20], the authors addressed the issue of NTL detection using a Maximum Overlap Decomposition and Packet Transform (MODWPT) and Random Under Sampling Boosting (RUSBoost) techniques. The RUSBoost method is effective in handling the imbalanced data. However, the authors do not perform optimization to select best parameter values to improve the classification process. Moreover, the random under sampling technique reduces the data size and results in under fitting the model. To address the issue of power losses in Brazil, Ramos et al. [21] designed a Binary Black Hole Algorithm (BBHA) for NTL detection in Brazil. The accuracy comparison shows that BBHA outperforms other optimization techniques, i.e., Genetic Algoithm (GA) and Particle Swarm Optimization (PSO). However, no reliable evaluation metrics like precision and recall are used to validate the performance of the system. The reliable evaluation is very necessary in case of imbalanced binary classification problems.

Authors in [18] proposed a solution based on XGBoost and SVM for the detection of NTL in the smart grid. The aim of this study is to rank the list of consumers based on the smart meter data and extract features from the auxiliary dataset. The XGBoost is utilized that operates as an ensemble model and boosts the classification performance. However, the data pre-processing is not considered to refine the input data. The performance of machine learning algorithms is dependent on the quality of input data. In [22], the authors proposed a new technique to detect the NTL, which is based on Maximum Information Coefficient (MIC) and Fast Search by Finding of Density Peaks (FSFDP). The refined data are achieved by the MIC method, while FSFDP is used for classification. However, it needs an additional cost of hardware installation. The summary of existing work related to supervised learning techniques is given in Table 2. It gives the information about contritions and limitations of the existing work done in ETD using the supervised learning techniques.

Table 2. Performance of supervised machine learning techniques in the literature. MODWPT: Maximum Overlap Packet Transform; LSTM: Long Short Term Memory; CNN: Convolutional Neural Network; XGBoost: Extreme Gradient Boosting Technique; SGCC: State Grid Corporation of China; MLP: Multi Layer Perceptron; SMOTE: Synthetic Minority Over-sampling Technique; SVM: Support Vector Machine; BBHA: Binary Black Hole Algorithm.

Dataset	Supervised Techniques	Data Balancing	Contributions	Limitations
SGCC [19]	LSTM, CNN	SMOTE	The CNN is utilized for feature extraction, while LSTM uses the refined features to classify the data into honest consumers and electricity thieves.	The overfitting problem is not considered, which is caused by the addition of duplicate information through SMOTE
Endesa [16]	MLP and LSTM	Not handled	Detect the NTL by combining the auxiliary data through MLP and electricity consumption data through LSTM	The imbalanced data are not balanced before classification
Honduras [20]	MODWPT, RUSBoost	National grid of Brazil	The MODWPT gives the refined input and RUSBoost method balances the labels in the data before classification	The random under sampling technique reduces the data size and results in underfitting the model
Brazilian utility [21]	BBHA	Not handled	Use of binary black hole optimization technique to identify the NTL	No reliable evaluation is performed to validate the performance of the system
Endesa [18]	SVM, XGBoost	RUS	The XGBoost is utilized that operates as an ensemble method and boosts the classification performance	The data pre-processing is not considered to refine the input data
Irish data [22]	MIC, FSFD	Not handled	The refined data are achieved by MIC method, while FSFDP is used for classification.	This model has a high cost of hardware installation
NAB [23]	LSTM-GMM	Not handled	The authors enhanced the internal structure of LSTM to solve the gradient vanishing problem	The model is complex and its execution time is high
EISA [24]	CNN, RF	SMOTE	The generalized performance is achieved by using the decision trees along with CNN	The SMOTE generate synthetic data, which causes overfitting issues

Ding et al. [23] solve the gradient vanishing problem by enhancing the internal structure of LSTM to detect the electricity theft. This approach is based on LSTM and Gaussian Mixture Model (GMM). This model achieved excellent results. i.e., precision 90.1% and recall 91.9%. However, this model has high execution time. In [24], the authors utilize the CNN model for detecting the electricity theft. In [18] CNN, the classification through fully connected layers leads towards the degradation of generalization. Therefore, the authors used Random Forest (RF) for final classification. Moreover, the imbalanced data are handled using SMOTE. The generalized performance is achieved by using the decision trees along with CNN. However, the SMOTE generates synthetic data, which causes the overfitting problem. Authors in [25] used a gradient Boosting theft detector for NTL detection. This technique improves its performance by learning from an ensemble of decision trees, which shows the effectiveness of the model. The simulation result shows that a gradient Boosting theft detector is superior to other machine learning techniques.

The performances of the existing Electricity Theft Detection (ETD) methods are reasonable. However, these methods have some limitations, which are given below.

1. Conventional ETD includes the manual methods, i.e., humanly checking the meter readings and examining the direct hooking of power transmission lines. However, these methods require the additional cost for hiring the inspection teams.
2. The game theory based solutions have a low detection rate and high False Positive Rate (FPR) [26].
3. The state based solution is expensive because it requires an additional cost for hardware implementation [27].
4. The major problem in ETD using machine learning techniques is handling the unbalanced data. In traditional models, this problem is left untreated. Some authors (as mentioned in Table 2) use the RUS and SMOTE methods, which cause the loss of information and overfitting problem, respectively.
5. In most cases, the available data contain erroneous values, which reduce the classification accuracy [28].
6. The traditional machine learning techniques like Logistic regression (LR) and Support Vector Machine (SVM) have poor classification performance for massive data [28].

1.3. Contributions

The flowchart of proposed methodology for ETD is also given in Figure 1. The mapping of problems addressed and our proposed approach is given in Table 3. In the proposed methodology, the electricity data are pre-processed using interpolation of missing values, three sigma rule and normalization methods to compute the missing values and remove the outliers in the data. An Adasyn algorithm is proposed for handling the imbalanced dataset. Afterwards, the balanced data are fed into the Visual Geometry Group (VGG-16) module for features extraction. A VGG-16 detects abnormal patterns in electricity consumption data. Finally, the extracted features are passed to the Firefly Algorithm based Extreme Gradient Boosting (FA-XGBoost) module for classification. The main applications of this paper are listed below.

- The proposed approach provides the solution for the problem present in the power sector, such as to wastage of electrical power due to electricity theft.
- This model can efficiently be applied by the utility companies using the real electricity consumption data to identify the electricity thieves and reduce the energy wastage.
- The proposed approach can be used against the all types of consumers who steal the electricity.

The key contributions of this paper are:

- A comprehensive data pre-processing is performed using interpolation, three sigma rule, and normalization methods to deal with missing values and outliers in the dataset. The data pre-processing step gives the refined input, which improves the performance of the classifier.

- A class balancing technique, Adasyn, is proposed to address the problem of imbalance data. The benefit of using Adasyn is two-fold. Firstly, it improves the learning performance of classifier to be more focused on theft cases that are harder to learn. Secondly, it prevents the model from being biased.
- We have introduced a new technique VGG-16 to solve the problem of overfitting to improve the classification performance. This technique is never being used before in ETD domain, and it has improved the accuracy of the classification model. The VGG-16 efficiently extracts useful information from data to truly represent electricity theft cases.
- XGBoost is applied to predict final classification, which improves the performance by combining multiple weak learners to make a strong learner.
- Along with XGBoost , an optimization technique, the Firefly Algorithm (FA) is utilized for efficient parameter optimization of the classifier.
- We conduct extensive simulations on real electricity consumption data set and for comparative analysis, precision, recall, F1-score, Matthews Correlation Coefficient (MCC), Receiving Operating Characteristics Area Under Curve (ROC-AUC), and Precision Recall Area Under Curve (PR-AUC) are used as performance metrics.

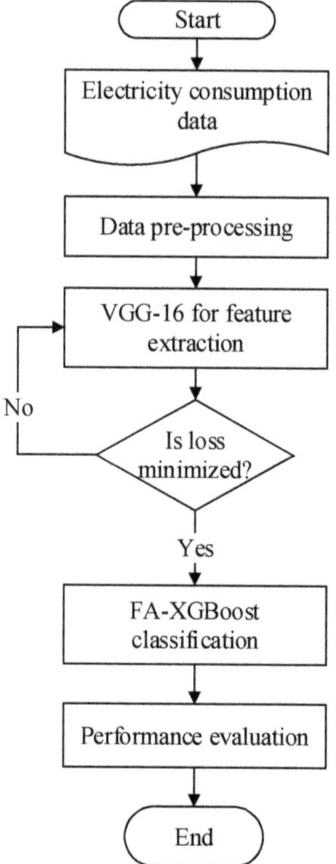

Figure 1. Flowchart of the proposed model for ETD.

Table 3. Mapping of problems addressed and proposed solution.

Limitation Number	Limitation Identified	Solution Number	Proposed Solution
L.1	Missing values and outliers	S.1	Pre-processing
L.2	Imbalanced data	S.2	Adasyn
L.3	Overfitting	S.3	VGG-16
L.4	Weak classification	S.4	FA-XGBoost
L.5	Reliable Evaluation	S.5	Precision, Recall, F1-Score, MCC, ROC-AUC, PR-AUC

1.4. Organization of Paper

The remaining paper is categorized as follows. Section 2 shows the proposed methodology. Section 3 provides the simulation results. Finally, this paper is concluded in Section 4.

2. Proposed System Model

Our proposed solution for ETD is presented in Figure 2. The proposed system model mainly consists of five parts: data pre-processing, data balancing, feature extraction, classification, and validation. Initially, electricity data are pre-processed using interpolation of missing values, three sigma rule, and normalization methods. Secondly, the pre-processed data are passed to the next model for data balancing. An Adasyn algorithm is used to balance the data. Thirdly, a VGG-16 is used to extract the important features from time series data and finally the important features are given to FA-XGBoost for classification. For comparative analysis, we use various performance metrics, i.e., precision, recall, F1-score, ROC-AUC, and PR-AUC to validate the effectiveness of our proposed model.

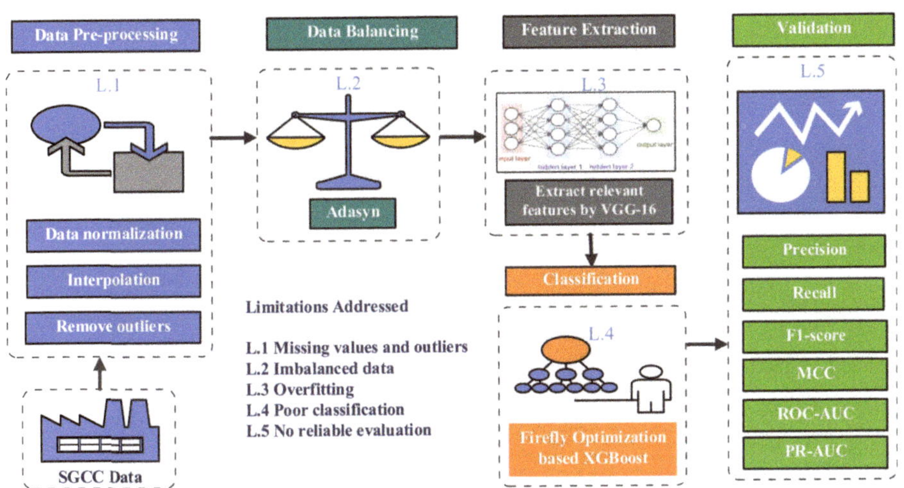

Figure 2. Proposed system model.

2.1. Overview of Proposed Methodology

The proposed methodolgy for ETD is described in the following subsections.

2.1.1. Information of Collected Data

The proposed system is tested using a high resolution real smart meter data, which is released by a State Grid Corporation of China (SGCC) [29]. These data are time series, i.e., recorded at regular intervals of time. The input dimensions or features are 1032. The duration of collected data are three years. It consists of electricity consumption data of 42,372 consumers. The released data also provide the ground truth according to which 9% of the total consumers are electricity thieves. This detail is given in Table 4. The daily electricity consumption pattern of the electricity thieves and honest consumers of over one month is given in Figure 3.

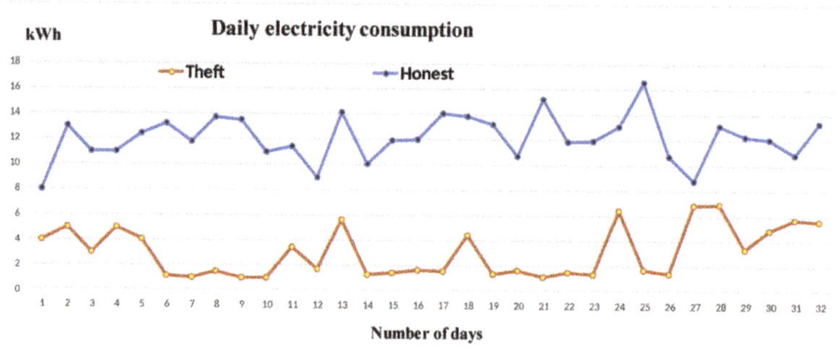

Figure 3. Electricity consumption pattern of thieves and honest consumers

In the electricity consumption data, the honest consumers have different consumption patterns than the electricity thieves. The electricity thieves have irregular patterns of energy consumption and their amount of energy consumption is also low due to meter tempering. In contrast, the honest consumers have regular periodicity in their consumption pattern. The machine learning algorithms use the smart meter data to track the anomalous consumption pattern of consumers to identify the electricity thieves.

Table 4. Description of Data.

Description	Values
Duration of collected data	2014–2016
Data type	Time series
Dimension	1034
Samples	42,372
Resolution	High resolution real time smart meter data
Number of fraudulent consumers	3800
Number of honest consumers	38,530
Total consumers	42,372

2.1.2. Data Pre-Processing

In this paper, data pre-processing is performed to achieve better results in ETD. We exploit interpolation method [30] to recover the missing information using Equation (1):

$$f(x_i) = \begin{cases} \frac{(x_{i+1}+x_{i-1})}{2} & \text{if } x_i \in NaN, x_{i-1} \text{ and } x_{i+1} \notin NaN \\ 0 & \text{if } x_i \in NaN, x_{i-1} \text{ or } x_{i+1} \in NaN \\ x_i & \text{if } x_i \notin NaN. \end{cases} \tag{1}$$

where x_i is the attribute of the electricity consumption data and NaN represents the non-numeric value.

Afterwards, we use the three sigma rule to remove outliers from the raw data. These outliers show the peak electricity consumption that occurs during non-working days. We restore these values using Equation (2) according to the "Three sigma rule of thumb" [31],

$$f(x_i) = \begin{cases} avg(x) + 2std(x) & \text{if } x_i > avg(x) + 2std(x) \\ x_i & \text{else.} \end{cases} \tag{2}$$

In Equation (2), *std (x)* is standard deviation and *avg (x)* is the average value of x. This method is effective in handling the outliers.

Along with interpolation and the three sigma rule, we also used Min-Max scaling method to normalize the data between the range 0 and 1. It is important because neural networks show poor performance on inconsistent data [32]. The data normalization improves the training process of deep learning models by assigning a common scale to the data. The following equation is used to normalize the data [33]:

$$A' = \frac{A - Min(A)}{Max(A) - Min(A)}, \tag{3}$$

where A' is the normalized value. The performance of machine learning algorithms depends on quality of input data. Data pre-processing enhances the data quality and performance of these models.

2.1.3. Data Balancing

In this section, we deal with imbalance dataset. The dataset collected from SGCC has a larger number of normal electricity consumers than thieves. This data imbalance is a major problem in ETD, which needs to be resolved; otherwise, the classifier will be biased towards the majority class and can result in performance degradation [34]. Various Random Under Sampling (RUS) and Random Over Sampling (ROS) techniques are used in the literature to solve this problem.

In the RUS technique, the data samples from majority class are made equal to the minority class [35]. This technique reduces the size of dataset, which is computationally beneficial. However, this technique is not preferred. As it reduces the dataset, this gives the model less data to train on. In contrast, ROS replicates the minority class instances in order to balance the data. However, due to the replication of minority instances unintelligently, the model leads towards the overfitting problem. Another method used for the data balancing technique is SMOTE [36,37]. In this technique, the minority class instances are increased by finding the n-nearest neighbor samples in the same class, i.e., the theft class. The example of synthetic data generation is represented in Figure 4. It draws a line between the neighbor of the minority class instances and creates new points on the lines, which are the synthetic data samples. Synthetic generation of minority instances avoids the overfitting problem which occurs due to ROS technique; however, synthetic generation of NTL instances do not reflect real world theft cases. In addition, SMOTEBoost [38] creates synthetic examples from the minority class samples, which indirectly change the updating weights and compensate for the imbalanced distributions.

Motivated by SMOTE [37] and SMOTEBoost [38], which are helpful for handling the imbalanced data set, we use the Adasyn method [39] to balance the dataset. It is an enhanced version of SMOTE. With a minor modification, after creating n-nearest neighbors samples, it adds random values that are linearly correlated to the parent samples and have a little more variance. This modification generates more realistic data samples.

The Adasyn algorithm initially finds out the number of synthetic data samples g that need to be created to increase the minority class instances. It can be calculated using the following equation [39]:

$$g = (m_j - m_i)\beta, \tag{4}$$

where m_j and m_i are the numbers of majority and minority classes samples, respectively. $\beta \in [0,1]$ is a constraint used to set the balance level of minority class to the majority class. Next, we calculate the ratio r_i by finding K nearest neighbors, which is based on the Euclidean distance given in Equation (5) and mentioned in [39] as:

$$r_i = \delta i / k, \ i = 1. \tag{5}$$

In the above equation, δi represents the synthetic samples and i represents the number of samples of the majority class in the k nearest neighbours; therefore, $r_i \in [0, 1]$. Finally, the number of synthetic data samples g_i are found by Equation (6) mentioned in [39] as:

$$g_i = r_i * g. \tag{6}$$

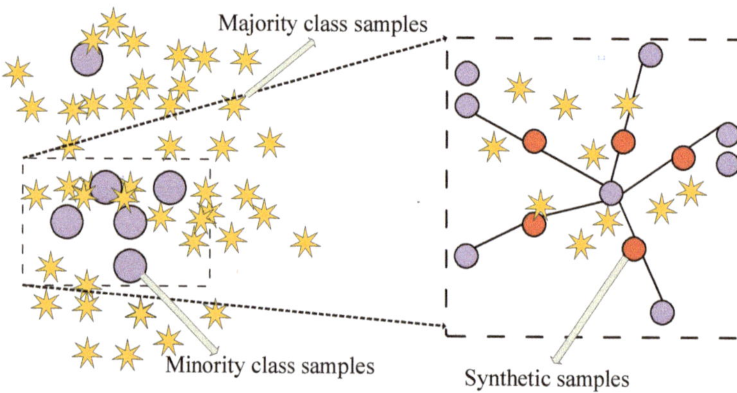

Figure 4. Generating of synthetic data through SMOTE.

The benefits of using Adasyn is two-fold; it improves the learning performance of the classifier to be more focused on theft cases that are harder to learn and prevents the model from being biased. The pseudo code of the Adasyn algorithm [40] is given in Algorithm 1.

Algorithm 1: Adasyn Algorithm

Input: Initial dataset X and desired balanced level β
Output: Synthetic dataset X_o
Initialize m_i as minority class samples
Initialize m_j as majority class samples
Synthesized total samples as $g = (m_j - m_i)\beta$
   for each $X_i \in m_i$ **do**
     find the K nearest neighbors of m_i
     $r_i = \delta i / k, \ i = 1$
  end for
   for each $x_i \in m_i$ **do**
     select the synthetic samples
     $g_i = r_i * g$
  end for
  return X_o

2.1.4. Feature Extraction Using VGG-16

VGG-16 is an enhanced version of CNN with 16 layers presented by the Visual Geometry Group [41]. It surpasses AlexNet by replacing large filters with small sized 3×3 filters [42]. It is used for feature extraction and transfer learning [43,44]. In this paper, VGG-16 is used for feature extraction where the representation spaces constructed by all filters of a layer are visualized in more comprehensive ways. All activations of a layer are used to extract the relevant features through a deconvolution network.

The architecture of VGG-16 [44] is shown in Figure 5. It consists of the pooling layers and convolutional layers. The operations of all layers are summed up by three fully connected layers at the end. The softmax is used as the activation function in the final dense layer.

The multiple pooling layers used in the VGG-16 module are better at extracting the high level features from the input data. We can visualize what features each filter captures by learning the input image that maximizes the activation of that filter. The convolutional operation is performed by sliding the kernel over the entire input, which produces a feature map. The final output from the convolution layer is integrated after multiple operations of feature mapping by the kernel function, which is given in [19] as:

$$y = x \times F \rightarrow y|i| = \sum_{j=-\infty}^{+\infty} \times [i-j]F[j]. \tag{7}$$

In Equation (7), x is input and F is the filter, which is also called the kernel. The input image is initially random while the loss is calculated as the activation of a particular filter. Relu [19] is used as an activation function to introduce nonlinearity to the model:

$$Relu(x) = max(0, x). \tag{8}$$

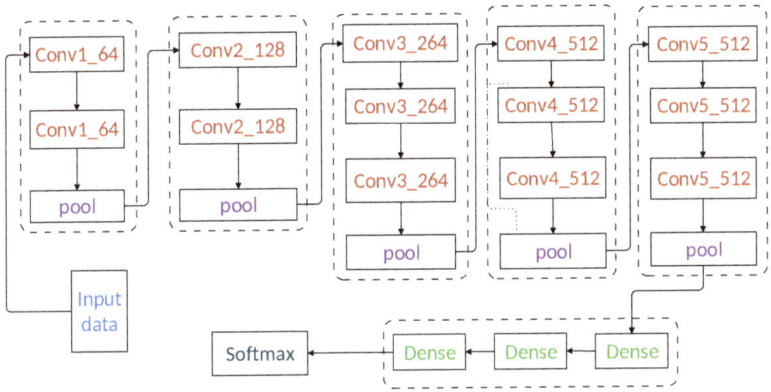

Figure 5. Architecture of VGG-16.

After the operations of pooling layers, three dense layers are used to visualize the important features. To avoid the overfitting problem, the dropout is set to 0.01 and the learning rate is 0.001. This method can be extended to the final dense layer having softmax as an activation function, which is defined in [19] as:

$$P(y = j|\varphi^{(i)}) = \frac{\rceil \varphi^{(i)}}{\sum_{j=0}^{k} \rceil \varphi_k^{(i)}}. \tag{9}$$

If the feature matrix and the weight matrix are denoted by X and W, then φ in the above equation is computed as:

$$\varphi = \sum_{i=1}^{l} W_i X_i = W^T X. \tag{10}$$

The hyper-parameters values of VGG-16 along with their description are given in Table 5. The hyper-parameters are batch size, learning rate, dropout rate, optimizer, and the number of epochs. These parameters play a key role in optimal performance of the VGG-16 module.

Table 5. Hyper-parameter values of VGG-16.

Hyper-Parameters	Values	Description
Batch size	130	It is training samples in each iteration
Leaning rate	0.001	It is a tuning parameter
Dropout	0.01	To avoid overfitting problem in neural networks.
Optimizer	Adam	It is adaptive learning rate.
Epochs	10	It is the number of iterations for training the algorithm

2.1.5. FA-XGBoost Based Classification

The XGBoost is one of the most popular machine learning methods [45]. On the Kaggle platform in 2015, the XGBoost as a classifier won 17 out of 29 competitions [45]. The extracted high level features given by VGG-16 become the inputs of the FA-XGBoost model. The FA-XGBoost library implements the gradient boosting decision tree algorithm. The ensemble model of XGBoost for classification is given in Figure 6. It shows that the XGBoost algorithm combines multiple weak models and makes a strong model to improve the final results. The final prediction is taken by voting of the majority of weak models.

There is a strong connection between hyper-parameters and outcome of a classifier [46]. Therefore, optimization is very important for accurate prediction. The hyper-parameters of XGBoost are learning rate and the number of estimators.

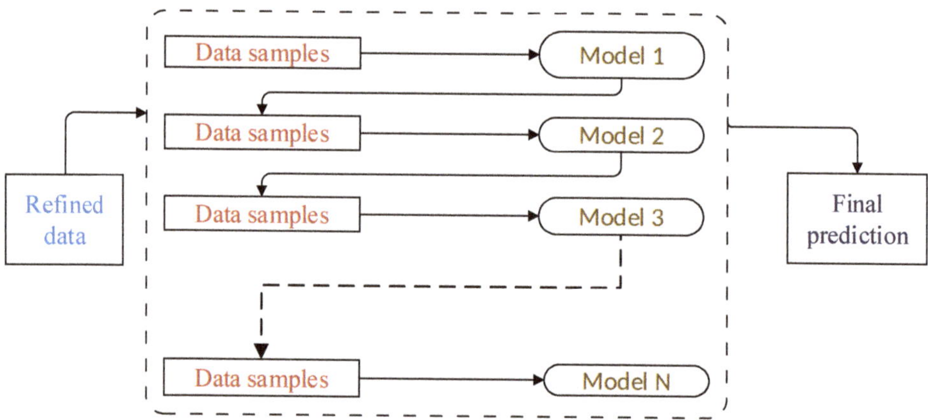

Figure 6. Ensemble model of XGBoost for classification.

The FA (developed by Yang [47]) is used in this paper to optimize the hyper-parameters of an XGBoost classifier. It is a nature inspired meta-heuristic algorithm based on flashing behavior of fireflies. The pseudo code of the FA-XGBoost [47] is given in Algorithm 2. The FA is based on three rules [48]:

1. Fireflies are uni-sexual in nature, so one firefly will be attracted to another regardless of whether the Firefly is male or female.

2. The attractiveness is proportional to light intensity of each firefly; thus, for any two flashing fireflies, the less bright firefly will be attracted by the brightest firefly. Attractiveness is calculated using Equation (11), which is mentioned in [49] as:

$$\beta(r) = \beta_0 e^{-\gamma r^2}. \tag{11}$$

In the above equation, $\beta(r)$ shows the attractiveness as a function of distance r, while β_0 represents attractiveness at zero distance. $e^{\gamma r^2}$ is the value of rate of light absorption in the air.

3. As distance between fireflies increases, the attractiveness decreases. The distance r_{ij} between two fireflies i and j can be calculated using Euclidean distance as:

$$r_{ij} = ||x_i - x_j|| = \sqrt{(\sum_{k=1}^{d} (x_{i,k} - x_{j,k})^2}, \tag{12}$$

where $x_{i,k}$ and $x_{j,k}$ are the t_{th} components of the position of fireflies i and j, respectively, while d is the number of dimensions. If no firefly is found brighter in the initialized population, then it moves in a random direction. The random movement towards the most brighter firefly is calculated using Equation (13), which is mentioned in [49] as:

$$x_i^{t+1} = x_i^t + \beta_0 e^{\gamma r_{ij}2} * (x_j^t - x_i^t) + \alpha * (rand - 1/2). \tag{13}$$

In the above equation, *rand* represents the random number, t is the number of iterations, while α controls the size of random walk.

Algorithm 2: FA-XGBoost

1: Set the objective function by $y = (1,2,3 ... n)$
2: Initialize the population of Fireflies by y_i (i = 1,2,3 ... n)
3: Define γ as the rate of light absorption in the air
4: Define I as the light intensity of a firefly
5: Maximum number of iteration is m and t is current iteration
6: **while** (t < m)
7: **for** i = 1:
8:   **for** j = 1:
9:     **if** $(I_i > I_j)$ **then**
10:      Move Firefly i towards j
11:     **end if** ;
12:     Attractiveness varies with distance r as given in Equation (8)
13:     Adjust the light intensity I to find new solutions
14:     Choose the best solution by random fly
15:   **end for** j
16:   **end for** i
17: Rank the Fireflies on the basis of minimum cost function
18: Choose the current best solution
19: **end while**
20: Return the best values of performance metrics

The XGBoost model for classification is given in Figure 6. It is based on ensemble learning in which several weak classifiers are combined to make a strong classifier. On each iteration, the classification rate of each learner is computed. The predicted value y_i after k iterations is computed using Equation (14):

$$\overline{y}_i = \sum_{k=1}^{k} [f_k(x_i)], \tag{14}$$

where $f_k(x_i)$ is the input function. The loss function $\mathcal{L}\ \phi$ is calculated by taking the difference between the actual y_i and predicted result $\bar{y}_i$ as given is Equation (15):

$$\mathcal{L}(\phi) = \sum_{i=1}^{l} [l(\bar{y}_i, y_i)]. \tag{15}$$

The objective of XGBoost is to minimize the loss function given Equations (16)–(18), which is computed by taking the summation of loss l of the multiple weak learners:

$$min\mathcal{L}(f_t) = min(f_t) \sum_{i=1}^{l} [l(\bar{y}_i, y_i)], \tag{16}$$

$$= min(f_t) \sum_{i=1}^{l} [l(y_i, \sum_{i=1}^{l} [l(f_k(x_i))])], \tag{17}$$

$$= min(f_t) \sum_{i=1}^{l} [l(y_i^{t-1} \bar{y}_i, f_t(x_i))]. \tag{18}$$

The instances of electricity theft that are misclassified by the learner are given more weight in the next iteration. The weights are adjusted by using the penalizes function $\Omega(f)$, which can be calculated using the following equation:

$$\Omega(f) = \gamma T + 1/2\lambda ||w||^2. \tag{19}$$

In Equation (19), γ and λ are the hyper-parameters, T is the number of tree node, and W is the vector of the nodes. When the penalized function is added to the loss, it minimizes the objective function and helps to smooth the final learnt weight:

$$\mathcal{L}(\phi) = min(f_t) \sum_{i=1}^{l} [l(\bar{y}_i, y_i)] + \sum_{i=1}^{l} \Omega f_k. \tag{20}$$

The final classification is performed by taking the mean of individual models. The prediction of each individual decision tree is weak and prone to overfitting. However, combining several decision trees in an ensemble method gives better results. For comparative analysis, various performance metrics are used, i.e., F1-score, precision, recall, and ROC curve to validate the effectiveness of our proposed model. They are discussed in detail in the simulation section.

3. Experiments and Results

In this section, the experimental results are discussed in detail.

3.1. Loss Function

For accurate prediction, the proposed model aims to reduce the loss function. The widely used logarithmic loss function is cross entropy. As we are doing binary classification, the loss function we are using is a binary cross entropy. It is calculated using the following equation [50]:

$$f(x) = 1/N \sum_{i=1}^{N} -(y_i log(p(y_i)) + (1 - y_i) log(1 - p(y_i))). \tag{21}$$

In Equation (21), N is the total number of consumers samples, $p(y_i)$ is the probability of electricity theft, and y_i is the ground truth label.

3.2. Model Evaluation Metrics

The concern of ETD in supervised learning is a class imbalance problem. In this problem, the number of honest customers varies remarkably from these fraudulent ones. Therefore, for evaluation, a simple accuracy measure is not reliable. In this paper, various performance metrics are considered. These evaluation metrics' values are determined from confusion matrix. The confusion matrix gives information about the following results:

- True positive (TP), the dishonest consumers accurately predicted as dishonest.
- True Negative (TN), the honest consumers accurately predicted as honest.
- False Positive (FP), the honest consumers predicted as thieves.
- False Negative (FN), the dishonest consumers predicted as honest consumers.

In this paper, we use precision, recall, F1-score, ROC-AUC, and *MCC* for evaluation of our system model. Precision is referred to as True Negative Rate (TNR); it shows the actual number of honest customers that are correctly identified by the classifier. It is formulated in [19] using Equation (22). Recall is referred to as True Positive Rate (TPR); it shows the actual number of positives that are correctly identified by classifier. It is formulated in [19] using Equation (23). Both precision and recall are not enough to show real assessment of a classifier. It is better to maximize precision and recall, which gives F1-score. It is a useful measure for binary classification problems where the distribution of classes is imbalanced. It is calculated by the weighed harmonic mean of precision and recall [19], which is given in Equation (24). Another suitable metric for ETD is the ROC-AUC. It shows a graphical representation of a model to evaluate its detection performance. The classifier having ROC-AUC close to 1 has better capability to separate two classes. However, ROC-AUC only summarizes the trade-off between the TPR and FPR of the model. *AUC* score is calculated by using Equation (25), mentioned in [30]. In Equation (25), *Rank* shows the number of samples, *M* is the number of positive class samples, and *N* is the number of negative class samples. Moreover, the PR-AUC are appropriate for imbalanced datasets. Its graphical representation is obtained by plotting the recall against the precision. The value of curve is in the range between 0 and 1. The classifier having ROC-AUC value close to 1 is considered a good classifier. In all performance matrices, *MCC* produces a high score only if the prediction obtained good results in all of the four confusion matrix values, i.e., *TP*, *TN*, *FP*, and *FN*. MCC score ranges between −1 to 1, whereas, close to 1 shows the accurate classification, 0 shows no class separation capability, and −1 shows the incorrect classification by model. It is calculated using Equation (26) mentioned in [19]. Accuracy is the number of correctly predicted data points out of all the data points. It is a widely used metric for classification problems in the data science community [51]. However, it is not considered as a reliable metric where the distribution of labels is imbalanced. It can be calculated by using Equation (27):

$$Precision = \frac{TP}{TP + FP}. \tag{22}$$

$$Recall = \frac{TP}{TP + FN}. \tag{23}$$

$$F1 = 2 * \frac{Precision * Recall}{Precision + Recall}. \tag{24}$$

$$AUC = \frac{\sum Rank_{i \in positive\ class} - \frac{M(1+M)}{2}}{M * N}. \tag{25}$$

$$MCC = \frac{TP * TN - FR * FN}{\sqrt{((TP + FP)(TP + FN)(TN + FP)(TN + FN))}}. \tag{26}$$

$$Accuracy = \frac{TP * TN}{TP + TN + FP + FN}. \tag{27}$$

3.3. Benchmark Models and Their Configuration

In this section, we describe the conventional models, which are widely used as classifiers for ETD. The range of hyper-parameter values is defined, and we select optimal values for each base model.

3.3.1. SVM Model

It is a popular classifier and widely used for ETD [18]. The hyper-parameters of SVM are γ and regularization parameter C, which are important in selecting an optimal hyperplane. The values of these parameters are given in Table 6. The optimal values are selected from the given range of values.

Table 6. Hyper-parameter values of SVM.

Hyper-Parameters	Range of Values	Selected Value
γ	1,3,5	3
C	0.001, 0.01,	0.01

3.3.2. LR Model

It is a supervised learning algorithm, which is used as benchmark model in this paper. Its hyper-parameters along with their range and selected values are given in Table 7. During implementation, we choose optimal values for accurate classification.

Table 7. Hyper-parameter values of LR.

Hyper-Parameters	Range of Values	Selected Value
R	0.001, 0.01, 0.1	0.001
C	l1 norm, l2 norm	l2 norm

3.3.3. RUSBoost Model

It is widely used for classification problems where the distribution of labels is imbalanced. The hyper-parameters of RUSBoost are learning rate and number of estimators. The best values are selected from the range of different values as given in Table 8.

Table 8. Hyper-parameter values of RUSBoost.

Hyper-Parameters	Range of Values	Selected Value
Learning rate	0.2, 0.5, 1	1
Estimator	150, 200, 300	200

3.3.4. CNN Model

Along with conventional machine learning algorithms, we also used CNN as a deep learning model for comparison. It is a feed-forward neural network and is mostly used for complex classification problems. We choose best values of CNN during model validation, which are given in Table 9.

Table 9. Hyper-parameter values of CNN.

Hyper-Parameters	Range of Values	Selected Value
Epochs	10,15,30	10
Batch size	50, 80,130	50
Dropout	0.01, 0.1, 0.2	0.2

3.4. Proposed Model Results

In this section, we present the performance of our proposed model on raw data and transformed data. Initially, the missing values in the data set are filled with interpolation and three sigma rule methods. In addition, the data are normalized using the Min-Max scaling method. Figure 7a,b shows the unbalanced and balanced distribution of labels, respectively. The thieves are represented by '0', while '1' shows honest customers. The x-axis represents observation values for the first sample, and the y-axis represents the observation values for the second sample. Each point on the plot represents a single observation.

The Adasyn algorithm is used to address the class imbalance problem. This algorithm intelligently balances the number of instances of electricity thieves and honest consumers. The distribution of two classes after applying the Adaysn algorithm is also shown in Figure 7b. The minority class instances are increased, and it shows the equal distribution of the two labels in the data.

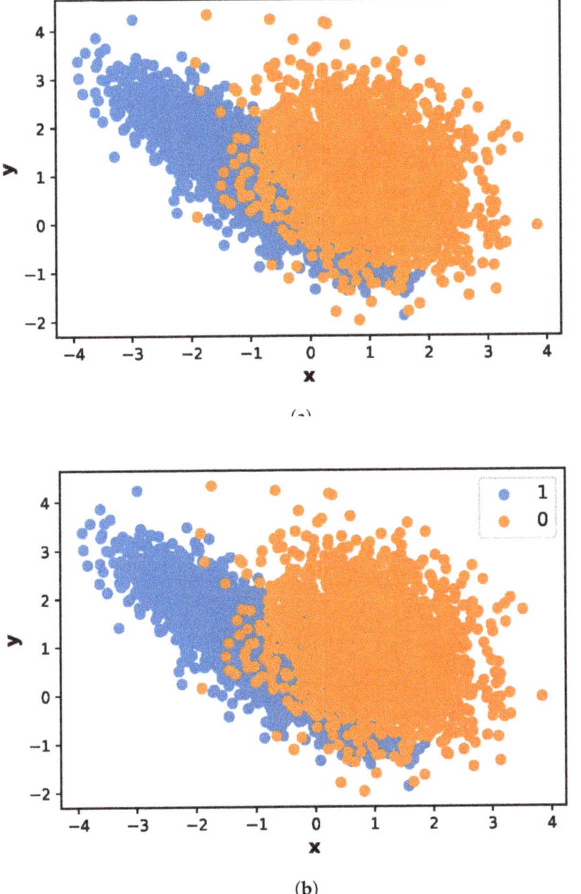

Figure 7. (**a**) visualization of unbalanced data; (**b**) visualization of balanced data.

Due to imbalanced data, the classifiers get biased and result in high FPR. In order to show the effectiveness of balanced data, we make a comparison that is shown in Table 10. Before applying Adaysn, the model could not classify effectively as evident from the scores mentioned in Table 10 and Figure 8. On imbalance data, the classifier achieves 60% precision, 62.1% recall, 59.01% F1-score, and the 63.2% ROC curve. The SMOTE improves the performance of the FA-XGBoost classifier. It achieves 79.1% precision, 80% recall, 78.7% F1-score, and 78% ROC curve. Adasyn improves the ability of the model by using synthetic data intelligently. The results of Adasyn are far better as compared to the results of the unbalanced data and SMOTE. On the imbalance dataset, the classifier becomes biased by considering the real electricity thieves as honest consumers. The Adasyn method improved the performance of the FA-XGBoost classifier. It achieves 93% precision, 97% recall, 93.7% F1-score, and the 95.9% ROC curve.

Table 10. Model performance before and after Adasyn.

Performance Metrics	Imbalaced Data	SMOTE	Adasyn
Precision	60	79.1	93
Recall	62.1	80	97
F1-score	59.01	78.7	93.7
ROC-AUC	63.2	78	95.9

Our goal in ETD is to maximize the TP and TN and reduce the FP and FN. Table 11 shows that our proposed model has achieved good values of the confusion matrix. The high TP and TN values show that our model has truly identified the electricity thieves and honest consumers, respectively.

Table 11. Confusion matrix values of the ETD model. TN: True Negative; FP: False Positive; FN: False Negative; TP: True Positive.

Confusion Matrix	Predicted No	Predicted Yes
Actual No	$TN = 9306$	$FP = 1296$
Actual Yes	$FN = 948$	$TP = 8996$

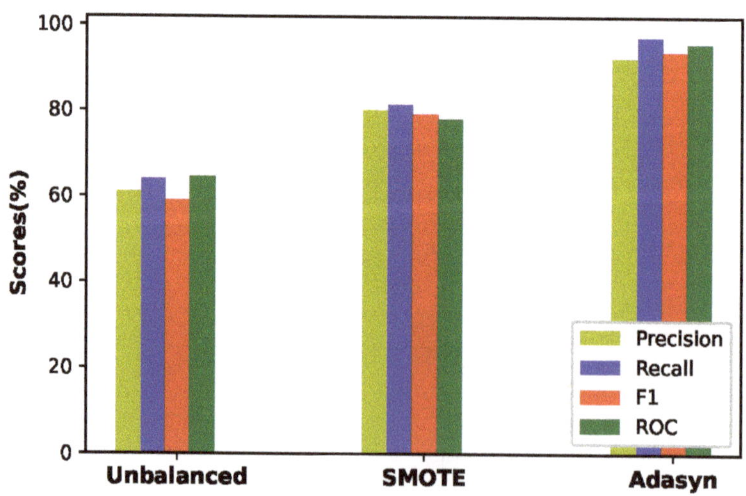

Figure 8. Performance comparison of balanced and unbalanced datasets

In this paper, FA-XGBoost is used as a classifier. Its performance is primarily dependent on the selection of hyper-parameter values. Initially, we randomly apply the XGBoost without tuning its hyper-parameter values. Still, it achieves better performance than the state-of-the-art models, i.e., 86.5% ROC-AUC. To enhance the classification performance, we utilize the Firefly algorithm to choose the optimal hyper-parameter values of XGBoost. The results of ROC-AUC of FA-XGBoost in Figure 9 shows the better result, i.e., 95.6% ROC-AUC.

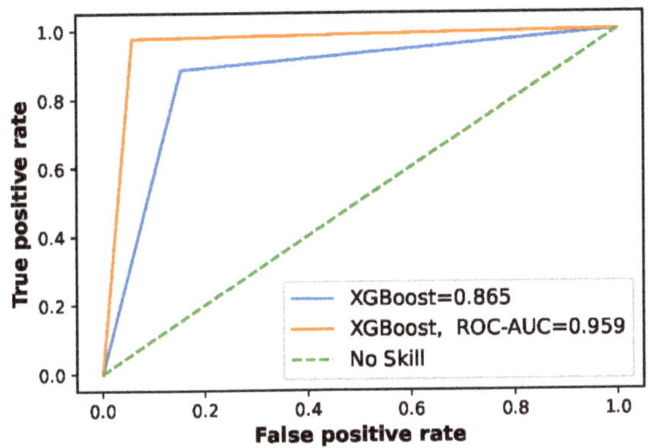

Figure 9. Performance comparison of parameter tuning.

3.5. Convergence Analysis

As expected, using VGG-16 for feature extraction improves the performance for ETD. Figure 10 shows the accuracy and loss of VGG-16 module. When we choose a smaller epoch value to optimize the training procedure, it is enough to let our model to learn from high dimensional data. However, it causes overfitting when we choose a larger epoch value. As the number of epochs increases up to four, the training and testing losses decrease and the accuracy increases significantly. It shows the better prediction capability of the model. The mapping of addressed problems to the validation results is presented in Table 12. There is no direct validation for pre-processing methods. The class imbalance problem is efficiently solved through the Adasyn method that is validated in Figure 8. The overfitting problem is solved by achieving a higher generalized performance through the VGG-16 module. In addition, the Firefly based XGBoost classifier enhanced the classification accuracy as shown in Figure 9.

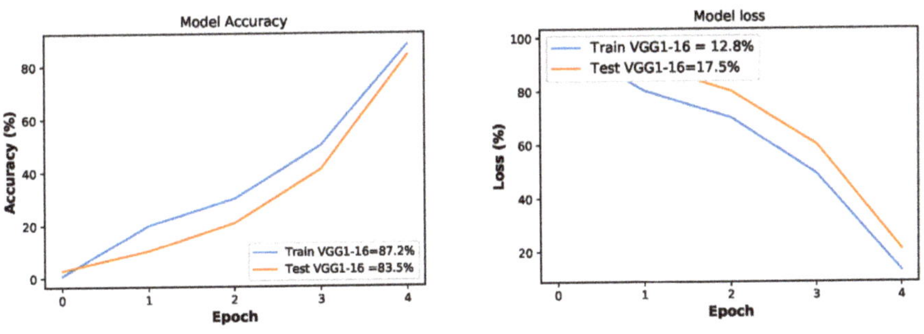

Figure 10. Accuracy and loss of VGG-16

Table 12. Mapping of problems addressed and validation results.

Limitation Number	Limitation Identified	Solution Number	Validation Results
L.1	Missing values and outliers	S.1	No direct validation
L.2	Imbalanced data	S.2	Adasyn algorithm effectively handles imbalance data as shown in Figure 8
L.3	Overfitting	S.3	Figure 10 shows a generalized performance of our proposed model
L.4	Poor classification	S.4	Firefly based XGBoost classifier achieved excellent results in terms of all performance metrics as mentioned in Figure 9
L.5	No reliable Evaluation	S.5	Figure 11 shows the performance of our proposed model in terms of several performance metrics

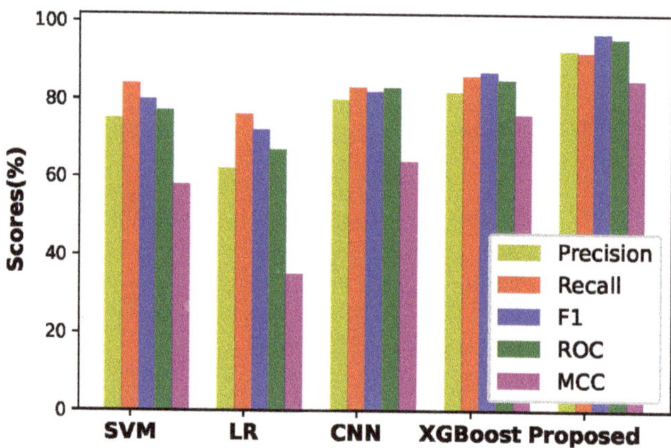

Figure 11. Performance metrics comparison with benchmark schemes.

3.6. Comparison with Benchmark Models

In order to show the comparison of our proposed model with benchmark schemes, we trained LR, SVM, CNN, and RUSBoost using the same dataset. These are basic models for classification problems. The configuration of these models is discussed in Section 3.4.

Figure 12 shows ROC-AUC of SVM, CNN, LSTM-RUSBoost, and LR models. We obtained the results by using the same dataset and setting optimal hyper-parameters of these models. It can be seen that RUSBoost outperforms the benchmark schemes by achieving 86.5% ROC-AUC. It balances the data by a random under sampling method. It performs classification through the adaptive boosting technique. This technique is better for unbalanced binary classification problems.

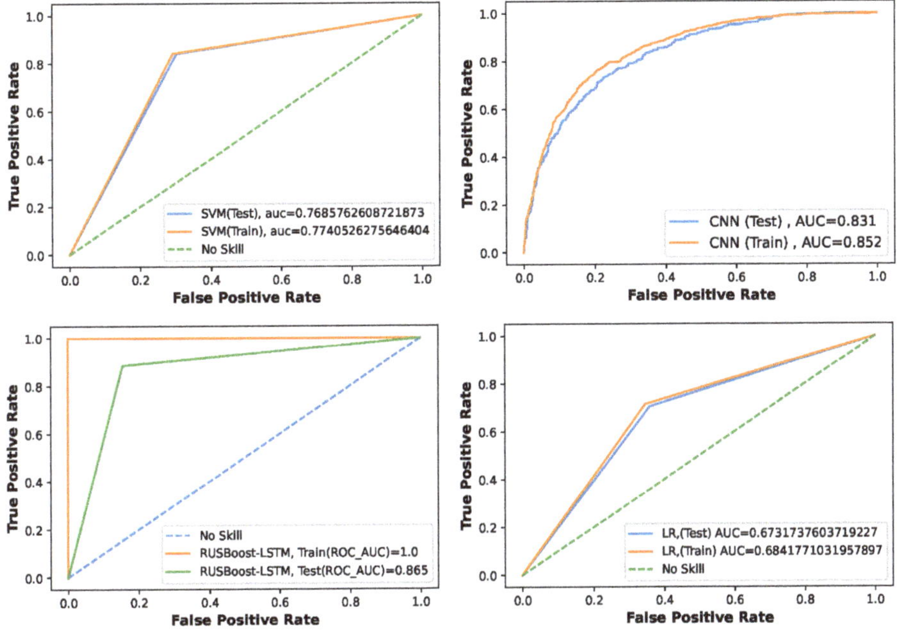

Figure 12. ROC-AUC of benchmark models.

However, LR performs worst among the classifiers, securing just 67.3 % ROC-AUC. It is due to the fact that LR is based on the concept of probability and uses the principle of neural networks, which do not capture the long-term dependencies from the large time series data. Moreover, it becomes biased during the identification of real electricity theft cases due to the training on the majority class samples. Hence, LR can not perform accurate classification on the large imbalanced dataset.

The performance of SVM is also not satisfactory, securing 76.8% ROC-AUC. It classifies the data by creating a hyperplane. However, in a complex binary classification problem, it becomes difficult for SVM to set optimal values for creating a hyperplane. Hence, this method is also not suitable for ETD. In contrast, the CNN slightly performs better than SVM by securing 83.1% ROC-AUC. The CNN is a learning deep model having multiple stacks of hidden layers that extract the hidden patterns from the electricity consumption data and identify the real electricity thieves. However, it has over fitting issues due to the dense layers. It fails to achieve a generalized performance.

Figure 11 presents the performance comparison of our proposed scheme with benchmark models. It is worth noting that our proposed model out performs the other models in terms of precision, recall, ROC-AUC, and F1-score. The Firefly based XGBoost achieves 95.9% ROC-AUC, 92.6% precision, 97% recall, and 93.7% F1-score.

The ROC-AUC and PR-AUC of benchmark schemes and our proposed model is shown in Figure 13. It can be seen that deep learning models like CNN perform better for the classification of high dimensional data. The CNN achieves 83.1% ROC-AUC on the test dataset. These models automatically extract the features from the data while traditional machine learning algorithms require the separate techniques for refining the data. They capture long-term dependencies and improve the performance as the dataset increases. Moreover, the traditional machine learning algorithm like SVM and LR are not efficient for the classification using a larger dataset. The SVM and LR get 76% ROC-AUC and 67.3% ROC-AUC, respectively.

In this paper, we also evaluate our proposed model on the PR-AUC. Our proposed model also covers more area under the PR-AUC than the benchmark models as shown in Figure 13. The results of

ROC-AUC and PR-AUC show that our proposed model is superior to other classifiers. The overall summary of the benchmark model and the proposed model is presented in Table 13. As it can be observed, FA-XGBoost outperforms the rest of the classifiers in terms of all the performance metrics. The high values of precision, recall, F1, and ROC-AUC show that our model has truly identified the number of honest consumers and electricity thieves.

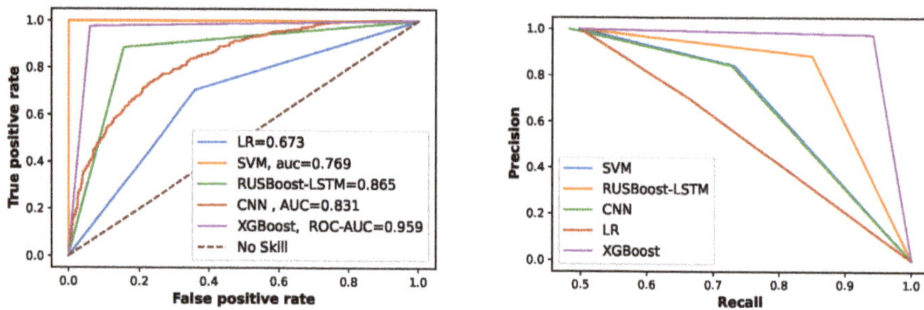

Figure 13. ROC-AUC and PR-AUC comparison

In this paper, our focus was more on accurate identification of electricity thieves. However, our proposed model has high execution time. It takes 25 min to run the model.

Table 13. Summary of results.

Models	Accuracy	Precision	Recall	F1-Score	ROC	MCC
CNN	0.812	0.805	0.862	0.845	0.813	61.5
SVM	0.772	0.765	0.883	0.819	0.769	56.3
LR	0.676	0.645	0.772	0.701	0.673	35.6
RUSBoost	0.869	0.85	0.896	0.871	0.865	77.8
Proposed Model	0.95	0.930	0.9700	0.937	0.959	85.6

4. Conclusions and Future Work

In this paper, the proposed methodology is implemented for ETD using the real smart meter data. The different limitations in literature are addressed in this work. The conclusions were drawn and are summarised as follows.

Initially, the real smart meter data, which is collected from SGCC, have a number of missing values and outliers. For this reason, we performed a comprehensive data pre-processing which consists of interpolation, the three sigma rule, and normalization methods. In addition, the dataset has a small number of instances for electricity thieves, which makes the classification model biased due to its training on majority honest instances. We employed the Adaysn algorithm to address this problem. This technique has improved the performance of the FA-XGboost classifier, which has achieved F1-score, precision, and recall of 93.7%, 92.6%, and 97%, respectively. Afterwards, the model has overfitting issues due to training the model on a large time series data, a VGG-16 module is introduced in ETD, which extracts relevant features from the data. It achieved a higher generalized performance by securing accuracy of 87.2% and 83.5% on training and testing data, respectively. Finally, the XGBoost method is applied to classify data into honest and dishonest consumers. To enhance the performance of XGBoost method, an FA is used for parameters' optimization. This method improved the performance of XGBoost and achieved 95.9% ROC-AUC and outperforming the benchmarks: SVM, LR, and CNN. However, as the dataset increases, the execution time of our proposed model also increases. In the future, we will improve its performance by reducing the delay in detecting the electricity theft.

Author Contributions: Z.A.K. and M.A. proposed and implemented the main idea. N.J. and M.N.S. performed the mathematical modeling and wrote the simulation section. M.S. and J.-G.C. organized and refined the manuscript. All authors have read and agreed to the published version of the manuscript.

Funding: This work was supported by the Basic Science Research Program through the National Research Foundation (NRF) of Korea funded by the Ministry of Education under Grant 2018R1D1A1B07048948.

Conflicts of Interest: The authors declare no conflict of interest.

References

1. Gul, H.; Javaid, N.; Ullah, I.; Qamar, A.M.; Afzal, M.K.; Joshi, G.P. Detection of Non-Technical Losses using SOSTLink and Bidirectional Gated Recurrent Unit to Secure Smart Meters. *Appl. Sci.* **2020**, *10*, 3151.

2. Adil, M.; Javaid, N.; Qasim, U.; Ullah, I.; Shafiq, M.; Choi, J.-G. LSTM and Bat-Based RUSBoost Approach for Electricity Theft Detection. *Appl. Sci.* **2020**, *10*, 1–21.

3. Mujeeb, S.; Javaid, N. ESAENARX and DE-RELM: Novel schemes for big data predictive analytics of electricity load and price. *Sustain. Cities Soc.* **2019**, *51*, 101642.

4. Nazari-Heris, M.; Mirzaei, M.A.; Mohammadi-Ivatloo, B.; Marzb, M.; Asadi, S. Economic-environmental effect of power to gas technology in coupled electricity and gas systems with price-responsive shiftable loads. *J. Clean. Prod.* **2020**, *244*, 118769.

5. Marzb, M.; Azarinejadian, F.; Savaghebi, M.; Pouresmaeil, E.; Guerrero, J.M.; Lightbody, G. Smart transactive energy framework in grid-connected multiple home microgrids under independent and coalition operations. *Renew. Energy* **2018**, *126*, 95–106.

6. Jadidbonab, M.; Mohammadi-Ivatloo, B.; Marzb, M.; Siano, P. Short-term Self-Scheduling of Virtual Energy Hub Plant within Thermal Energy Market. *IEEE Trans. Ind. Electron.* **2020**, accepted.

7. Gholinejad, H.R.; Loni, A.; Adabi, J.; Marzb, M. A hierarchical energy management system for multiple home energy hubs in neighborhood grids. *J. Build. Eng.* **2020**, *28*, 101028.

8. Mirzaei, M.A.; Sadeghi-Yazdankhah, A.; Mohammadi-Ivatloo, B.; Marzb, M.; Shafie-khah, M.; Catalão, J.P. Integration of emerging resources in IGDT-based robust scheduling of combined power and natural gas systems considering flexible ramping products. *Energy* **2019**, *189*, 116195.

9. Biswas, P.P.; Cai, H.; Zhou, B.; Chen, B.; Mashima, D.; Zheng, V.W. Electricity Theft Pinpointing through Correlation Analysis of Master and Individual Meter Readings. *IEEE Trans. Smart Grid* **2019**, 1–12.

10. Lydia, M.; Kumar, G.E.P.; Levron, Y. Detection of Electricity Theft based on Compressed Sensing. In Proceedings of the 2019 5th International Conference on Advanced Computing and Communication Systems (ICACCS) IEEE, Coimbatore, India, 15–16 March 2019; pp. 995–1000.

11. Razavi, R.; Gharipour, A.; Fleury, M.; Akpan, I.J. A practical feature-engineering framework for electricity theft detection in smart grids. *Appl. Energy* **2019**, *238*, 481–494.

12. Depuru, S.S.S.R.; Wang, L.; Devabhaktuni, V. Support vector machine based data classification for detection of electricity theft. In Proceedings of the 2011 IEEE/PES Power Systems Conference and Exposition, Phoenix, AZ, USA, 20–23 March 2011; pp. 1–8.

13. Saeed, M.S.; Mustafa, M.W.; Sheikh, U.U.; Jumani, T.A.; Mirjat, N.H. Ensemble Bagged Tree Based Classification for Reducing Non-Technical Losses in Multan Electric Power Company of Pakistan. *Electronics* **2019**, *8*, 860.

14. Razavi, R.; Fleury, M. Socio-economic predictors of electricity theft in developing countries: An Indian case study. *Energy Sustain. Dev.* **2019**, *49*, 1–10.

15. McDaniel, P.; McLaughlin, S. Security and privacy challenges in the smart grid. *IEEE Secur. Priv.* **2009**, *7*, 75–77.

16. Buzau, M.M.; Tejedor-Aguilera, J.; Cruz-Romero, P.; Gomez-Exposito, A. Hybrid deep neural networks for detection of non-technical losses in electricity smart meters. *IEEE Trans. Power Syst.* **2019**, *35*, 1254–1263.

17. Jamil, A.; Alghamdi, T.A.; Khan, Z.A.; Javaid, S.; Haseeb, A.; Wadud, Z.; Javaid, N. An Innovative Home Energy Management Model with Coordination among Appliances using Game Theory. *Sustainability* **2019**, *11*, 6287.

18. Buzau, M.M.; Tejedor-Aguilera, J.; Cruz-Romero, P.; Gómez-Expósito, A. Detection of non-technical losses using smart meter data and supervised learning. *IEEE Trans. Smart Grid* **2018**, *10*, 2661–2670.
19. Hasan, M.; Toma, R.N.; Nahid, A.A.; Islam, M.M.; Kim, J.M. Electricity Theft Detection in Smart Grid Systems: A CNN-LSTM Based Approach. *Energies* **2019**, *12*, 3310.
20. Avila, N.F.; Figueroa, G.; Chu, C.C. NTL detection in electric distribution systems using the maximal overlap discrete wavelet-packet transform and random under sampling boosting. *IEEE Trans. Power Syst.* **2018**, *33*, 7171–7180.
21. Ramos, C.C.; Rodrigues, D.; de Souza, A.N.; Papa, J.P. On the study of commercial losses in Brazil: a binary black hole algorithm for theft characterization. *IEEE Trans. Smart Grid* **2016**, *9*, 676–683
22. Zheng, K.; Chen, Q.; Wang, Y.; Kang, C.; Xia, Q. A novel combined data-driven approach for electricity theft detection. *IEEE Trans. Ind. Inform.* **2019**, *15*, 1809–1819
23. Ding, N.; Ma, H.; Gao, H.; Ma, Y.; Tan, G. Real-time anomaly detection based on long short-Term memory and Gaussian Mixture Model. *Comput. Electr. Eng.* **2019**, *70*, 106458.
24. Li, S.; Han, Y.; Yao, X.; Yingchen, S.; Wang, J.; Zhao, Q. Electricity Theft Detection in Power Grids with Deep Learning and Random Forests. *J. Electr. Comput. Eng.* **2019**, *2019*, 1–12.
25. Punmiya, R.; Choe, S. Energy theft detection using gradient boosting theft detector with feature engineering-based preprocessing. *IEEE Trans. Smart Grid* **2019**, *10*, 2326–2329.
26. Amin, S.; Schwartz, G.A.; Cardenas, A.A.; Sastry, S.S. Gametheoretic models of electricity theft detection in smart utility networks: Providing new capabilities with advanced metering infrastructure. *IEEE Control. Syst. Mag.* **2015**, *35*, 66–81,
27. Leite, J.B.; Mantovani, J.R.S. Detecting and locating non-technical losses in modern distribution networks. *IEEE Trans. Smart Grid* **2016**, *9*, 1023–1032.
28. Wang, S.; Chen, H. A novel deep learning method for the classification of power quality disturbances using deep convolutional neural network. *Appl. Energy* **2019**, *235*, 1126–1140.
29. State Grid Corporation of China. Available online: https://www.sgcc.com.cn (accessed on 22 February 2020).
30. Zheng, Z.; Yang, Y.; Niu, X.; Dai, H.N.; Zhou, Y. Wide and deep convolutional neural networks for electricity-theft detection to secure smart grids. *IEEE Trans. Ind. Informat.* **2017**, *14*, 1606–1615.
31. Chola, V.; Banerjee, A.; Kumar, V. Anomaly detection: A survey. *Acm Comput. Surv. (Csur)* **2009**, *41*, 1–58.
32. Nam, H.; Kim, H.E. Batch-instance normalization for adaptively style-invariant neural networks. In *Advances in Neural Information Processing Systems*; The MIT Press: Cambridge, MA, USA, 2018; pp. 2558–2567.
33. Pandey, A.; Jain, A. Comparative analysis of KNN algorithm using various normalization techniques. *Int. J. Comput. Netw. Inf. Secur.* **2017**, *9*, 36–42.
34. Figueroa, G.; Chen, Y.S.; Avila, N.; Chu, C.C. Improved practices in machine learning algorithm for NTL detection with imbalanced data. In Proceedings of the 2017 IEEE Power Energy Society General Meeting, Chicago, IL, USA, 16–20 July 2017; pp. 1–5.
35. Hasanin, T.; Khoshgoftaar, T. The effects of random under sampling with simulated class imbalance for big data. In Proceedings of the 2018 IEEE International Conference on Information Reuse and Integration (IRI), Salt Lake City, UT, USA, 6–9 July 2018; pp. 70–79.
36. Qin, H.; Zhou, H.; Cao, J. Imbalanced Learning Algorithm based Intelligent Abnormal Electricity Consumption Detection. *Neurocomputing*, **2020**, *402*, pp. 112–123.
37. Qu, Z.; Li, H.; Wang, Y.; Zhang, J.; Abu-Siada, A.; Yao, Y. Detection of Electricity Theft Behavior Based on Improved Synthetic Minority Oversampling Technique and Random Forest Classifier. *Energies* **2020**, *13*, 1–20.
38. Pelayo, L.; Dick, S. Synthetic minority oversampling for function approximation problems. *Int. J. Intell. Syst.* **2019**, *34*, 2741–2768.
39. He, H.; Bai, Y.; Garcia, E.A.; Li, S. ADASYN: Adaptive synthetic sampling approach for imbalanced learning. In Proceedings of the 2008 IEEE International Joint Conference on Neural Networks (IEEE world Congress on Computational Intelligence), Hong Kong, China, 1–8 June 2008.
40. Xiang, Y. *Polarity Classification of Imbalanced Microblog Texts*; AIST: Tsukuba, Ibaraki, Japan, 2019; pp. 1–61.
41. Simonyan, K.; Zisserman, A. Very Deep Convolutional Networks for Large-Scale Image Recognition. *arXiv* **2014**, arXiv:1409.1556.

42. Yu, W.; Yang, K.; Bai, Y.; Xiao, T.; Yao, H.; Rui, Y. Visualizing and comparing AlexNet and VGG using deconvolutional layers. In Proceedings of the 33rd International Conference on Machine Learning, New York City, NY, USA, 19–24 June 2016; pp. 1–7.
43. Dixon, J.; Rahman, M. Modality Detection and Classification of Biomedical Images with Deep Transfer Learning and Feature Extraction. In Proceedings of the International Conference on Image Processing, Computer Vision, and Pattern Recognition (IPCV) The Steering Committee of The World Congress in Computer Science, Computer Engineering and Applied Computing (WorldComp), Las Vegas, NV, USA, 29 July–1 August 2019; pp. 55–58.
44. Cıbuk, M.; Budak, U.; Guo, Y.; Ince, M.C.; Sengur, A. Efficient deep features selections and classification for flower species recognition. *Measurement* **2019**, *137*, 7–13.
45. Chen, T.; Guestrin, C. XGBoost: A scalable tree boosting system. In Proceedings of the 22nd ACM SIGKDD Conference on Knowledge Discovery and Data Mining, San Francisco, CA, USA, 13–17 August 2016; pp. 785–794
46. Zahid, M.; Ahmed, F.; Javaid, N.; Abid Abbasi, R.; Zainab Kazmi, H.S.; Javaid, A.; Bilal, M.; Akbar, M.; Ilahi, M. Electricity Price and Load Forecasting using Enhanced Convolutional Neural Network and Enhanced Support Vector Regression in Smart Grids. *Electronics* **2019**, *8*, 122.
47. Yang, X.-S. Firefly Algorithm, Stochastic Test Functions and Design Optimization. *Int. Bio-Inspired Comput.* **2010**, *2*, 78–84
48. Yang, X.S. Chaos-enhanced firefly algorithm with automatic parameter tuning. In *Recent Algorithms and Applications in Swarm Intelligence Research*; Information Science Reference (IGI Global): Hershey, PA, USA, 2013; pp. 125–136.
49. Chen, K.; Zhou, Y.; Zhang, Z.; Dai, M.; Chao, Y.; Shi, J. Multilevel image segmentation based on an improved firefly algorithm. *Math. Probl. Eng.* **2016**, *2016*, 1–12
50. Janocha, K.; Czarnecki, W.M. On loss functions for deep neural networks in classification. *arXiv* **2017**, arXiv:1702.05659.
51. Zhu, W.; Zeng, N.; Wang, N. Sensitivity, specificity, accuracy, associated confidence interval and ROC analysis with practical SAS implementations. *Nesug Proc. Health Care Life Sci. Balt. Md.* **2010**, *19*, 67.

 © 2020 by the authors. Licensee MDPI, Basel, Switzerland. This article is an open access article distributed under the terms and conditions of the Creative Commons Attribution (CC BY) license (http://creativecommons.org/licenses/by/4.0/).

Article

Improving Residential Load Disaggregation for Sustainable Development of Energy via Principal Component Analysis

Arash Moradzadeh [1], Omid Sadeghian [1], Kazem Pourhossein [2], Behnam Mohammadi-Ivatloo [1,3,*] and Amjad Anvari-Moghaddam [1,4,*]

[1] Faculty of Electrical and Computer Engineering, University of Tabriz, Tabriz 5166616471, Iran; arash.moradzadeh@tabrizu.ac.ir (A.M.); omidsadeghian1991@gmail.com (O.S.)

[2] Department of Electrical Engineering, Tabriz Branch, Islamic Azad University, Tabriz 5157944533, Iran; k.pourhossein@iaut.ac.ir

[3] Institute of Research and Development, Duy Tan University, Da Nang 550000, Vietnam

[4] Department of Energy Technology, Aalborg University, 9220 Aalborg, Denmark

* Correspondence: bmohammadi@tabrizu.ac.ir (B.M.-I.); aam@et.aau.dk (A.A.-M.)

Received: 16 March 2020; Accepted: 11 April 2020; Published: 14 April 2020

Abstract: The useful planning and operation of the energy system requires a sustainability assessment of the system, in which the load model adopted is the most important factor in sustainability assessment. Having information about energy consumption patterns of the appliances allows consumers to manage their energy consumption efficiently. Non-intrusive load monitoring (NILM) is an effective tool to recognize power consumption patterns from the measured data in meters. In this paper, an unsupervised approach based on dimensionality reduction is applied to identify power consumption patterns of home electrical appliances. This approach can be utilized to classify household activities of daily life using data measured from home electrical smart meters. In the proposed method, the power consumption curves of the electrical appliances, as high-dimensional data, are mapped to a low-dimensional space by preserving the highest data variance via principal component analysis (PCA). In this paper, the reference energy disaggregation dataset (REDD) has been used to verify the proposed method. REDD is related to real-world measurements recorded at low-frequency. The presented results reveal the accuracy and efficiency of the proposed method in comparison to conventional procedures of NILM.

Keywords: load disaggregation; non-intrusive load monitoring (NILM); dimensionality reduction; principal component analysis (PCA)

---

1. Introduction

Energy is one of the most important aspects of industrial and economic development in all countries. Future energy systems must be equipped to provide sustainable, affordable, and reliable energy, and to provide consumers with the ability to guarantee sustainable development. Effective use and energy efficiency are essential for sustainable development [1]. Therefore, energy consumption monitoring processes and planning in conserving energy in buildings are considered to be energy management for sustainable development. Due to rising costs and environmental impacts of energy consumption, the importance of energy conservation and planning is growing significantly [2]. Today, policy efforts to reduce CO_2 emissions from energy sources are one of the major expert efforts of environmentalists. On the other hand, energy demand of consumers is increasing exponentially, and the energy demand is projected to double by 2030. Therefore, several researches have been conducted on the effective management of energy supply and demand [3,4]. Nowadays, worldwide smart electricity meters are widely installed and used in homes and other places. According to research, it is estimated that by the end of 2020, approximately 72% of European homes will have electricity smart meters installed [5,6]. With the advances in smart electricity metering technologies, consumers are aware of their energy consumption patterns over days, weeks, or months. Adding features such as power/energy consumption onto the surface of home appliances will make them "smart" users of energy. In addition to monitoring the energy consumption of the entire home, they can monitor the energy consumption of each device [7].

Load monitoring or energy disaggregation is a very effective and useful step in energy management. Obtaining this information about the active loads of a grid is very effective and useful to the energy management system. Monitoring the load level of home appliances can be measured by two types: intrusive load monitoring (ILM) and non-intrusive load monitoring (NILM). In the ILM method, a sub-meter is attached to each appliance. This method is expensive and inconvenient, because an ILM-based power system with several appliances to read and record data needs a magnitude of sensors, which incur a prohibitive extension cost, and it does not respect consumer privacy. In contrast, NILM, or load disaggregation, can analyze the aggregated power consumption data of the appliances' exclusive power consumption via the appliance power consumption patterns, with no need to have data recorder sensors (Figure 1) [8,9].

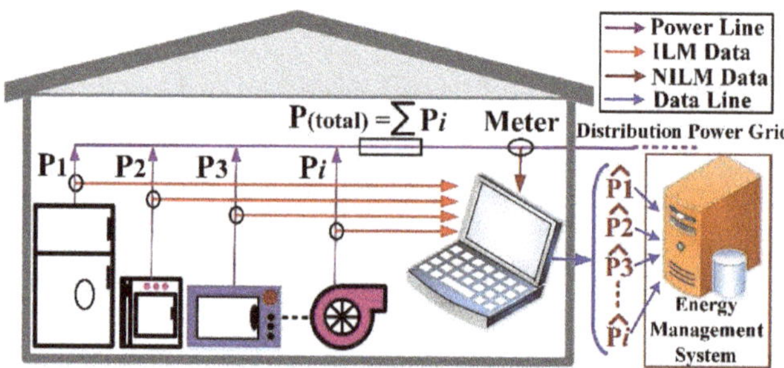

Figure 1. Principal structure of intrusive load monitoring (ILM) and non-intrusive load monitoring (NILM).

In NILM, all measured power consumption is processed in the smart meter. This process is continued until the required information about the time and amount of home electrical appliances consumption is calculated. Electrical smart meters have the capability to record the total customer energy consumption of the building. Using the information that electrical smart meters record from home power consumption can have several useful benefits in the areas of energy, trade, and economics, such as improving short and long term forecasts of demand profiles, load forecasting, providing consumers with detailed feedback on their energy consumption, designing demand management plans, and measuring and validating energy efficiency plans of buildings. Therefore, the development of techniques to improve load disaggregation problems, which can recognize the individual appliance's signal signatures through reading the total power consumption, has emerged as an interesting research topic in academic and industrial fields [10,11].

Data collection, event discovery, pattern recognition, and appliance identification are the four principal steps of an NILM system. Identifying electrical appliances operating at the same time in a home is the core of the NILM system [12].

There are many methods to improve the NILM system problems. Some of these methods are based on numerical indices, classical methods, and optimization. In some studies, different approaches for load disaggregation based on hidden Markov models (HMM), are used to model each appliance. In [13], segmented integer quadratic constraint programming is used to solve the load disaggregation problem. In [14], the load disaggregation problem is solved via increasable factorial approximate maximum posteriori. In [15], the event-based load disaggregation method is suggested, in which multiple signatures including distortion, active, and reactive powers are used. The information coding perspective of the load disaggregation is proposed in [16], in which appliances with similar power draws are recognized. In [17], the segmented integer quadratic programming problem is suggested to improve the NILM problem. Recognition of the simultaneous on and off state of multiple devices is dealt with a Cepstrum smoothing-based load disaggregation in [18]. Optimization-based methods to solve the NILM problem are proposed in [9,19–21].

On the other hand, data mining methods are widely used to solve energy management problems. Some works have been done on this basis to solve the NILM program. These methods are usually divided into two types: supervised and unsupervised. The main difference between these two methods is in learning the features that are in the essence of data. Unsupervised methods do not need to learn these features.

Supervised applications such as artificial neural networks, support vector machine applications, deep learning, feature learning, etc., use the training dataset of each appliance to identify and extract the features and build a feature dictionary [22–28]. In [23], a deep long short-term memory (LSTM) recurrent network is used to classify the types of electrical appliances into a set. A convolutional neural network (CNN) for recognizing multi-state appliances is suggested in [24], in which low-frequency power measurements are used. In [24], a support vector machine (SVM) is used to improve NILM problems, so that the K-means is considered to reduce the SVM training set size. A deep convolutional neural network is used in [26] to implement a practical data reinforcement technique with the need of sub-metering for new unseen houses, which makes a post-processing technique to solve the NILM problem. In [27], load disaggregation based on deep learning methods is proposed, in which deep dictionary learning and deep transform learning techniques are used. The transform learning method is also proposed in [28] for solving the NILM problem.

Unsupervised methods [29–31] collect features through power consumption data sets. In [32], the graph-based signal processing (GSP) load disaggregation is developed without the need for training. NILM based on unsupervised learning is proposed in [33], in which the fuzzy clustering algorithm called entropy index constraints competitive agglomeration (EICCA) is improved and utilized for solving the load disaggregation problem.

In this paper, a transparent unsupervised approach based on dimensional reduction is used to improve the residential load disaggregation problem via a visual and transparent process. Here, the power consumption curves of home electrical appliances are acting as a vector in high-dimensional space. The high-dimensional power consumption curves, related to household electrical appliances, are diminished to low-dimensional ones via principal component analysis (PCA) to disaggregate them. The proposed method does not require the training of specific networks by the use of training data to identify their characteristics, so some probabilities, based on inaccurate learning, will reduce the accuracy of the problem. The proposed method uses a feature space to transfer data from a high-dimensional space to a low-dimensional one. Because every household electrical appliance has its own consumption pattern, it is possible to obtain the inherent characteristics and patterns of each appliance by extracting eigenvalues and eigenvectors of the consumption curve of each appliance.

To apply the proposed method, the low-frequency data of power consumption readings at the meter, related to the REDD dataset [34], is utilized. This data shows the power consumption of several home electrical appliances in the real world. To obtain the best results in this paper, transient state information for each appliance are considered, because selecting the operating state of each appliance has a great impact on the aggregation operation.

The rest of this paper is structured as follows: PCA is elucidated in Section 2. Section 3 describes in detail the case study. How to apply PCA on data, experimental results, and load disaggregation results via proposed method are presented in Section 4. Finally, Section 5 concludes the paper.

2. Principal Component Analysis

In statistical analysis, principal component analysis (PCA) was introduced by Hotelling as a tool for dimension reduction of data in 1933. PCA is a convenient and useful method to compress images, reduce dimensions in high-dimensional data, and a common application for pattern recognition and feature extraction of big data [35,36]. The fundamental idea of PCA is to find an orthogonal linear model, which designs the high-dimensional data on a low-dimensional space known as the principal component (PC), while maximizing the variance of the data and minimizing the mean squared reconstruction error [37,38]. Achieving this idea first requires the calculation of the covariance matrix (CM) and then the obtaining of the eigenvalues and eigenvectors. In this paper, the PCA is used to identify eigenvalues and eigenvectors of power consumption curves of home electrical appliances, and to re-display them in a low-dimensional space. Let us suppose each database has power consumption curves of home electrical appliances with a column vector F_i, the length of which consists of n eigenvectors that are in the power consumption curves of home electrical appliances inside the original space. For m items of F_i vectors related to power consumption curves, F-matrix with the size of $n \times m$ could be defined [39]:

$$F = [F_1, F_2, F_3, \ldots, F_m]$$

(1)

The F data matrix can be transformed into a low dimensional space using PCA:

$$P = H^T F$$

(2)

where eigenvectors of data matrix F have formed the columns of scheme matrix H, and H^T is the transpose of the matrix H.

The steps of PCA are as follows [39]:

- approximation of the CM,
- eigen-dissociation of the CM and selecting the k highest eigenvalues,
- building the feature matrix I via respective eigenvectors, and
- mapping the main power consumption curves to the k-dimensional vector space by applying the I.

Considering m items of power consumption vectors, a CM is obtained from the following equation [40]:

$$CM = \frac{1}{m} \sum_{i=1}^{m} F_i . F_i^T$$

(3)

where T represents the transmission of the vectors. Solving the following eigenvalues equation is required to conduct a specific analysis of the CM [39,40]:

$$\lambda I = CM.I$$

(4)

where I and λ show the identification matrix and eigenvalues, respectively. The total variance of the main matrix (dataset) elements for the average zero is equivalent to the sum of the eigenvalues [38]. After the transmutation, the variance of the ith element equalize λ_i. To discover the adequate number of PCs to discriminate home electrical appliances, the accumulative contributory ratio (ACR) can be a useful parameter [41]. If the obtained eigenvalues are sorted in descending order, the ACR related to the first k PCs is explained as

$$\gamma_k = \frac{\sum_{i=1}^{k} \lambda_i}{\sum_{i=1}^{n} \lambda_i}$$

(5)

Having obtained the CM and computed the eigenvalues, we arrange their eigenvectors in descending order. To create the feature space, finding the first k PCs in which their γ_k exceeds 0.85 is necessary [41]. After finding these *k* principal components and placing their eigenvectors in a matrix, the feature matrix I is formed. As the final result of PCA, matrix P is obtained from Equation (6).

$$P = I.CM \tag{6}$$

Every electrical appliance has its own unique consumption pattern, but in most NILM problem solving techniques, some of the features of the power consumption curves are lost. The PCA method, by using its ability to detect the intrinsic structure and nature of data, can disaggregate the share of any electrical appliances' power consumption of the total home power consumption. Figure 2 illustrates the flowchart of the proposed method in this paper for load disaggregation. Figure 3 shows the basic principal diagram of the work done in this paper, step by step.

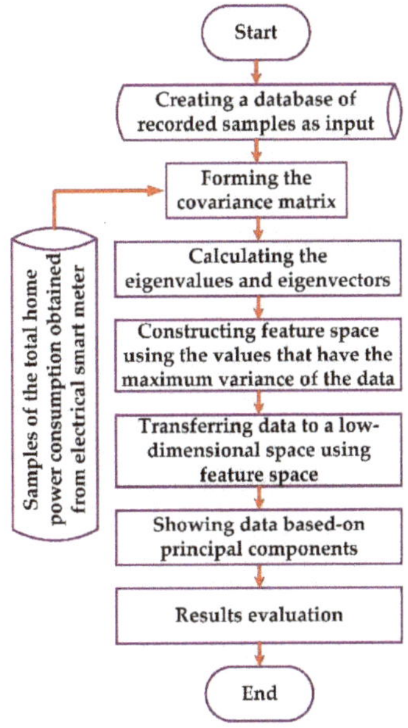

Figure 2. Flowchart of the proposed procedure.

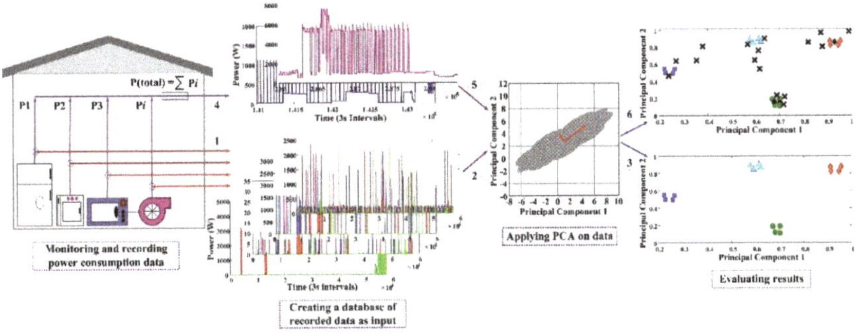

Figure 3. The basic principal diagram of the work done in this paper, step by step.

3. Case Study

In this paper, experiments were performed on the REDD dataset. The REDD dataset contains low-frequency data for 6 homes in Massachusetts, USA, including the total power consumption of the home, and the power consumption of each individual electrical device in the home [34]. Given that the main grid is sampled at 1 Hz, we used a 3 s interval to match these readings with the main grid, on both the main and plug levels (1/3 Hz), for on-line disaggregation. Because this data is relevant to real-world use, it has been used in most studies in NILM fields. To apply the proposed method, data from three houses including REDD house 1, REDD house 2, and REDD house 3 were used. Table 1 presents the types of household electrical appliances that the proposed method was able to identify.

Table 1. The names of the electrical appliances considered from reference energy disaggregation dataset (REDD) houses.

REDD Houses	Appliances
House 1	Wall oven, refrigerator, dishwasher, kitchen outlets, lighting, washer dryer, microwave, bathroom ground fault interrupters (GFI), electric heat, stove, different
House 2	Kitchen outlets, lighting, stove, washer dryer, microwave, refrigerator, dishwasher, garbage, different
House 3	Electronics, lighting, refrigerator, unknown, dishwasher, furnace, washer dryer, microwave, smoke alarms, garbage, bathroom GFI, kitchen outlets, different

4. Experimental Results

It is necessary to identify and extract the features and consumption patterns of the appliances, to load/energy disaggregate and to assess the consumption of each electrical appliance from the total power consumption of the whole house. In this paper, extraction of features and consumption patterns of household electrical appliances is done using PCA.

Using the proposed method requires a database as input. We used the power consumption curves of the electrical appliances presented in Table I as inputs. In this database, the power consumption curve of the two-day (2880 min) operation of each appliance was considered as a sample of each appliance. Four samples from each appliance (power consumption for the first eight days of each house) were considered as inputs. Figure 4 illustrates the data considered for the power consumption of the appliances in REDD House 1, as the network input.

After collecting the database, the steps were performed as presented in the flowchart of Figure 1. For more accuracy, load disaggregation was conducted based on dimension reduction. This method maintains the highest variance and eigenvalue for each power consumption curve in the principal component. The five highest values of the calculated eigenvalues, and the computed ACRs for them, are given in Table 2. It is visible that k exceeded 0.90 using two PCs. Thus, the power consumption curves of home electrical appliances in high-dimensional space could be reduced to vectors in two-dimensional space by sustaining the highest variance. Figure 5 shows the results of the separation of the power consumption pattern of each electrical appliance in the studied homes of the REDD dataset via PCA. It can be seen that the proposed method was able to disaggregate the power consumption patterns of electrical appliances in two-dimensional space by extracting the power consumption curve features.

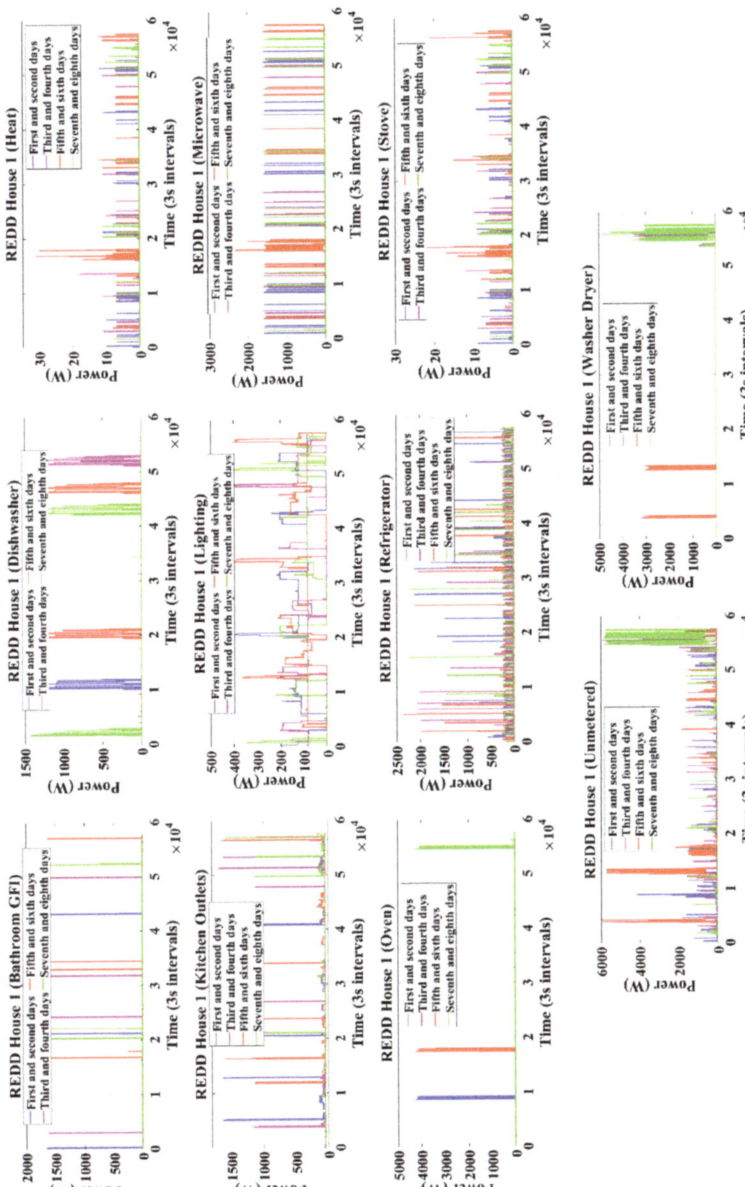

Figure 4. Data considered for the power consumption of the appliances in REDD House 1, as the network input.

Table 2. The five highest eigenvalues in descending order and their accumulative contributory ratios (ACRs).

	λ_1	λ_2	λ_3	λ_4	λ_5
Eigenvalue	535.24	208.28	65.38	21.07	5.90
ACR	0.713	0.908	0.954	0.987	0.991

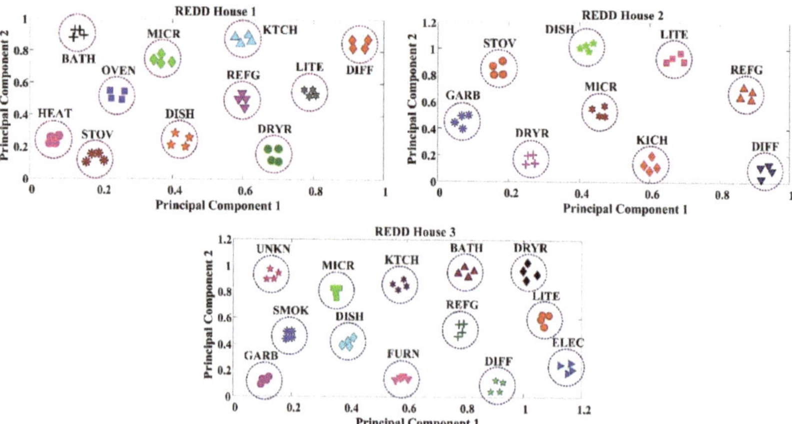

Figure 5. Separation of the power consumption pattern of each electrical appliance in the studied homes of REDD dataset via principal component analysis (PCA).

Now, to test and monitor the accuracy and efficiency of the proposed method, new samples of the power consumption of each electrical appliance were needed. To do this, new samples of the power consumption of each electrical appliance were considered over a two-day period. The PCA method was used to extract the features of this data, and the test results for the electrical appliances of each house by new samples are shown in Figure 6. In this figure the black color was used to represent each new sample of each electrical appliance.

From the above figures, it is clearly visible that the proposed method fully recognized the power consumption patterns of the new samples of each of the home appliances and identified them from the previous samples. However, the basic principle in the NILM is the disaggregation and detection of the power consumption of each electrical appliance from the total power consumption of the home. In this regard, samples of whole-house power consumption taken from the home's electricity smart meter were considered, each of which represented two hours. This data was used as input for PCA. Figure 7 shows plotted samples of the total home power consumption of each dataset. Dimension reduction via PCA was applied to the new samples that were obtained from the total home consumption. Figure 8 shows the test results of the proposed method using the new data.

The results of load disaggregation for the intended data revealed the performance and accuracy of the proposed method. It was found that PCA was able to efficiently display and distribute the nature and pattern of power consumption for household electrical appliances, in a two-dimensional space. Identifying the intrinsic behavior of any electrical appliance with regard to its power consumption could accurately perform the disaggregation of the total power consumption of an entire house in different hours.

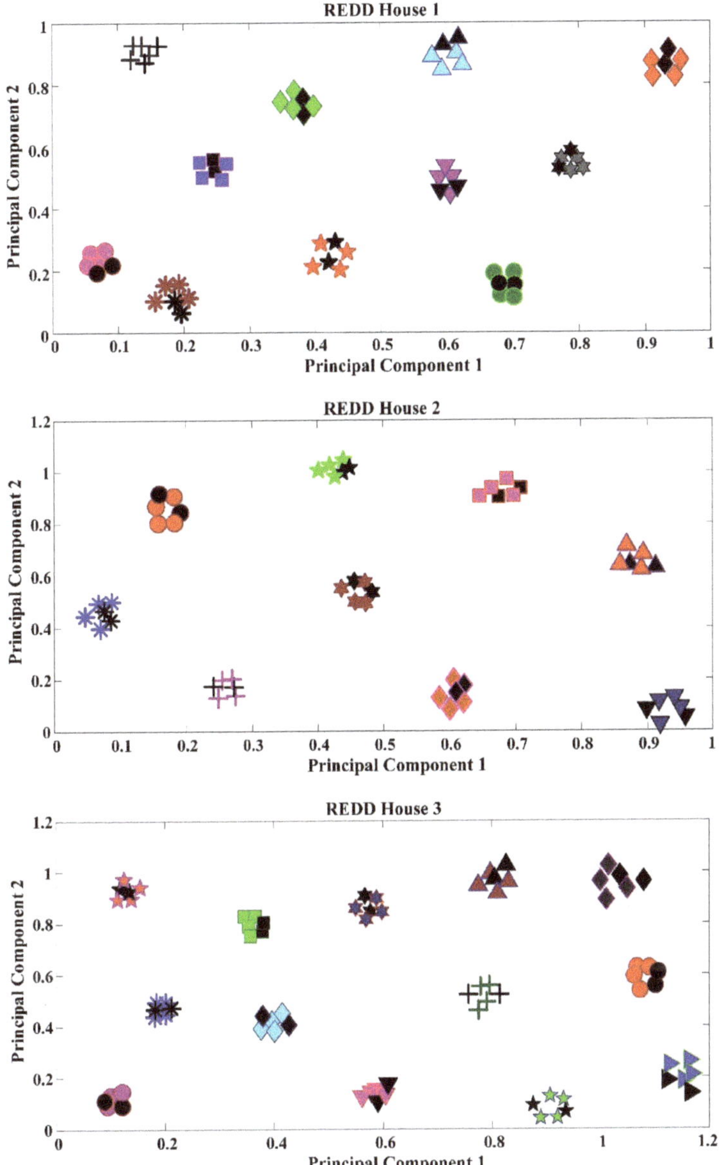

Figure 6. Test results for the electrical appliances of each house by new samples.

A comparison of the results should be made to show the accuracy and efficiency of the proposed method compared to other methods [12]. This is done by calculating the F-score as follows:

$$F - score = \frac{TI}{TI + FI} \tag{7}$$

where the F-Score is the accuracy evaluation metric for the predicted results, TI represents the number of samples that were truly identified, and FI represents the number of samples that were falsely identified.

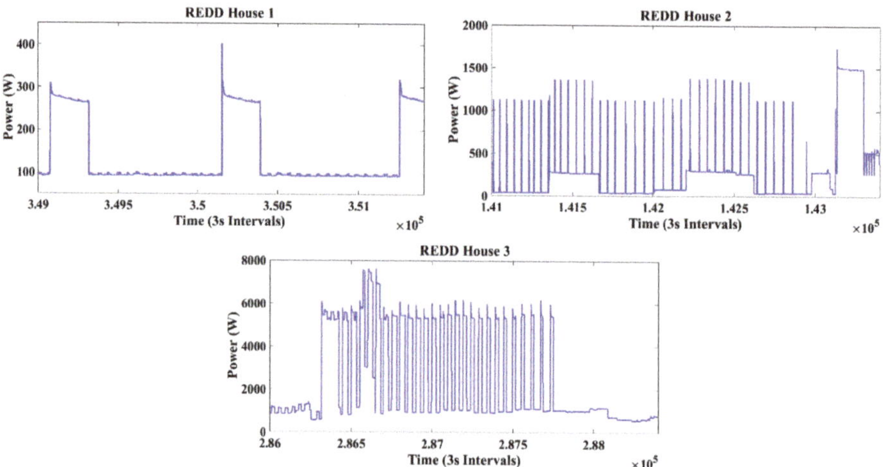

Figure 7. Samples of the total home power consumption of REDD Houses.

Direct comparisons of results should be done with extreme caution; this was done by ensuring that the database used in all cases were similar. Because this paper used REDD data, comparisons were made with the works that previously used this data, and the results of these comparisons are stated in Table 3.

Table 3. Performance comparison of the proposed method with other unsupervised solutions for REDD data.

Appliance Identification Method	Remarks	F-Score
Proposed Method	Using all appliances from REDD houses 1, 2, and 3	94.68%
Basic NILM [42]	Using all appliances from REDD	79.7%
Supervised GSP [43]	Using 5 appliances selected from the REDD	64%
Unsupervised GSP [32]	Using 5 appliances selected from the REDD	72.2%
Unsupervised HMM [44]	Using 7 appliances selected from the REDD	62.2%
Unsupervised dynamic time warping (DTW) [45]	Using 9 appliances selected from the REDD	68.6%
Supervised decision-tree (DT) [45]	Using 9 appliances selected from the REDD	76.4%
Viterbi algorithm [46]	Using 9 appliances selected from the REDD	88.1%

Since unsupervised solutions have no information (target) about the features of input data (power consumption curves), they must have the ability to enable the extraction high features of data so that they can perform the detection operation well. The presented results in Table 3, show the accuracy and efficiency of the proposed method compared to other unsupervised methods, in identifying power consumption patterns of household electrical appliances.

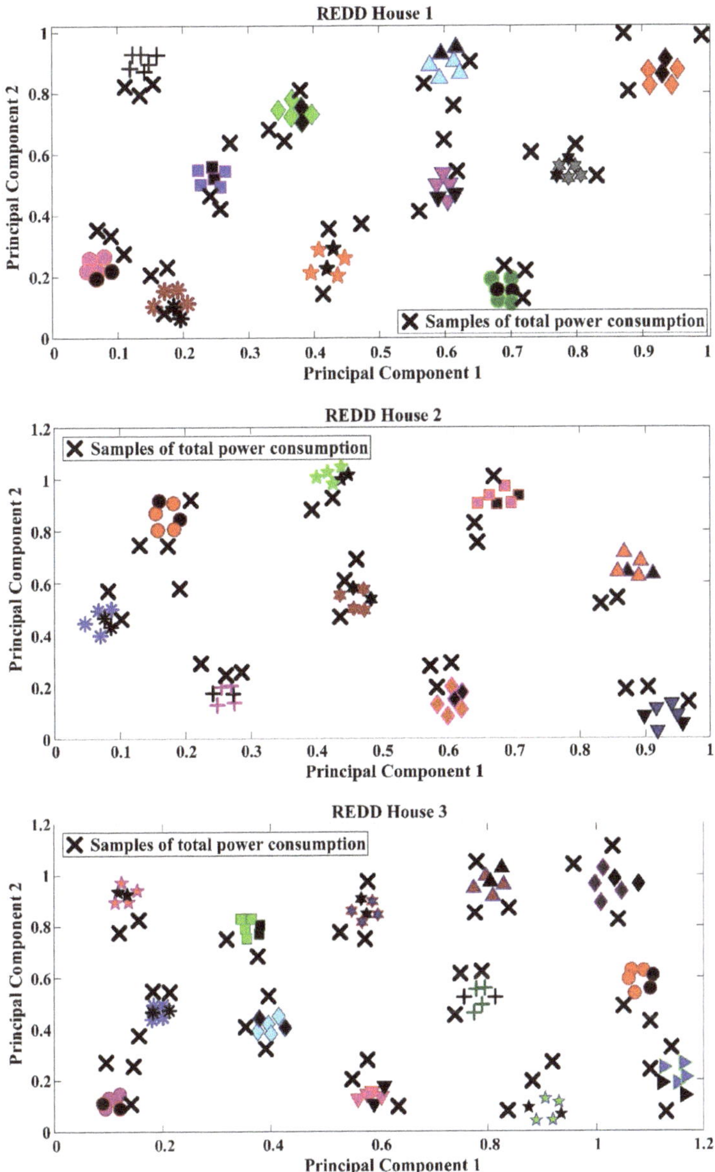

Figure 8. Test results of the proposed method using the samples of the total home power consumption.

5. Conclusions

To address non-intrusive load monitoring in an efficient and transparent manner, the pattern recognition of power consumption time series is very helpful. In this paper, principal component analysis, as an unsupervised approach, was used to extract useful features from power consumption data to detect consumer type. This approach displays high-dimensional data in a low-dimensional space by preserving maximum information of the initial data. Extraction of features and recognition of the consumption patterns of each electrical appliance in the load disaggregation make it possible for the consumers to be aware of their own power consumption pattern in any time period. Low-frequency sampled data from the REDD was used to test the proposed method. Power consumption signatures received from each home electrical appliance at different times were considered as input data for PCA. By applying the proposed method to the input data power, consumption patterns of each electrical appliance in a two-dimensional space was transparently observed. Subsequently, PCA was applied to the samples of the total home power consumption for load disaggregation. The power consumption of the household electrical appliances was estimated from the total power consumption of the home at different times. Clarity and transparency in displaying power consumption patterns of different electrical appliances in a low-dimensional space, makes the proposed method desirable.

Author Contributions: A.M.: Writing—original draft, software, and methodology; O.S.: Investigation and validation; K.P.: Formal analysis, resources, writing—review and editing; B.M.-I.: Conceptualization, data curation, writing—review and editing; A.A.-M.: Supervision, writing—review and editing. All authors have read and agreed to the published version of the manuscript.

Funding: This research received no external funding.

Conflicts of Interest: The authors declare no conflict of interest.

References

1. Anvari-Moghaddam, A.; Mohammadi-ivatloo, B.; Asadi, S.; Guldstrand Larsen, K.; Shahidehpour, M. Sustainable Energy Systems Planning, Integration, and Management. *Appl. Sci.* **2019**, *9*, 4451. [CrossRef]
2. Liu, Y.; Wang, X.; Zhao, L.; Liu, Y. Admittance-based load signature construction for non-intrusive appliance load monitoring. *Energy Build.* **2018**, *171*, 209–219. [CrossRef]
3. Nazari-Heris, M.; Mirzaei, M.A.; Mohammadi-Ivatloo, B.; Marzband, M.; Asadi, S. Economic-environmental effect of power to gas technology in coupled electricity and gas systems with price-responsive shiftable loads. *J. Clean. Prod.* **2020**, *244*, 118769. [CrossRef]
4. Miyasawa, A.; Fujimoto, Y.; Hayashi, Y. Energy disaggregation based on smart metering data via semi-binary nonnegative matrix factorization. *Energy Build.* **2019**, *183*, 547–558. [CrossRef]
5. Devlin, M.A.; Hayes, B.P. Non-Intrusive Load Monitoring and Classification of Activities of Daily Living Using Residential Smart Meter Data. *IEEE Trans. Consum. Electron.* **2019**, *65*, 339–348. [CrossRef]
6. He, K.; Stankovic, L.; Liao, J.; Stankovic, V. Non-Intrusive Load Disaggregation Using Graph Signal Processing. *IEEE Trans. Smart Grid* **2018**, *9*, 1739–1747. [CrossRef]
7. D'Incecco, M.; Squartini, S.; Zhong, M. Transfer Learning for Non-Intrusive Load Monitoring. *IEEE Trans. Smart Grid* **2020**, *11*, 1419–1429. [CrossRef]
8. Aiad, M.; Lee, P.H. Unsupervised approach for load disaggregation with devices interactions. *Energy Build.* **2016**, *116*, 96–103. [CrossRef]
9. Zeinal-Kheiri, S.; Shotorbani, A.M.; Mohammadi-Ivatloo, B. Residential Load Disaggregation Considering State Transitions. *IEEE Trans. Ind. Inform.* **2020**, *16*, 743–753. [CrossRef]
10. Quek, Y.T.; Woo, W.L.; Logenthiran, T. Load Disaggregation Using One-Directional Convolutional Stacked Long Short-Term Memory Recurrent Neural Network. *IEEE Syst. J.* **2020**, *14*, 1395–1404. [CrossRef]
11. Wang, Y.; Pandharipande, A.; Fuhrmann, P. Energy Data Analytics for Nonintrusive Lighting Asset Monitoring and Energy Disaggregation. *IEEE Sens. J.* **2018**, *18*, 2934–2943. [CrossRef]
12. Morais, L.R.; Castro, A.R.G. Competitive Autoassociative Neural Networks for Electrical Appliance Identification for Non-Intrusive Load Monitoring. *IEEE Access* **2019**, *7*, 111746–111755. [CrossRef]
13. Kong, W.; Dong, Z.Y.; Ma, J.; Hill, D.J.; Zhao, J.; Luo, F. An Extensible Approach for Non-Intrusive Load Disaggregation with Smart Meter Data. *IEEE Trans. Smart Grid* **2018**, *9*, 3362–3372. [CrossRef]
14. Ji, T.Y.; Liu, L.; Wang, T.S.; Lin, W.B.; Li, M.S.; Wu, Q.H. Non-Intrusive Load Monitoring Using Additive Factorial Approximate Maximum a Posteriori Based on Iterative Fuzzy c-Means. *IEEE Trans. Smart Grid* **2019**, *10*, 6667–6677. [CrossRef]
15. Alcala, J.; Urena, J.; Hernandez, A.; Gualda, D. Event-Based Energy Disaggregation Algorithm for Activity Monitoring from a Single-Point Sensor. *IEEE Trans. Instrum. Meas.* **2017**, *66*, 2615–2626. [CrossRef]
16. Pöchacker, M.; Egarter, D.; Elmenreich, W. Proficiency of power values for load disaggregation. *IEEE Trans. Instrum. Meas.* **2016**, *65*, 46–55. [CrossRef]

17. Kong, W.; Dong, Z.Y.; Hill, D.J.; Luo, F.; Xu, Y. Improving Nonintrusive Load Monitoring Efficiency via a Hybrid Programing Method. *IEEE Trans. Ind. Inform.* **2016**, *12*, 2148–2157. [CrossRef]

18. Kong, S.; Kim, Y.; Ko, R.; Joo, S.K. Home appliance load disaggregation using cepstrum-smoothing-based method. *IEEE Trans. Consum. Electron.* **2015**, *61*, 24–30. [CrossRef]

19. De Baets, L.; Ruyssinck, J.; Develder, C.; Dhaene, T.; Deschrijver, D. On the Bayesian optimization and robustness of event detection methods in NILM. *Energy Build.* **2017**, *145*, 57–66. [CrossRef]

20. Piga, D.; Cominola, A.; Giuliani, M.; Castelletti, A.; Rizzoli, A.E. Sparse Optimization for Automated Energy End Use Disaggregation. *IEEE Trans. Control Syst. Technol.* **2016**, *24*, 1044–1051. [CrossRef]

21. Machlev, R.; Levron, Y.; Beck, Y. Modified Cross-Entropy Method for Classification of Events in NILM Systems. *IEEE Trans. Smart Grid* **2018**, *10*, 4962–4973. [CrossRef]

22. Sadeghianpourhamami, N.; Ruyssinck, J.; Deschrijver, D.; Dhaene, T.; Develder, C. Comprehensive feature selection for appliance classification in NILM. *Energy Build.* **2017**, *151*, 98–106. [CrossRef]

23. Mauch, L.; Yang, B. A new approach for supervised power disaggregation by using a deep recurrent LSTM network. In Proceedings of the 2015 IEEE Global Conference on Signal and Information Processing, GlobalSIP, Orlando, FL, USA, 14–16 December 2015; pp. 63–67.

24. Zhang, C.; Zhong, M.; Wang, Z.; Goddard, N.; Sutton, C. Sequence-to-point learning with neural networks for non-intrusive load monitoring. In Proceedings of the 32nd AAAI Conference on Artificial Intelligence, AAAI 2018, Louisiana, NO, USA, 2–7 February 2018; pp. 2604–2611.

25. Altrabalsi, H.; Stankovic, L.; Liao, J.; Stankovic, V. A low-complexity energy disaggregation method: Performance and robustness. In Proceedings of the IEEE Symposium on Computational Intelligence Applications in Smart Grid, CIASG, Orlando, FL, USA, 9–12 December 2014; Volume 2015-Janua.

26. Kong, W.; Dong, Z.Y.; Wang, B.; Zhao, J.; Huang, J. A practical solution for non-intrusive type II load monitoring based on deep learning and post-processing. *IEEE Trans. Smart Grid* **2020**, *11*, 148–160. [CrossRef]

27. Singhal, V.; Maggu, J.; Majumdar, A. Simultaneous Detection of Multiple Appliances from Smart-Meter Measurements via Multi-Label Consistent Deep Dictionary Learning and Deep Transform Learning. *IEEE Trans. Smart Grid* **2019**, *10*, 2969–2978. [CrossRef]

28. Gaur, M.; Majumdar, A. Disaggregating transform learning for non-intrusive load monitoring. *IEEE Access* **2018**, *6*, 46256–46265. [CrossRef]

29. Kim, H.; Marwah, M.; Arlitt, M.; Lyon, G.; Han, J. Unsupervised disaggregation of low frequency power measurements. In Proceedings of the 11th SIAM International Conference on Data Mining, SDM 2011, Mesa, AZ, USA, 28–30 April 2011; pp. 747–758.

30. Parson, O.; Ghosh, S.; Weal, M.; Rogers, A. An unsupervised training method for non-intrusive appliance load monitoring. *Artif. Intell.* **2014**, *217*, 1–19. [CrossRef]

31. Gonçalves, H.; Ocneanu, A.; Bergés, M.; Fan, R.H. Unsupervised disaggregation of appliances using aggregated consumption data. *SustKDD Workshop* **2011**.

32. Zhao, B.; Stankovic, L.; Stankovic, V. On a Training-Less Solution for Non-Intrusive Appliance Load Monitoring Using Graph Signal Processing. *IEEE Access* **2016**, *4*, 1784–1799. [CrossRef]

33. Liu, Q.; Kamoto, K.M.; Liu, X.; Sun, M.; Linge, N. Low-Complexity Non-Intrusive Load Monitoring Using Unsupervised Learning and Generalized Appliance Models. *IEEE Trans. Consum. Electron.* **2019**, *65*, 28–37. [CrossRef]

34. Kolter, J.Z.; Johnson, M.J. REDD: A Public Data Set for Energy Disaggregation Research. *SustKDD Workshop* **2011**, *25*, 59–62.

35. Witten, I.H.; Frank, E.; Mark, A. *Hall Data Mining: Practical Machine Learning*; Elsevier: Amsterdam, The Netherlands, 2011; ISBN 9780123748560.

36. Moore, B.C. Principal Component Analysis in Linear Systems: Controllability, Observability, and Model Reduction. *IEEE Trans. Autom. Control* **1981**, *26*, 17–32. [CrossRef]

37. Zhang, X.; Kano, M.; Li, Y. Principal Polynomial Analysis for Fault Detection and Diagnosis of Industrial Processes. *IEEE Access* **2018**, *6*, 52298–52307. [CrossRef]

38. Honeine, P. Online kernel principal component analysis: A reduced-order model. *IEEE Trans. Pattern Anal. Mach. Intell.* **2012**, *34*, 1814–1826. [CrossRef]

39. Pourhossein, K.; Gharehpetian, G.B.; Rahimpour, E.; Araabi, B.N. A probabilistic feature to determine type and extent of winding mechanical defects in power transformers. *Electr. Power Syst. Res.* **2012**, *82*, 1–10. [CrossRef]

40. Theodoridis, S.; Pikrakis, A.; Koutroumbas, K.; Cavouras, D. *Introduction to Pattern Recognition: A Matlab Approach*; Academic Press: Cambridge, MA, USA, 2010; ISBN 9780123744869.
41. BinFeng, Y.; FeiLu, L.; Dan, H. Research on edge identification of a defect using pulsed eddy current based on principal component analysis. *NDT E Int.* **2007**, *40*, 294–299. [CrossRef]
42. Dinesh, C.; Nettasinghe, B.W.; Godaliyadda, R.I.; Ekanayake, M.P.B.; Ekanayake, J.; Wijayakulasooriya, J.V. Residential Appliance Identification Based on Spectral Information of Low Frequency Smart Meter Measurements. *IEEE Trans. Smart Grid* **2016**, *7*, 2781–2792. [CrossRef]
43. Stankovic, V.; Liao, J.; Stankovic, L. A graph-based signal processing approach for low-rate energy disaggregation. In Proceedings of the IEEE SSCI 2014—2014 IEEE Symposium Series on Computational Intelligence—CIES 2014: 2014 IEEE Symposium on Computational Intelligence for Engineering Solutions, Orlando, FL, USA, 9–12 December 2014; pp. 81–87.
44. Parson, O.; Ghosh, S.; Weal, M.; Rogers, A. Non-intrusive load monitoring using prior models of general appliance types. In Proceedings of the National Conference on Artificial Intelligence, Toronto, ON, Canada, 22–26 July 2012; Volume 1, pp. 356–362.
45. Liao, J.; Elafoudi, G.; Stankovic, L.; Stankovic, V. Power Disaggregation for Low-sampling Rate Data. In Proceedings of the 2nd International Non-intrusive Appliance Load Monitoring Workshop, Austin, TX, USA, 3 June 2014; pp. 1–4.
46. Zeifman, M. Disaggregation of home energy display data using probabilistic approach. *IEEE Trans. Consum. Electron.* **2012**, *58*, 23–31. [CrossRef]

 © 2020 by the authors. Licensee MDPI, Basel, Switzerland. This article is an open access article distributed under the terms and conditions of the Creative Commons Attribution (CC BY) license (http://creativecommons.org/licenses/by/4.0/).

MDPI

St. Alban-Anlage 66

4052 Basel

Switzerland

Tel. +41 61 683 77 34

Fax +41 61 302 89 18

www.mdpi.com

Sustainability Editorial Office

E-mail: sustainability@mdpi.com

www.mdpi.com/journal/sustainability

www.ingramcontent.com/pod-product-compliance
Lightning Source LLC
LaVergne TN
LVHW070154170726
843515LV00007B/1916